NATURE MATHEMATIZED

VOLUME I

THE UNIVERSITY OF WESTERN ONTARIO SERIES IN PHILOSOPHY OF SCIENCE

A SERIES OF BOOKS

IN PHILOSOPHY OF SCIENCE, METHODOLOGY,

EPISTEMOLOGY, LOGIC, HISTORY OF SCIENCE,

AND RELATED FIELDS

VOLUME 20

NATURE MATHEMATIZED

Historical and Philosophical Case Studies in Classical Modern Natural Philosophy

Papers Deriving from the Third International Conference on the History and Philosophy of Science, Montreal, Canada, 1980

VOLUME I

Edited by

WILLIAM R. SHEA

Department of Philosophy, McGill University

D. REIDEL PUBLISHING COMPANY

DORDRECHT : HOLLAND / BOSTON : U.S.A.

LONDON : ENGLAND

Library of Congress Cataloging in Publication Data

International Conference on the History of Philosophy of Science (3rd : 1980 : Montréal, Québec)
Nature mathematized.

(The University of Western Ontario series in philosophy of science ; v. 20)
Includes indexes.
1. Science–Philosophy–Congresses. 2. Science–History–Congresses. I. Shea, William R. II. Title. III. Series.
Q174.I563 1980a 501 82–9846
ISBN-13: 978-94-009-6959-9 e-ISBN-13: 978-94-009-6957-5
DOI: 978-94-009-6957-5

Published by D. Reidel Publishing Company,
P.O. Box 17, 3300 AA Dordrecht, Holland

Sold and distributed in the U.S.A. and Canada
by Kluwer Boston Inc.,
190 Old Derby Street, Hingham, MA 02043, U.S.A.

In all other countries, sold and distributed
by Kluwer Academic Publishers Group
P.O. Box 322, 3300 AH Dordrecht, Holland

D. Reidel Publishing Company is a member of the Kluwer Group

TABLE OF CONTENTS

VOLUME I

TABLE OF CONTENTS

VOLUME II

PART V

PREFACE

These remarks preface two volumes consisting of the proceedings of the Third International Conference on the History and Philosophy of Science of the International Union of History and Philosophy of Science. The conference was held under the auspices of the Union, The Social Sciences and Humanities Research Council of Canada, and the Canadian Society for History and Philosophy of Science. The meetings took place in Montreal, Canada, 25–29 August 1980, with Concordia University as host institution.

The program of the conference was arranged by a Joint Commission of the International Union of History and Philosophy of Science consisting of Robert E. Butts (Canada), John Murdoch (U.S.A.), Vladimir Kirsanov (U.S.S.R.), and Paul Weingartner (Austria). The Local Arrangements Committee consisted of Stanley G. French, Chair (Concordia), Michel Paradis, treasurer (McGill), François Duchesneau (Université de Montréal), Robert Nadeau (Université du Québec à Montréal), and William Shea (McGill University). Both committees are indebted to Dr. G. R. Paterson, then President of the Canadian Society for History and Philosophy of Science, who shared his expertise in many ways. Dr. French and his staff worked diligently and efficiently on behalf of all participants. The city of Montreal was, as always, the subtle mixture of extravagance, charm, warmth and excitement that retains her status as the jewel of Canadian cities.

The funding of major international conferences is always a problem. This conference was exceptional in that financial support came forward from many sources. Contributions from the Division of History of Science and the Division of Logic, Methodology and Philosophy of Science of the International Union of History and Philosophy of Science were matched by a seed grant from the Social Sciences and Humanities Research Council of Canada. The seeds bore fruit in the form of grants from Concordia University, McGill University, Université de Montréal, University of Calgary, University of Guelph, University of Lethbridge, The University of Western Ontario, the Hannah Institute for the History of Medicine, and The University of Western Ontario Series in Philosophy of Science. On behalf of the IUHPS, the members of the Joint Commission thank all sponsors for their generous support, and for their cooperative demonstration of the fact that there is, indeed, strength in numbers!

William R. Shea (ed.), Nature Mathematized, ix–x.

The conference organizers owe many additional debts which we here gratefully acknowledge. We thank the many people who worked for Dr. French, but especially Stephanie Manuel, who organized most of the details of local arrangements, Shona French, a gracious and helpful lady during a long week, and Susan Hudson, who designed the conference posters and the conference program booklet, always useful items but in this case elevated to a new standard of aesthetic attraction. We thank all of those who were our hosts – for drinks between sessions, for lunch or dinner, especially we thank William Shea and Mario Bunge for a marvellous luncheon at McGill, Dr. and Mrs. Robert V. V. Nicholls, host and hostess at a glittering reception in the storied Ritz-Carlton Hotel, and the administration of Concordia University, for a second welcoming reception in the Faculty Club of Concordia.

Comparison of the tables of contents of the two conference volumes with the final program of the conference will reveal that not all papers delivered at the meetings were made available for publication. In some cases, material appearing in the proceedings is largely new (Albury, Moore), or very extensively revised (Duchesneau, Lennon, Okruhlik). The conference contribution of Jonathan Hodge grew to monograph length in revision; it is our hope that it might appear separately in a short time. The papers by Joly and Clavelin were delivered at the conference in French. We thank Vida Bruce for the translation of Joly's paper, and Violaine Arès and Stella-Marie Baza for the translation of Clavelin's paper. In both cases the translations were corrected and approved by the authors.

We are especially grateful to Michael Ruse and William Shea for their willing acceptance of the chores of editing the volumes, and for writing the special introductions to each of the volumes. Mrs. Nel Jones, WOS editor at Reidel, was patient as usual, and as usual enormously helpful. We are very grateful. Judith York, editorial assistant in the WOS office at The University of Western Ontario, edited the difficult papers, and organized the final typescripts. We owe her many thanks. She was a cog in the wheels at Western, as was Pat Orphan, who did all of the thousand little things connected with getting the project launched, things usually ignored or forgotten. We remember and take notice, with sincere thanks.

On behalf of editors and organizers,

October, 1981 ROBERT E. BUTTS

PROGRAM

THIRD INTERNATIONAL CONFERENCE ON HISTORY AND PHILOSOPHY OF SCIENCE, CONCORDIA UNIVERSITY, MONTREAL, CANADA, 25–29 AUGUST, 1980

CONCEPTUAL AND EMPIRICAL PROBLEMS IN ANCIENT GREEK MEDICINE AND NATURAL PHILOSOPHY

Paul Potter (Canada), 'Hippocratic Nosology: the Identification and Classification of Diseases'.
John Wright (Canada), Comments on Potter's paper.
Robert Joly (Belgium), 'Hippocrate et l'Ecole de Cos: entre le mythe et l'hypercritique'.
Jaap Mansfeld (The Netherlands), Comments on Joly's paper.

In the Chair: Maurice Lebel (Canada).

HISTORY AND PHILOSOPHY OF PSYCHIATRY

William Bynum (U.K.), 'Themes in British Psychiatry, J. C. Pritchard (1785–1848) to Henry Maudsley (1835–1918)'.
Stefano Poggi (Italy), Comments on Bynum's paper.
Morris Eagle (Canada), 'Anatomy of the Self in Psychoanalytic Theory'.
Michael Moore (U.S.A.), Comments on Eagle's paper.
David Gruender (U.S.A.), Comments on Eagle's paper.

In the Chair: Eva Lester (Canada).

MATHEMATICS IN THE DEVELOPMENT OF SCIENTIFIC THEORIES

Paul Weingartner (Austria), 'The Ideal of the Mathematization of All Sciences and of 'More geometrico' in Descartes and Leibniz'.
François Duchesneau (Canada), Comments on Weingartner's paper.
Thomas Lennon (Canada), Comments on Weingartner's paper.
V. Vizgin (U.S.S.R.), 'The Genesis of the Tensor Geometrical Conception of Gravitation'.

In the Chair: Joseph Pitt (U.S.A.).

William R. Shea (ed.), Nature Mathematized, xi–xiii.

HISTORY AND PHILOSOPHY OF 19TH CENTURY BIOLOGY

Camille Limoges (Canada), 'L'économie politique d'une tendance hégémonique en histoire naturelle: le cuviérisme en France au XIXe siècle'.

W. R. Albury (Australia), Comments on Limoges' paper.

M. Jaroshevsky (U.S.S.R.), 'The Problem of Voluntary Motion Determination in Russian Psycho-Physiology'.

Jonathan Hodge (U.K.), 'Darwin on Natural Selection: his Methods and his Methodology'.

Michael Ruse (Canada), Comments on Hodge's paper.

John Beatty (U.S.A.), 'What's in a Word? The Problem of Coming to Terms in the Darwinian Revolution'.

David Hull (U.S.A.), Comments on Beatty's paper.

A. Lindenmayer (The Netherlands), Comments on Beatty's paper.

In the Chair: Mario Bunge (Canada).

MATHEMATICS AND PHYSICS IN THE 17TH AND 18TH CENTURIES

Maurice Clavelin (France), 'Aspects conceptuels et techniques de la géometrisation Galiléene du mouvement de graves'.

William Shea (Canada), Comments on Clavelin's paper.

J. E. McGuire (U.S.A.), 'Space, Geometry and Infinity: Newton and Descartes on the Indefiniteness of Extension'.

John North (The Netherlands), 'Finite and Otherwise. Some Seventeenth Century Views'.

Peter Harman (U.K.), 'Force and Inertia: Euler and Kant's *Metaphysical Foundations of Natural Science*'.

Kathleen Okruhlik (Canada), Comments on Harman's paper.

V. Kirsanov (U.S.S.R.), 'Non-Mechanistic Ideas in Physics and Philosophy: from Newton to Kant'.

V. Kartsev (U.S.S.R.), 'Hypothetical Experiments of Academician Petrov and Electrical Experiments of the XVIII Century'.

In the Chair: Roger Angel (Canada).

Communications not read in sessions:

MATHEMATICS IN THE DEVELOPMENT OF SCIENTIFIC THEORIES

Karel Berka (Czechoslovokia), 'The Ideal of Mathematization in B. Bolzano'.

MATHEMATICS AND PHYSICS IN THE 17TH AND 18TH CENTURIES

Ashot Grigorian (U.S.S.R.), 'Measure, Proportion and the Mathematical Structure of Galileo's Mechanics'. (Copied for distribution).

INTRODUCTION

WILLIAM R. SHEA

DO HISTORIANS AND PHILOSOPHERS OF SCIENCE SHARE THE SAME HERITAGE?

INTRODUCTION

There is an obvious and trivial sense in which the answer to this question can only be in the affirmative. Historians and philosophers of science study the same phenomenon and hence share the same field of enquiry. But the interesting question, of course, is whether they meet on that common ground, not merely as individuals politely exchanging comments about the weather but as scholars who have something to learn from one another. In other words, do historians and philosophers of science profit from an exchange of ideas about their different ways of looking at science?

Most people would assume that such a dialogue is not only possible but eminently desirable. Experience, however, teaches us that cooperation is more honoured in the breach than in the observance. There may be a simple reason for this. When one recalls that a fruitful dialogue can only take place between mature partners who are conscious of their own value and willing to learn from others, the relative youth of both history and philosophy of science explains why they have not sought each other out. Adolescent sciences, like teen-agers, are more intent on doing their own thing than on enriching their perspectives by confronting them with those of others. Many historians are so busy getting their facts straight and chartering a course through the tangled skein of historical anecdoctes that they have little time for the broader issues that interest the philosopher. On the other hand, some philosophers of science are so occupied in establishing a viable program of their own that they have no truck with the vagaries of history. Who cares after all whether Galileo was born on the same day Michelangelo died, as was once believed, or three days earlier as recent historians have shown? In a sensitive and penetrating essay, Robert E. Butts discusses how, as a result of this isolation, history of science has often locked itself within the narrow confines of a strictly expository accumulation of the details of past science, while the philosophy of science frequently advocated " a model of science as psychology-free, value-free, history-free, free, indeed, of almost everything but the tools of logic and maturing semantic."[1]

It could hardly have been different. Historians felt, rightly, that the data

William R. Shea (ed.), Nature Mathematized, 3–20.

needed to be supplied before a more fruitful discussion of the nature of science could be undertaken. The same kind of professional constraint acted on the philosophers who saw that they must first attend to the genuinely philosophical aspects of science, namely argument and analysis. What was unfortunate was the incidental but deep cleft that developed between the two disciplines. An anecdote will illustrate the climate in the early forties. When Carnap spent a term at Harvard, I. B. Cohen, then a young historian of science, asked him whether he would speak to their history of science club. He did not refuse at once, writes Cohen, "but before giving me his answer he asked me to walk home with him to tell him why I found the history of science interesting and worthy of study. We spent a good bit of the afternoon together; in the end, he thanked me, and remarked that he felt he could *not* speak to our group. I had convinced him, he said, that he really was as unhistorically minded a person as one could imagine. He had nothing whatever to say, he concluded, about the study of the history of scientific ideas that could possibly be of interest to historians!"[2]

Philosophers and historians did not merely salute one another from divided heights; they came to entertain deep suspicions about their respective motives. Historians of science accused philosophers of distorting actual scientific practice to make it fit their philosophical predilections. A larger world view was attained but only at the price of historical accuracy! Philosophers of science retorted that historians were trying to foist upon them an ideology of science based on the crude logical fallacy that the genesis of an idea is relevant to its validity.

This is not to say that there has been no communication between the two groups or that some individuals have not worked successfully in both fields. One need only run through a list of some of the most distinguished figures of the nineteenth century to come across such names as William Whewell, Charles Sanders Peirce, Ernst Mach, Pierre Duhem and Arthur Hannequin. More recently, the philosopher Alexandre Koyré caused a major historiographical revolution in the history of science, and it was the historian Thomas S. Kuhn who located what might be the limits of historical understanding in interpreting science. His work has given rise, directly or indirectly, to a vigorous debate among historians and philosophers of science.[3]

THE RELEVANCE OF HISTORY

Two schools or trends of thought may be said to prevail among contemporary philosophers of science. In the first we find philosophers such as

Hanson, Laudan, Shapere and Toulmin who examine historical cases in the sciences in order to uncover the structure of scientific reasoning and define the growth of scientific concepts as they actually emerged. Their research rests on the following methodological consideration: a sound philosophy of science is based on the study of both past and present scientific activities, and is defended or justified by appealing to both past or present scientific activities. The other school, of which Carnap, Quine, and Suppes are distinguished representatives, consider these methodological points irrelevant. They are not interested in the historical genesis but the rational structure of theories. Whereas the first group, which I shall refer to as the 'History-minded', wish to present a faithful picture of science as it came into being and to display its building blocks, the second group believes that we should not lose time unravelling the swaddling clothes of great ideas. In this sense we may call them 'Rationality-minded' since their aim is to provide rigour and accuracy where there is too often historical vagueness or confusion. Carnap, for instance, advocated a process of 'explication' whereby concepts such as theory, probability, model or explanation, that are often used in an imprecise way, are to be replaced by more exact and clearer concepts.[4] Substituting the notion of hypothetico-deductive system for the general concept of theory, for example, was said to bring greater clarity by explicitly indicating which propositions are to be treated as axioms, which as theorems, and how the latter are to be derived from the former. The hypothetico-deductive system is formulated by using 'theoretical' and 'observational' terms from an agreed-upon vocabulary, but it is empirically meaningful since each term is either observational or related to observational terms via 'correspondence' rules. It has explanatory value because of its deductive order as well as its empirical content. Formulations of a particular theory, such as Newton's mechanics, are to be recast in a more precise and elegant axiomatic system of the sort that P. Suppes and others have supplied.[5] Roughly speaking, the 'History-minded' want to 'get their facts straight'; the 'Rationality-minded' want to "straighten out the facts".

Despite appearances to the contrary, the 'Rationality-minded' cannot carry out their program unless they turn to the history of science. They advocate replacing a concept or system in use on the ground that, as Carnap puts it, there are certain 'critical contexts' in which its use "involves confusions or even inconsistencies" and leads to unanswerable questions and paradoxes.[6] But unless the defects of the present concept or system are identified and recognized as such, we are in no position to say whether the new concept or system avoids these difficulties. In other words, in order to know what is wrong with the present system we have to know something about it. What strikes us as obscure at first blush, may, upon better acquaintance, turn out to be perfectly satisfactory. Any student of physics knows that this is often the case. Even Carnap says that we must, at the outset 'clarify the *explicandum*',[7] although he is, admittedly, in no great hurry to practice what he preaches. We cannot identify the real deficiencies in such a particular model unless we know what counts as a model for scientists, and what purpose they are meant to serve. It may turn out that there are several different concepts of model in use. All could be vague and inconsistent, but it is equally possible that they are all free of internal contradictions. No amount of armchair philosophy can settle this: unless we look at the actual models we will never know. Just as listening to the engine of a motor car idling is not quite the same thing as watching it perform on the motorway, so looking at the definitions of a model is no substitute for seeing it at work within a given system. But if only an examination of the models as they are used in the sciences will tell us what is wrong with them, *a fortiori* we can never know whether the model of the 'Rationality-minded' avoids these difficulties. They may reply that what counts is the demonstrable logic and consistency of the replacement, but this would leave us again with a merely subjective reason for abandoning the current use. Hence, if he is to get us to follow him, the "Rationality-minded" must at some point show us the flaw in the old theory and this may involve some appeal, however covert, to the history of sciences. A completely a-historical philosophy of science is a myth.

But general considerations, like pious wishes, have little value in settling an issue that concerns particular instances. The usefulness, indeed the necessity, of paying attention to the history of concepts used in the sciences can be illustrated with the aid of a few famous cases. I shall examine two: The Galilean principle of inertia and Newton's second law of motion, before I comment briefly on Newton's use of 'hypotheses', and on some translations of Galileo.

Gelileo stands in an uncertain position in the light of recent scholarship. Is he to be considered as the first of the moderns or the last of a long lineage of medieval and Renaissance natural philosophers? Until fairly recently, the first interpretation seemed not only the most plausible but the only one that deserved consideration. Ernst Mach taught that Galileo, virtually single-handed, founded the science of dynamics, created the modern notions of force and acceleration and discovered "the so-called law of inertia, according to which a body not under the influence of forces, i.e., of special circumstances that change motion, will retain forever its velocity (and direction)"[8]

It is indeed surprising that a gifted scientist and philosopher like Mach should have been prey to such a simplistic historical fallacy as to assume that the past was essentially different from the present. To make Galileo the *fons et origo* of modern science is a dangerous anachronism, especially when it leads to the interpretation of his methodology (and hence his philosophy of science) as the key that unlocked the door to all the marvels of modern physics. The Galileo revealed to us by historical research was a creative scientist not a *deus ex machina*. His greatness does not consist in the mere good luck that he was born before anyone else had invented the science of dynamics, but that he was alone among his contemporaries to weave the strands of ancient and medieval science in a novel and daring pattern. Galileo did not merely pick up ideas which were in the air and produce what is sometimes called a 'synthesis' of them; he transformed these ideas and imbued them with new meaning. He went further than anyone of his generation in creating a new science of motion, but it is not to do him honour to attribute to him achievements that were beyond his ken. The principle of inertia is a case in point. The principle itself is of interest to both the historian and the philosopher of science because it shows that the new physics of the seventeenth century entailed a radical departure from the traditional concepts of motion and rest since Aristotle. As Alexandre Koyré has so brilliantly shown, the realization that uniform rectilinear motion was a state and not a process was as revolutionary in its implications as it was baffling to common sense. It should come as no surprise that it did not spring Minerva-like from a single scientific head, however great.[9]

The word 'inertia' in its technical sense was not introduced by Galileo but by Kepler who conceived matter as characterized by an 'inertness', a 'sluggishness', an 'inclinatio ad quietem'. Thus any sample of matter will stay at rest or come to rest if it is not actually moved by a mover or a motor. An important consequence of this view is that a body will come to rest not only *whenever* but also *wherever* a motive force stops acting on it. Hence the entire Aristotelian scheme of natural places is set aside. Without a motive force, the planets would come to rest, and Kepler argues that this force must originate from the sun itself.

If we turn to Galileo, we find in his *Discorsi e dimostrazioni matematiche intorno a due nuove science* of 1638 the following statements: "Attendere insuper licet, quod velocitatis gradus, quicunque in mobili reperiatur, est in illo suapte natura indelebiliter impressus, dum externae causae accelerationis aut retardationis tollantur, quod in solo horizontale contingit";[10] "Mobile quoddam super planum horizontale proiectum mente concipio, omni secluso impedimento; iam constat, ex his quae fusius alibi dicta sunt, illius motum aequabilem et perpetuum super ipso plano futurum esse, si planum in infinitum extendatur".[11]

Mach apparently took these statements to be identical with the Newtonian law of inertia. But, on closer inspection, one sees that Galileo had not travelled that far. In Galileo's physics (and this is no longer the case in Newton's) a body has to be supported on a 'horizontal plane'. In the real physical world, as opposed to the mathematical world of pure thought, there is no infinitely extended horizontal plane. Indeed, motion in a straight line is merely transitory: what endures is perfect circular motion. The limitation of a supporting plane and the implied condition that all horizontal planes are in fact circular characterize Galileo's perspective. The Newtonian analysis of planetary motion as compounded of a linear inertial component and a descent toward the centre is absent from his writings. His study of motion, as is clear from the quotations above, is limited to an elevated physical plane. Galileo's position becomes even clearer when we consider another passage of the *Discorsi* where he appears to state the principle of inertia. He observes that with respect to the Earth's surface, "ben possiamo prendere un minuto di un grado del cerchio massimo come se fusse una linea retta",[12] and he refers to the practice of Archimedes, who "piglia come principio vero, l'ago della bilancia o stadera essere una linea retta in ogni suo punto equalmente distante dal centro commune de i gravi, e le corde alle quali sono appesi i gravi esser tra di loro parallele". The reason is that "nel nostre pratiche gli strumenti nostri e le distanze le quali vengono da noi adoperate, son cosi piccole in comparazione della nostra gran lontananza dal centro del globo terrestre".[13]

Galileo never broke loose from the bondage of circularity since it allowed him to claim that bodies on a rotating earth would behave exactly like bodies on a stationary one because their circular motion was natural. In his *Letters on the Sunspots* of 1613, he wrote: "mi par di osservare che i corpi naturali abbino naturale inclinazione a qualche moto . . . a qualche altro movimento hanno repugnanza . . ad alcuni movimenti si trovano indifferenti". He concluded: "e però, rimossi tutti gl'impedimenti esterni, un grave nella superficia sferica e concentrica alla Terra sarà indifferente alla quiete ed a i movimienti verso qualunque parte dell'orizonte, ed in quello stato si conserverà nel qual una volta sarà stato posto; cioè se sarà messo in stato di quiete, quello conserverà, e se sarà posto in movimento, v.g. verso occidente, nell'istesso si manterrà".[14]

The statement that motion is a state sounds daring since motion had traditionally been considered a process that required a motive force, but it is clear in the light of our foregoing discussion that Galileo was not flying in the face of tradition but merely restating the common belief in the perennial nature of uniform circular motion.

Descartes may have come across the concept of a *state* of motion in Galileo but he endowed the concept with a richness that is altogether implicit in Galileo's use of the word. What is merely a passing remark becomes a general principle of nature, the "prima lex naturae" that governs the whole interpretation of the physical world. Descartes adds, in a "second law of nature", that for motion to be a state it must necessarily be not only uniform but also rectilinear. This second and separate law of rectilinearity emerges from a radical transformation of Galileo's own example of what he conceived to be a state of motion: "e così una nave, per essempio, avendo una sol volta ricevuto qualche impeto per il mar tranquillo, si moverebbe continuamente intorno al nostro globo senza cessar mai".[15] Galileo made no distinction between a plane and a portion of an enormous spherical surface as large as the earth. The profound difference between the Galilean and the Cartesian concept is brought home when we look at the work of a gifted philosopher of nature, John Wallis, who shares the ambiguity of Galileo's position. In his *Mechanica* of 1670, Wallis repeatedly introduces apparently linear terrestrial inertial motion only to specify that it holds so long as it is understood that the horizontal plane is coincident with the surface of a sphere concentric with the earth.

Newton probably encountered the concept of a state of motion in Descartes' *Principia Philosophiae*. The continuing dialogue with Descartes can be seen in the very title of Newton's major work where the *Principia Philosophiae* becomes the *Philosophiae Naturalis Principia Mathematica*. The transformation of the title is but a sign of the profound change that the Cartesian law of inertia underwent in Newton's hands. Descartes did not clearly grasp that the conservation of motion depends on a vector concept of motion and he found it necessary to introduce a second law of nature to state that all motion, in and of itself, is rectilinear. Descartes also needed a metaphysical underpinning and he explicitly derives the conservation of motion from the immutability of God. Clearly a significant change had to occur to get Newton's single law of inertia that follows directly from the nature of matter and is not a direct consequence of the attributes of the Creator.

There is a continuous and intelligible path from Galileo to Newton, but we must not commit the fallacy of ascribing to the earlier scientist insights that were arrived at by working out implications of his assertions that he himself did not even perceive let alone contemplate or analyse. Major transformations are required to convert Galileo's concepts into Newton's first law of motion, for inertia must be: (1) recognized as a fundamental law of nature, (2) seen to imply rectilinearity, (3) generalized from motion on the earth to any unsupported motion in empty space, and (4) associated with mass as quantity of matter. The first three steps were taken by Descartes, the fourth by Newton alone. The first two are the most significant since they indicate the real difference between reading Galileo before 1687 and after that date. For once Newton had published his *Principia* with its plain statement of the principle of inertia, and had declared that Galileo had known and used this law, no one could ever again read Galileo without seeing Newtonian implications or at least Newtonian possibilities. But to go on and say that, after Newton, the gropings of his predecessors become irrelevant is to be beguiled into considering the problem at an inappropriate level of abstraction, indeed, it is to assume, naively, that the Newtonian synthesis is not only a privileged but a definitive moment in our interpretation of nature.

Galileo's example shows us how far from obvious or self-evident the law of inertia actually is. It came into being after a long gestation and as the result of the transformation of an idea thrown out by Galileo, elaborated by Descartes, and modified by Newton. Hence, any philosophy of science that rests on an easy and absolute identification of 'Newton's physics' and 'Galileo's physics' can only be a gross misrepresentation of the true nature of scientific enquiry.

But even Newton's physics can be anachronistically interpreted, and my second example of the relevance of history of science for the philosophy of science is Newton's second law of motion, familiar in the form $F = ma$ of our physics textbooks. If one turns to the *Principia* however, one finds neither this notation nor its equivalent formulation. The second law is stated expressly by Newton as follows: "Mutationem motus proportionalem esse vi motrici impressae et fieri secundum lineam rectam qua vis illa imprimitur". By force (*vis*) Newton has in mind what we would call *impulse*, and he tells us that the "alteration of motion", measured by the product of mass and velocity is proportional to such a 'force'. He then explains: "Si vis aliqua motum quemvis generet; dupla duplum, tripla triplum generabit, sive simul et semel, sive gradatim et successive impressa fuerit".[16] How shall we reconcile this statement apparently so straightforward with the familiar $F = ma$?. There is no obvious way of doing this. Only a study of Newton's actual procedure can shed light on the problem.

In our modern notation, impulse is expressed as $F.\Delta t$ whereby the time

t in which the force acts is made explicit. This is done because we distinguish between impulse and force. But Newton did not conceive force as primary, and impulse as a derived concept obtained by multiplying the force by the time interval in which it acts. In his discussion of the second law, he says that a 'force' (or 'impulse') will produce the same change in momentum whether it acts all at once or gradually and successively. The end result will be the same whether a single impulse or blow is applied or whether the single impulse is broken up into a succession of lesser ones. In the limit, as the interval between blows decrease, we will have a continuous force acting. Thus, Newton uses both mv and ma as a measure of force. Where the distinction of force and total force (F and $\int Fdt$) implies dimensional incompatibility to us, the "force" common to both implied identity to Newton. Hence, the danger of identifying more recent and clearer formulations of the second law with Newton's own understanding of it.

An equally cautionary tale is to be found in Newton's celebrated "*Hypotheses non fingo*" that appeared in the General Scholium that Newton wrote for the second edition of the Principia of 1713. Newton makes the point: "phaenomena caelorum et maris nostri per vim gravitatis exposui, sed causam gravitatis nondum assignavi". He then says: "Rationem vero harum gravitatis proprietatum ex phaenomenis nondum potui deducere, et hypotheses non fingo." He proceeds to explain that "Quicquid enim ex phenomenis non deducitur, hypothesis vocanda est; et hypotheses seu metaphysicae, seu physicae, seu qualitatum occultarum, seu mechanicae, in philosophia experimentali locum non habent".[17]

Attempts to elucidate this passage illustrate the different approaches of historians and philosophers of science. A well-known philosopher, William Kneale offers, in his *Probability and Induction*, an explanation of the passage as it stands. He notices "that Newton speaks in a very curious way of *deducing* propositions from phenomena. This expression occurs in other places, and we must assume that Newton uses it deliberately; but it obviously cannot mean what is ordinarily called deduction, and I can only conclude that Newton meant that the propositions which interested him were derived from observations in a very strict way. Apart, however, from the peculiarity of its phraseology, the passage is fairly clear."[18] While acknowledging the linguistic (?) clarity of the passage, Kneale finds the doctrine "very puzzling", and he argues that it does not square with Newton's practice. The law of universal gravitation could not conceivably be established by the discovery in experience of instances falling under it, and Kneale leaves us with the melancholy suggestion "that for Newton the word 'hypothesis' had some of the unfortunate associations which the word 'speculation' now has for natural scientists".[19]

I. B. Cohen, one of the leading Newtonian scholars, approaches the problem in a different fashion. First, he observes that what Newton says about 'hypotheses' must be taken with great caution since Book II of the *Principia* contains a plainly labelled 'Hypothesis' in every edition Secondly, the second edition (1713), Book III, to which is appended the General scholium, contains two hypotheses (Hypothesis I and Hypothesis II). Thirdly, in the first edition (1687) in which there was no General Scholium as a conclusion, we find a whole series of hypotheses, labelled 'Hypothesis I', 'Hypothesis II', and so on. In the second edition, in which the General Scholium with its 'Hypotheses non fingo' was inserted, some of these hypotheses were rechristened 'Regulae philosophandi' while others were called 'Phaenomena'.[20]

The philosopher of science, whose main concern is logical analysis, is struck at once by the curious way in which Newton speaks of *deducing* propositions from phenomena, and he spontaneously addresses himself to Newton's unusual logical procedure. The historian of science could have saved the philosopher much trouble, or at least protected him from unwarranted speculation, by suggesting that he should look carefully at the meaning of 'Phaenomena' since what were 'hypotheses' in the first edition of the *Principia* become 'phaenomena' in the second.

Another illustration of the importance of attending to the actual words of a scientist is to be found in the standard English translation of Galileo's *Discorsi*, which would have him say that he "discovered *by experiment* some properties of motion which are worth knowing and which have not hitherto been either observed or demonstrated."[21]

The words 'by experiment', as Alexandre Koyré pointed out, are absent from the original text, and were added by Crew and de Salvio who obviously felt, like good empiricists, that they merely made explicit what Galileo would have spelled out had he been writing for a twentieth century audience. The result, of course, is to alter the very thrust of Galileo's argument.[22]

This attempt – conscious or unconscious – to mould Galileo into the image of the experimental scientist began in his own century. This can be seen in a passage of the *Dialogo* in Thomas Salusbury's translation of 1667. The context is the discussion concerning the path of an object let fall from the mast of a moving ship. Salviati, Galileo's spokesman, leads the Aristotelian Simplicio to argue that this experiment can be useful in determining whether a ship is moving or not. "Voi dite: Perchè, quando la nave sta ferma, il sasso cade al piè dell'albero, e quando ell'è in moto cade lontano dal piede, adunque, per il converso, dal cadere il sasso al piede si inferisce la nave star ferma, e dal caderne lontano s'argumenta la nave muoversi". The significance of the 'experiment' is its applicability to the larger problem of the motion of the earth: "perchè quello che occorre della nave deve parimente accader della Terra". Simplicio himself has not performed the experiemtn aboard a ship, "ma ben credo che quelli autori che la producono, l'abbiano dilligentemente osservata". Salviati makes little of this appeal to authority, and claims that "chiunque la farà, troverà l'experianza mostrar tutto 'l contrario di quel che viene scritto: cioè mostrerà che la pietra casca sempre nel medesimo luogo della nave, stia ella ferma o muovasi con qualsivoglia velocità". When Salviati admits, however, that he has not made the experiment, Simplicio retorts: "Che dunque voi non n'avete fatte cento, non che una prova, e l'affermate cosi francamente per sicura? Io ritorno nella mia incredulità, e nella medesima sicurezza che l'esperienza sia stata fatta da gli autori principali che se ne servono, a che ella mostri quel che essi affermano". Salviati's immediate reply is translated as follows by Salusbury: "I am assured that the effect will ensue as I tell you; for it is necessary that it should". The original Italian reads: "Io *senza esperienza* son sicuro che l'effetto seguirà come vi dico".[23] Salusbury, perhaps unwittingly, left out the crucial "senze esperienze"!

I. B. Cohen, who first spotted this interesting omission, points out that Salusbury, writing at the time of the founding of the Royal Society, saw Galileo as a scientist for whom only experience counted.[24] Two and a half centuries later Crew and de Salvio, implicitly subscribing to the fashionable positivistic interpretation of correct science, made Galileo think as they believed he must have. Such attempts to impose an empiricist philosophy on great scientists is not an uncommon occurrence. Dalton is usually credited with painstaking precision whereas, in fact, his atomic theory was made possible by the adjustments of certain experimental results.[25] Einstein is sometimes said to have devised his Special Theory of Relativity to explain the Michelson-Morley experiment when there is no mention of this experiment in his original paper, which is primarily concerned with purely theoretical considerations arising from the classical electromagnetic theory.[26]

THE USES OF PHILOSOPHY

If philosophers of science can learn something from historians of science, at least to the extent of being told where their quarry might be found, historians can also derive profit from a continuing dialogue with philosophers of science. I wish to suggest three ways in which the philosopher of science may prove useful to the historian, namely by providing an interpretative framework, by analysing arguments and elucidating concepts, and by clarifying the meaning of theories.

If history is to rise above a mere collection of anecdotes, it must be written from some point of view and with some unifying theme. It is here that the philosopher has a contribution to make by supplying either a general philosophical stance, for instance, positivism of the kind found in Mach, or some special and distinctive perspective, such as Kuhn's view about paradigms, normal science and revolutions. The two examples of the falsification of Galileo's texts by well-meaning translators discussed above reflect a philosophical aspect of the distortion of history, and make it abundantly clear that no one can completely escape the climate of intellectual opinion prevalent in his day. Unfortunately, historians only too often employ such frameworks without thinking about them. They are left to operate as tacit assumptions, and are dangerous because they are not drawn out into the open and scrutinized for what they really are.

No one would gainsay that it is of intrinsic interest to discover whether an argument actually employed by a scientist of the past is cogent, but some might deny that this is history of science. The historian, it could be said, should worry over what the argument is, not over whether it is any good. But this would be a narrow and ultimately stultifying approach. One of the most interesting questions in intellectual history is the determination of the value of arguments at the time they were formulated. It is a taks that requires the skills of both the philosopher and the historian of science, since we have to assess not only the validity of the logical procedure but the nature of the evidence at hand. In this domain philosophical analysis can beautifully complement the historian's craft.

But aid and comfort can be furnished in yet another way. The philosopher can be useful to the historian by providing a careful analysis not only of the arguments but of the concepts used in the sciences.

This raises an important issue. It is sometimes claimed that historians of science should immerse themselves in the writings of scientists of previous ages, and this is indeed necessary if they are to understand what they were actually up to. But it would be a futile exercise if their program of total immersion led them to lose their bearings in the world in which they actually live. Immersion is only profitable if it leads to eventual emergence into the contemporary setting with an enhanced ability to translate the past into terms meaningful for a present-day audience. The historian aims at recapturing the past not in order to live in the past but in order to interpret it to those who cannot read its lesson at first hand. Were the historian to divest himself of his twentieth century frame of reference to the point of acquiring the full panoply of, say, Aristotelian thought, he would no longer be an historian but a living intellectual fossil.

Any effort to reconstruct the past must be accompanied by a critical examination of what, in the light of hindsight, we know to have actually been the case. For instance, in investigating the models of Maxwell, Kelvin, FitzGerald, Helmholtz and others, it is important to recognize the nature and thrust of the methodological assumptions that guided nineteenth century physicists. In his paper "On Physical Lines of Force", published in 1861, Maxwell proposed a model of the electromagnetic field with the aid of certain assumptions, for example, that electromagnetic phenomena are due to the existence of matter under certain conditions of motion or pressure in every part of the magnetic field and not to action at a distance, and that there is inequality of pressure in a magnetic field which is produced by the existence of vortices. What he does not discuss is the ontological status of these assumptions, in plainer words, the reality that he ascribed to them. Was he saying that the electromagnetic field was really composed of the elements he described? Was he merely drawing an analogy with a mechanical system? Or, rather, was he showing what the electromagnetic field would be like if it operated on purely mechanical principles without claiming that this was necessarily the case?

One need only raise these questions to realize that they are important if we are to understand what Maxwell was actually doing. The philosopher of science may be in a position to help the historian become aware of the various ways in which the term 'model' is used. Peter Achinstein proposes to distinguish between three kinds of models that are frequently used in sciences. Following his suggestive terminology, I shall refer to them as representational, theoretical, and imaginary models.[27]

Representational models are three-dimensional physical representations of objects such that, by considering them, we are able to know some facts about the original. The simplest kinds of these models are tinkertoys models of a molecule or solar systems found in museums. They may be bigger or smaller than the original. They may also represent only those characteristics that a scientist is interested in. In this case, they may serve as an analogue for the original as, for instance, when Maxwell represented the electric field by describing an imaginary incompressible fluid flowing through tubes of variable section. The analogous properties here are electrostatic force and the velocity of the imaginary fluid, which both vary as the square of the distance from their sources, and the potential of the electric field and the pressure of the fluid, both inversely proportional to the distance. But whether the properties of the model are reproduction or analogous of properties of the original, a model in the present sense is an object distinct from the one it represents. This is not the case of the next category.

Theoretical models, like the billiard-ball model of a gas, Bohr's model of the atom, the corpuscular model of light or the shell model of the atomic nucleus, do not refer to a physical object that is distinct from the one of which it is a model, but to a set of assumptions about the object that is itself under scrutiny. For instance, the billiard-ball model is a set of assumptions according to which molecules in a gas exert only contact forces on one another, travel in straight lines except at the instant of collision, are small in size compared to average molecular distances, and so on. These theoretical models can be further characterized. First, they describe an object or system by attributing to it an inner structure or a mechanism that is intended to account for certain features of the object or system. In the case of the billiard-ball model, a molecular structure is ascribed to gases in order to explain observed relationships of pressure, volume, temperature, entropy, etc., of gases. Second, they are treated as useful approximations not exhaustive explanations. The billiard-ball model assumes that the only intramolecular forces are contact forces and thus ignores non-contact attractive and repulsive forces. This is useful in allowing a number of important relationships to be derived and in suggesting how the kinetic theory might be expanded. Thirdly, a theoretical model is set in the broader context of a more comprehensive theory. In the billiard-ball model, the behaviour of the molecules all obey Newton's laws.

The third group of models, imaginary ones, refer to a set of assumptions about a system which are supposed to show what the system would be like if it were to satisfy certain conditions but for which no factual claims are made. An example is Poincaré's model of a non-Euclidean world in which a number of assumptions are made such as that the temperature is greatest at the centre and gradually decreases as one moves towards the circumference where it is absolute zero, that bodies contract as they recede from the centre and, as they move, achieve instant thermal equilibrium with their environment. This model satisfies the postulates of Lobachevskian geometry but Poincaré does not claim that such a physical world exists or that if a Lobachevskian world occurred it would necessarily be the one he describes. Such imaginary models serve the purpose of showing that certain suppositions that may otherwise be thought self-contradictory, e.g. postulates of non-Euclidean geometry, are at least consistent.

Armed with these distinctions, the historian of science can probe deeper into the status of Maxwell's mechanical assumptions. Until this is known it will be impossible to proceed to the analysis of Maxwell's arguments. It is crucial to know whether Maxwell was actually ascribing the mechanical structure he described to the electromagnetic field or whether this was simply intended as a description of an analogue or as a description of a possible mechanism if the field were purely mechanical. Unless we know what his model was intended to do, there is no way we can assess the validity of his reasoning.

One must guard against the danger of anachronism. Nineteenth-century physicists did not explicitly distinguish three uses of the term model, but this does not mean that we cannot derive enlightenment from looking at their work with clearer concepts. If we consider the various models that were proposed in the nineteenth-century, we readily see that it helps to bear these distinctions in mind. It is reasonably clear that in his 1861 paper Maxwell was proposing an imaginary model of the electromagnetic field. He was saying what this field could be like if it were purely mechanical, but he was not claiming that it is actually like this or even that it is purely mechanical. In the case of Kelvin's celebrated mechanical contraptions what was being offered were representational models, while FitzGerald's proposition of 1899 according to which aether is a fluid can be classified as a theoretical model.

Finally, to the clarification of the concepts used by scientists, the philosopher of science can add a modest attempt to rescue the historian from much contemporary fuzzy-thinking about the *meaning* of theories. Although Kuhn and Feyerabend were right in stressing that the framework of a given hypothesis (the paradigm or disciplinary matrix) determines to a large extent what questions can be raised and what views can be suggested about a particular problem, they failed to explain how different theories can be contrasted and appraised. On their view, one is practically driven to describe scientific change in revolutionary terms, to speak, for instance, of the overthrow of Aristotelian mechanics or of the victory of phlogiston. As a result, theories become "incommensurable" and change can no longer be rendered intelligible in rational terms. This relativism is not, however, the outcome of an investigation of actual science and its history, but merely a logical consequence of a narrow presupposition about the *meaning* of scientific terms. Kuhn and Feyerabend held that if the terms do not retain precisely the same meaning over the history of their incorporation into more general theories, then these theories cannot be compared, and the similarities they exhibit must be considered at the best as superficial, and at the worst as deceptive and misleading.

This claim rests on the assumption that two expressions or sets of expression must either have exactly the same meaning or must be completely different. The only possibility left open by this rigid dichotomy of meanings is that history of science, since it is not a process of development by accumulation, must be a completely non-cumulative process of replacement.

The inherent weakness of this position turns out to be its retention of a positivistic concept of meaning. If anything, the revolution is not radical enough! Kuhn and Feyerabend, in spite of their spirited attack on the positivist view that theories are parasitic on 'observations', nevertheless approach their problems with that distinction in mind. They have applied the old classification to a new purpose rather than invented new ways of dealing with old problems. They have merely inverted the respective roles of the two members of the classical distinction: it is now the "theory" that determines the meaning and acceptability of the "observation", rather than the other way round. Observations are now so imbedded in a particular theory that they lose any identity of their own, and they cease to be comparable. But this does not solve the problem of meaning: it simply replaces the theory of meaning invariance with the doctrine of incommensurable meanings. An alternative is to consider meanings as similar or analogous: comparable in some respects while differing in others. For instance, the medieval theory of impetus can be compared with Aristotelian mechanics on the one hand, and with the principle of inertia on the other. The impetus theory shares with Aristotelian physics the view that a constant force is needed to keep a body in motion, but it tends toward the modern view by making this force internal and incorporeal rather than external and corporeal. It can be argued that the impetus theory encouraged a fresh approach to traditional problems by removing long-standing conceptual barriers. The Aristotelians rejected outright any suggestion that the earth rotated because the physical assumptions of their system entailed that a strong wind would be set up in the direction opposite to the earth's motion. But the impetus theory made it possible to entertain the idea that the earth might move by affirming that the air would receive an impetus and would be carried around as though nothing has happened. Likewise, by internalizing the cause of motion it shifted attention to a new set of possibilities. Since air was no longer the force producing motion but merely an impediment, motion in a vacuum ceased to be impossible, and the ground was cleared for thinking about the idealized case of a body moving in the absence of impeding forces. Furthermore, by treating all cases of motion, whether terrestrial or celestial, natural or constrained, in terms of one kind of cause, namely, impetus, it paved the way for a unified account of all motion and provided an alternative approach to the traditional Aristotelian division.

The difficulty in this interpretation lies in the concept of similarity or degrees of likeness of meanings. It is here that historians and philosophers of science can work together to ascertain both the historical relevance and the epistemological significance of shifts in meaning. The papers contained in this volume offer a rich variety of historical and conceptual analyses of important developments in the history of natural philosophy and science. They illustrate that the historians and the philosophers of science can indeed complement one another.

McGill University

NOTES

[1] Robert E. Butts, 'Methodology and the Functional Identity of Science and Philosophy', in J. Hintikka, D. Gruender and E. Agassi, *Pisa Conference Proceedings*, D. Reidel Publ. Co., Dordrecht, Holland, 1980, Vol. II, pp. 253–270.
[2] I. Bernard Cohen, 'History and the Philosophy of Science', in Frederick Suppe (ed.), *The Structure of Scientific Theories*, University of Illinois Press, Urbana, 1974, p. 310, n. 10.
[3] See, for instance, I. L. Lakatos and A. Musgrave (ed.), *Criticism and the Growth of Knowledge*, Cambridge University Press, Cambridge, 1970.
[4] Rudolf Carnap, *Logical Foundations of Probability*, second edition, University of Chicago Press, Chicago, 1962, pp. 3 ff.
[5] See, for instance, J. C. C. McKinsey, A. C. Sugar, and Patrick Suppes, 'Axiomatic Foundations of Classical Mechanics', *Journal of Rational Mechanics and Analysis* 2 (1953), pp. 253–272.
[6] Paul A. Schilpp (ed.), *The Philosophy of Rudolf Carnap*, Open Court, La Salle, Ill., 1963, p. 935.
[7] Rudolf Carnap, *Logical Foundations of Probability*, Chicago University Press, Chicago, 1962[2], p. 3.
[8] Ernst Mach, *The Science of Mechanics*, trans. by Thomas J. McCormack, Open Court, La Salle, Ill., 1960, p. 169.
[9] Alexandre Koyré, *Galileo Studies*, translated by John Mepham, Humanities Press, New Jersey, 1978, pp. 129–236.
[10] Galileo, *Discorsi e dimostrazioni matematiche intorno a due nuove scienze* in *Le Opere di Galileo Galilei*, A Favaro (ed.), 20 vols. G. Barbèra, Florence, 1890–1909. Vol. VIII, p. 243.
[11] *Ibid.*, p. 268.
[12] *Ibid.*, p. 275.
[13] *Ibid.*, p. 274–275.
[14] Galileo *Istoria e dimostrazioni intorno alle macchie solari* in *Opere*, Vol. V, p. 134.

[15] *Ibid.*, pp. 134–135.

[16] Isaac Newton, ***Philosophiae Naturalis Principia Mathematica***, A. Koyré and I. B. Cohen (eds.), 2 vols. Harvard University Press, Cambridge, Mass., 1972, Vol. I, p. 54. See I. B. Cohen, 'Newton's Second Law and the Concept of Force in the *Principia*', *Texas Quarterly* 10 (1967), 125–157.

[17] *Ibid.*, Vol. II, p. 764.

[18] William Kneale, *Probability and Induction*, Clarendon Press, Oxford, 1966, p. 99.

[19] *Ibid.*, p. 100.

[20] See I. B. Cohen's excellent essay 'Hypotheses in Newton's Philosophy', *Physis* 8 (1977), pp. 163–184.

[21] Galileo, *Two New Sciences*, trans. by Henry Grew and Alfonso de Salvio. New York: Dover, no date, p. 213. The original reads: "symptomatum tamen, quae complura et scitu digna insunt in eo, aduc inobservata, necdum indemonstrata, comperio" (*Le Opere di Galileo Galilei*, Vol. VIII, p. 190).

[22] Alexandre Koyré, 'Traduttore-traditore, à propos de Copernic et de Galilée', *Isis* 34 (1943), pp. 209–210.

[23] Galileo, *Dialogo sopra i due massimi sistemi del mondo* in *Opere*, Vol. VII, pp. 169–171.

[24] I. B. Cohen, 'History and the Philosophy of Science', p. 340.

[25] Arnold Thackray, *John Dalton*, Harvard University Press, Cambridge, 1972.

[26] Gerald Holton, *Thematic Origins of Scientific Throught*, Harvard University Press, Cambridge, 1974, pp. 261–352.

[27] Peter Achinstein, *The Concepts of Science*, The Johns Hopkins Press, Blatimore, 1968, pp. 209–225.

PART I

MAURICE CLAVELIN

CONCEPTUAL AND TECHNICAL ASPECTS OF THE GALILEAN GEOMETRIZATION OF THE MOTION OF HEAVY BODIES

This paper is inspired by a specific conception of the history of science. This latter, like any historical description, must in the first place study facts: establishing when and how a particular result is obtained, discerning heritage and genuine creation and thus preparing an accurate assessment of the successive contributions which gave rise to the major scientific theories – such is clearly its first task. But there also exists another way of looking at the history of science: more concerned with internal logic than with chronological details, it aims above all at bringing out which, at the level of conceptual initiatives, opened the way to the new doctrines, allowed them to take form, and thereby to acquire this systematicity and deep consistency without which no intellectual construction can really endure. So construed – and without neglecting the facts – it mainly focuses on the changes – be they abrupt or progressive – thanks to which an ever more precise representation of reality is built. In other words one will find, in the following pages, neither discussions of unpublished documents, nor chronological conjectures, nor even a comparative study. We begin by an observation: the creation by Galileo, within a few decades, of a mathematical theory dealing with the natural motion of heavy bodies. This theory does not appear in a conceptual void and in no way is free from technical constraints. Galileo finds himself facing a severely weakened but not totally ruined conception which he must initially combat and, later, replace piece by piece. At the same time, he uses a very specific type of mathematics – basically Euclidean geometry – guided (or held back) by requirements which were rendered obsolete by subsequent development in mathematics. Keeping these different aspects in mind, I should like to restore, in their internal logic and originality, some of the initiatives which made Galileo the creator of an entirely new discourse about nature.

FROM MOTION AS A PROCESS TO MOTION AS A STATE

(1) One cannot overemphasize the fact that this motion which Galileo deals with – which he tries to make into the object of a mathematical theory – is from the outset the most familiar physical phenomenon there is: the motion

William R. Shea (ed.), Nature Mathematized, 23–50.

of heavy bodies on the surface of the earth. But, it is also the one – and both are related – appearently explained with maximum depth and clarity by natural philosophy for close to two thousand years before Galileo. According to this interpretation (of which I shall mention only the main points), inherited from Aristotle and his successors, the natural motion downwards is due neither to action at a distance nor to an internal property of matter. What makes a body heavy is the presence of a formal quality precisely termed *gravity* which is characteristic of the element Earth and instilled in it by the center of the World. Associated, in this way, to a specific kind of body, downward motion finds its ultimate explanation in the cosmological order: its role is to bring as close as possible to the center of the World, viz. their natural resting place (which happens to coincide with the center of the Earth), these heavy bodies which, for some reason or another, have moved away from it.

Against this completely finalized view, which makes a mere utopia of a geometrical analysis of natural motion, Galileo reacted very early indeed. From his years in Pisa, he tried to substitute an analysis inspired by Archimedean hydrostatics.[1] From an agent of cosmological reordering, gravity becomes a constituent property of matter, the cause both of its weight and its tendency to downward motion every time the conditions are fulfilled. The lack of any obstacle is of course the first of these conditions: but they also include a situation which can be analysed in terms of density, according to the model given by Archimedes in his treatise *On Floating Bodies*. The idea can be grasped in a few words. Whenever the density of a body is greater than that of the surrounding medium, the body will tend downward (in the absence of obstacles); if, on the other hand, the density of the medium exceeds its density, it will spontaneously tend upward. A formula allows us to sum up this conception. If by W_{sb} we represent the specific weight of any body and by W_{sm} the specific weight of the medium wherein it happens to be, then the new theory advocated by Galileo amounts to claiming that the body's speed will always be proportional to the difference $W_{sb} - W_{sm}$.[2]

(2) Compared with the Aristotelian doctrine, the progress is in no way negligible. By turning gravity, seen as a force inherent to bodies, into the unique agent of their natural motion when they are free to go downward, Galileo loosened the hold of cosmology, and the appeal to Archimedes opened the way to a quantitative analysis. The fact that up to the end of his career, he never ceased to view gravity as a force inherent in bodies, clearly testifies to the liberating role of this first transformation of traditional ideas.

"I say", he still writes in the 1638 *Discourses*, "that a heavy body has from nature an intrinsic principle of moving toward the common center of heavy objects (that is of our terrestrial globe) with a continually accelerated movement".[3] It is clear, however, that most of the work was still to be done. Whatever simplification had been made, we remain in a causal perspective, much more positive indeed, but in no way free from difficulties, nay from contradictions. Though we lack a precise chronology, Galileo probably realized it quickly, for by working out the implications of the new theory it is easy to note that, as far as compatibility with observations is concerned, it is no better than Aristotle's old hypothesis.[4] For suppose that the speeds of naturally falling bodies be actually proportional to their respective specific weights: the rarer the medium, the more visible the differences in speeds should appear, particularly between wood and lead. Now we just find out the opposite: the differences in speeds are much more important in a dense medium (like water) than in a rare medium (like air) in which they tend to vanish.[5]

Along with this clash between prediction and observation, more general reasons must have convinced Galileo that the dynamics of specific weights was no more favourable to the geometrisation than the traditional conception. First, it sheds no light on the peculiarity of the natural motion of heavy bodies, that is, the progressive increase in speed. How is this increase taking place? Is it continuous or discontinuous? Is it proportional to space or to time? So many questions preliminary to a geometrical analysis and about which the theory has nothing to say. That is not all. Incapable of contributing to geometrisation, this concrete dynamics still maintains the major obstacle which, since Aristotle, blocked the application of mathematics to the study of motion. To link directly motion and its modes to the specific gravity of each body is to continue *de facto* to see in it the necessarily transitional effect of a force, and in no way a *sui generis* state which the action of a force can indeed modify but which is nevertheless susceptible of existence independently of that action. Although related to a unique cause, motion thus keeps the status, conferred on it by Aristotle, of a simple process (between a *terminus a quo* and a *terminus ad quem*). Now how can one claim to apply the methods and constructions of geometry, which by definition imply the idea of a stable and unique model, to an object without permanence or independence, entirely reliant on factors which must be reevaluated each time? Only a decisive step by which the motion would at last lose its status of a *process* for that of a *state* could render physically plausible the project of a geometrisation of the natural motion of the heavy bodies.

(3) This step Galileo precisely succeeded in making at the very dawn of the seventeenth century. I will content myself with recalling its essential aspect, emphasizing mainly the fact that Galileo achieved it by exploiting with an extraordinary ingenuity his conception of gravity as a natural motive force inherent to bodies. Let us imagine an inclined plane and suppose that a ball can roll on it without friction.[6] Placed on the surface of this inclined plane, the ball will show a tendency to go down the plane as well as a resistance to any upward traction. Next, imagine that the angle of incline diminishes gradually: the tendency which pushes the ball downward will decrease simultaneously with its resistance in the opposite direction. Now, contrary to natural media such as water or air, the inclined plane does not affect the specific weight of the bodies at all: only the tendency to descend, that is *gravity as a motive force*, is actually modified. A distinction thus becomes necessary between the weight and the natural motive force or, better still, between a properly gravific function and a motor function of weight. The expression 'moment of descent' (*momento del discendere*), which will henceforth designate the greater or lesser downward tendency of a moving body, will fix this distinction and, thus, make of the motor function of weight a conceptually autonomous phenomenon, characterized by its variability, and open to a direct study. This first step being made, let us suppress any incline and let us turn our plane into a "horizontal" surface on which the moving body will neither approach – nor withdraw from – the common center of heavy bodies. On such a surface – a spherical surface as a matter of fact – gravity will not show its motor function anymore, so that the moving body "will remain indifferent between the propensity and the resistance to the motion",[7] one and the other being as it were neutralized. At this very moment – and for the first time in the history of scientific thought – an entirely new problem was raised: how is one to understand the behaviour of a moving body if there is left in it neither tendency nor resistance to motion?

Galileo will answer this question in two successive stages. Is a minimal force, short of which no effect would ensue, necessary to set a body moving on such a plane? The answer is unambiguously given in *Le Meccaniche* (*On Mechanics*), i.e., as early as 1595–97: ". . . it is perfectly clear that on an exactly balanced surface the ball would remain indifferent and questioning between motion and rest, so that even the least force would be sufficient to move it, just as on the other hand any little resistance, such as that merely of the air that surrounds it, would be capable of holding it still".[8] With this in mind, the essential stage can be tackled: if a moving body on a plane neither moves from nor comes closer to the common center of the heavy bodies, what will it do once it has been set into motion by a force, as small as one likes? Will it stop after a certain time or will the absence of any resistance – we also discard that of air – logically entail a quite different conclusion? With admirable intellectual daring, Galileo then introduces in its first formulation the idea of the conservation of acquired motion: on a horizontal plane, i.e., concentric to the common center of the heavy bodies, where all cause of acceleration or deceleration has disappeared, a motion, once started, will go on indefinitely.[9] It is useless to insist here on the conceptual transformation that such an idea implies. The important point for our study is that this motion, of which Galileo has just asserted indefinite conservation, though it constitutes a borderline case of the natural motion of the heavy bodies, also stems from it by a *continuous transition*, and thus corresponds to a *perfectly significant phenomenon from a physical point of view*. Hence, a conceptualization had been set up in which motion, at least when uniform, receives this reality, viz. this full existence as an independent spatio-temporal object, without which its geometrisation could only appear as an unwarranted philosophical speculation. The way to the first geometrical study of motion was prepared for good.

GEOMETRISATION OF UNIFORM MOTION

(4) To geometrize uniform motion is, by definition, to bring to light and expound by means of a geometrical type of analysis its essential properties. Now, to be geometrical – and this is clear since Archimedes – the analysis of a physical problem must fulfil two conditions: first it must proceed deductively according to the model determined by Euclid, and second it must use for its reasonings the very tool of the geometrical proof, viz. the method of proportions. To geometrize uniform motion – once the essential definitions and assumptions are given – is thus tantamount to showing that the method of proportions can be used for its study, or maybe better: that the magnitudes necessary to carry this study to a successful conclusion satisfy the requirements of the method and, in return, can benefit from all its resources. This task is accomplished by Galileo with the help of Theorems I and II which, while formulating the first propositions on uniform motion, also establish the possibility of using the demonstrative tool of geometry in a theory with motion as its object.

Directly inspired by Archimedes, Theorem I first proves the geometrical legitimacy of proportions two terms of which are spaces and the two others times.

If a moveable equably carried [*latum*] with the same speed passes through two spaces, the times of motion will be to one another as the spaces passed through.

Let the moveable equably carried with the same speed pass through two spaces, *AB* and *BC*; and let the time of motion through *AB* be *DE*, while the time of motion through *BC* is *EF*; I say that space *AB* is to space *BC* as time *DE* is to time *EF*. [10]

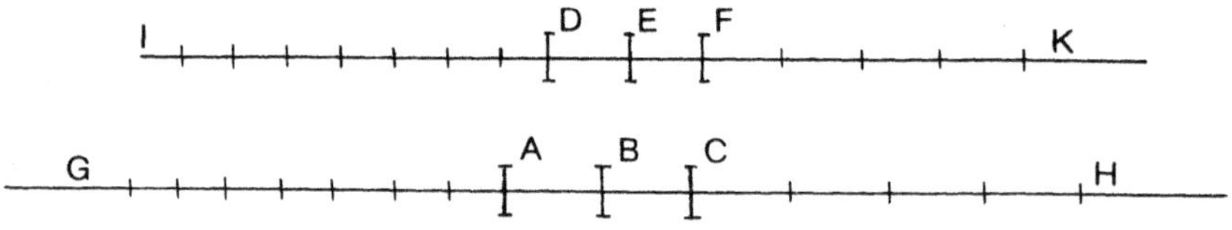

Since time as such is not a geometrical magnitude, this can only mean one thing, in good Euclidean orthodoxy, that the conditions according to which four magnitudes can be said to be proportional are verified in the present case. Definition 5 of Book V of the *Elements* gives a famous exposition of these conditions:

Magnitudes are said to be in the same ratio, the first to the second and the third to the fourth, when, if any equimultiples whatever be taken of the first and third, and any equimultiples whatever of the second and fourth, the former equimultiples alike exceed, are alike equal to, or alike fall short of, the latter equimultiples respectively taken in corresponding order.

Transcribed in modern notation, this definition can easily be put in the following form:

$$\text{(A)} \qquad \frac{a}{b} = \frac{c}{d} \equiv_{\text{def}} (m, n)\ (m, n, \in N)\ [(ma \gtreqless nb) \rightarrow (mc \gtreqless nd)]$$

This formula, a biconditional, allows us to derive:

$$(m, n)\ (m, n, \in N)\ [(ma \gtreqless nb) \rightarrow (mc \gtreqless nd)] \rightarrow \frac{a}{b} = \frac{c}{d}$$

and taking $a = AB$, $b = BC$, $c = DE$, $d = EF$, then

$$\text{(B)} \qquad (m, n)\ (m, n, \in N)\ [(mAB \gtreqless nBC) \rightarrow (mDE \gtreqless nEF)] \rightarrow \frac{AB}{BC} = \frac{DE}{EF}.$$

What Galileo must establish is thus the antecedent of (B) – the consequent then following directly by *modus ponens*. How does he do this?

(5) Though geometry may construct some of its objects from motions, it nevertheless says nothing, as geometry, about motion and so new assumptions are essential if the conditions specified in the antecedent of (B) are to be satisfied. The definition at the beginning of the treatise of uniform motion provides the first of these suppositions. Galileo writes: "Equal or uniform motion I understand to be that of which the parts run through by the moveable in any equal times whatever are equal to one another".[11]

From this definition – noting, especially, the term 'any'[12] – we can easily extract a first relation:

$$(s_1 = s_2) \rightarrow (t_1 = t_2)$$

where s_1 is the space travelled during t_1 and s_2 the space travelled during t_2.

The relation, however, only covers the case of equality appearing in the antecedent of (B), but not the two cases of inequality. Two axioms that Galileo presents as obvious consequences of the definition will fill this gap:

AXIOM I: During the same equable motion, the space completed in a longer time is greater than the space completed in shorter time.[13]

That is, using the same notation as before:

$$(s_1 > s_2) \rightarrow (t_1 > t_2).$$

AXIOM II: The time in which a greater space is traversed in the same equable motion is longer than the time in which a smaller space is traversed.[14]

Or, using the same symbols:

$$(s_1 < s_2) \rightarrow (t_1 < t_2).$$

With these three suppositions (the definition and the two axioms), it becomes possible to show that the conditions stated in definition 5 of Book V of the *Elements* are fully satisfied, and hence, that the four magnitudes,

AB and *BC* on the one hand, and *DE* and *EF* on the other, constitute a system of proportional magnitudes:

Extend the spaces toward *G* and *H*, and the times toward *I* and *K*. In *AG* take any number of spaces [each] equal to *AB*, and in *DI* likewise as many times [each] equal to *DE*. Further, let there be taken in *CH* any multitude of spaces [each] equal to *CB*, and in *FK* that multitude of times [each] equal to *EF*. Space *BG* and time *EI* will now be equimultiples of space *BA* and time *ED* [respectively], according to whatever multiplication was taken. Similarly, space *HB* and time *KE* will be equimultiples of space *CB* and time *EF* in such multiplication. And since *DE* is the time of movement through *AB*, the whole of *EI* will be the time of the whole [space] *BG*, since this motion is assumed equable, and in *EI* there are as many equal times *DE* as there are equal spaces *BA* in *BG*; and similarly it is concluded that *KE* is the time of movement through *HB*. But since the motion is assumed equable, if the space *GB* is equal to *BH*, the time *IE* will be equal to time *EK*, while if *GB* is greater than *BH*, so will *IE* be greater than *EK*; and if less, less. Thus there are four magnitudes, *AB* first, *BC* second, *DE* third, and *EF* fourth; and of the first and third (that is, of space *AB* and time *DE*), equimultiples are taken according to any multiplication, [i.e.] the time *IE* and the space *GB*; and it has been demonstrated that these either both equal, both fall short of, or both exceed the time *EK* and the space *BH*, which are equimultiples of the second and fourth. Therefore the first has to the second (that is, space *AB* has to space *BC*) the same ratio as the third to the fourth (that is, time *DE* to time *EF*); which was to be demonstrated.[15]

With the first proposition of uniform motion, the possibility of reasoning in terms of space and time in conformity with the requirements of geometry is thus justified.

(6) Theorem II will extend this possibility to arguments concerning spaces and speeds. The proposition he establishes is stated as follows:

If a moveable passes through two spaces in equal times, these spaces will be to one another as the speeds. And if the spaces are as the speeds, the times will be equal.

Taking the [previous diagram], let there be two spaces, *AB* and *BC*, completed in equal times, space *AB* with speed *DE* and space *BC* with speed *EF*; I say that space *AB* is to space *BC* as speed *DE* is to speed *EF*.[16]

A modern reader may be somewhat surprised to see Galileo treat speed in the same way he treated space and time, as a prime dimension and, as it were, irreducible for the study of motion. It suffices, nevertheless, to recall the constraints of traditional geometry for any surprise to disappear. Indeed, the result of the lack of correspondence between number and continuous magnitude is that there can be relations only between homogeneous magnitudes – between spaces or between times, but, in all cases, not between spaces and times. Speed, of which the expression would require a ratio of type s/t is, therefore, not mathematically constructible and can only appear directly, in the form of an unanalyzed magnitude, in the theory that Galileo is setting up.[17] How does he succeed in introducing and justifying its usage in a geometric discourse? Such is the major question posed by this second theorem of uniform motion.

Contrary to what one might believe, the answer is not to be found in the mechanism of the demonstration. Indeed, technically speaking, this second proof merely repeats the scheme of the preceding one: to show that the four magnitudes, between which one suggests a certain relation of proportionality (two spaces and two speeds), strictly satisfy the requirements of definition 5 of Book V of the *Elements*. In reality, the crucial point is the manner in which Galileo prepares and renders this demonstration possible. That of Theorem I, as we have seen, depends on the initial definition of uniform motion, completed by two axioms. Two new axioms, directly joined to the first two, will play an identical role for Theorem II.

AXIOM III: The space traversed with greater speed is greater than the space traversed in the same time with lesser speed.[18]

That is, with obvious abbreviations:

$$(s_1 > s_2) \rightarrow (v_1 > v_2).$$

AXIOM IV: The speed with which more space is traversed in the same time is greater than the speed with which less space is traversed.[19]

Or, still with the same abbreviations:

$$(s_1 < s_2) \rightarrow (v_1 < v_2).$$

A simple glance reveals the manner in which Galileo manages to obtain these two axioms: he simply replaces time by speed in the first two axioms. For want of a still impossible mathematical definition, speed thus enters in the wake of time, as if a perfect reciprocity existed between the two, as if what holds for time held *eo ipso* for speed. And it is true that in the case of uniform motion time and speed can be in some way inverted. At equal speed a greater space is covered in a longer period of time, just as, for an equal time, a greater space is covered with a higher speed. This is a relation that everyone directly perceives, and on which Galileo relies to introduce, in the most favourable conditions, a motion vital for the continuation of his analysis. Moreover, it is striking to note that he does not feel the need to explicitly extend to speed the equality relation given by the definition of uniform motion, that is to pass from the formula '$(s_1 = s_2) \rightarrow (t_1 = t_2)$', valid for two distances covered with equal speed, to the formula '$(s_1 = s_2) \rightarrow (v_1 = v_2)$' valid for two distances covered in equal times. Essential to the demonstration of Theorem II, this extension is probably too obvious to be mentioned. So the remarkable situation in which Galileo finds himself to be vis-à-vis speed is now clear. Incapable of constructing it geometrically, he must assume it as a first dimension for the analysis of motion in the same way as space and time; the reciprocity which however relates it to time, in the case of uniform motion, allows him, by a simple intuitively guaranteed substitution, to introduce it in the framework of the argument and thereby to make up for the absence of an adequate mathematical tool.

This being the case, one must insist on the significance of Theorem II which marks the real beginning of the Galilean theory of uniform motion.[20] By establishing the legitimacy of the ratios between speeds, it provides the essential complement to Theorem I. Henceforth we know that the specific magnitudes necessary for the study of motion – time and speed – can be geometrized in the strict sense, that is, subjected directly to all operations that the method of proportions allows. At the same time, the construction of a theory is justified which, without being part of mathematics, is nevertheless able to define and characterize its object in such a way that one can use for its study the method as well as the true propositions of mathematics. If we add that, as far as obviousness is concerned, the definition and axioms by which the geometrical treatment of uniform motion is prepared are in no way second to the assumptions of the *Elements*, we will easily perceive the extent of the Galilean success.

The four following theorems, by the way, produce immediate proof of this. Theorems I and II have barely demonstrated the geometrical validity of the ratios between times and speeds by comparing them to ratios between distances, that Theorem III thoroughly makes the most of this result by putting together times and speeds in a new relation of proportionality: "Of movements through the same space at unequal speeds, the times and speeds are inversely proportional." Finally Theorems IV to VI, taking as settled that times and speeds obey the fundamental criterion of proportionality, apply without the slightest hesitation to the study of uniform motion, one of the most fertile notions of the theory of proportions, namely that of a *ratio composed of two ratios.*[21] True, there was no longer any obstacle to a systematic use, in the new theory of motion, of the various possibilities of geometry.

SECOND NEUTRALIZATION OF GRAVITY

(7) Going from uniform motion to naturally accelerated motion is nevertheless far from being immediate. Not only does the geometrisation of this motion require, from a technical point of view, the intervention of new suppositions, but for the enterprise itself to make sense, new conceptual progress is essential. Stating the principle of conservation on a plane concentric at the common center of the heavy bodies was sufficient to make of uniform motion a stable object, definable as a spatio-temporal entity, to which a geometrical study could apply. Now, important as it may be, this progress does not settle the case of naturally accelerated motion. Such a motion is indeed the direct product of the action of gravity which, acting from within the bodies, drives them as close as possible to the center of the Earth. Impossible to eliminate, gravity apparently also acts in a very different way according to the bodies, since observation reveals notable variations between their respective speeds of fall. The problem is therefore quite clear: if gravity, the intrinsic cause of naturally accelerated motion, shows itself differently according to the bodies' nature, how can one be sure, as is required for its geometrisation, that this motion really takes place *in the same manner for all bodies*? That in spite of the differences between the natural speeds of fall, it always and everywhere obeys the same laws? That therefore in any case a same geometrical model can apply? Only the demonstration that the action of gravity is identically exerted for *all bodies* and is not responsible for the differences in their speeds can supply this certainty and, at the same time, render *physically significant* a geometrical study of naturally accelerated motion.

(8) We had left Galileo (cf. above §2) at the moment when, after having tried to dynamically explain the natural motion of bodies by the excess of their specific gravity over that of the medium,[22] he realized the conflict between this hypothesis and his observations. The passage from Archimedean hydrostatics to naturally accelerated motion proved impossible. At least directly, since, as Galileo finally saw it, the error lay less in the withheld factors (specific gravities of the bodies and media) than in the way to combine them. For instead of the reductions which the specific weights of the

bodies are subjected to in the various media, let us consider the ratio between the specific weight of each of these media and the specific weight of the bodies, that is, abridged, W_{sm}/W_{sb}.[23] This ratio will express in what proportion the specific weight of a given body is reduced by that of the medium, and will characterize the lifting effect of the latter. Let us fix our ideas by saying, e.g., that the density of air is 1, that of water 800, that of wood 600, that of ebony 1000, and that of lead 10 000.[24] We will see immediately that, in air, the ratio for wood will be 1/600, which means that the weight of wood will be reduced by 1/600th. Similarly, the ratio for ebony will be 1/1000, and that of lead, 1/10 000. In water, on the other hand, the ratio will become equal to 800/600 for wood (which signifies a reversal of motion) whereas the respective values of the ratios for ebony and lead in water will be 800/1000 and 800/10 000. The crucial point is then that the variations in these ratios correspond exactly to the variations of behaviour (for wood) and of speed (for ebony and lead) revealed by observation. Deprived of a very small part of their weight in the air (even the most important proportional decrease is negligible in absolute value), wood, ebony, and lead fall with nearly the same speed; in water where ebony and lead are deprived, on the other hand, of parts of their weight which are very different in absolute value (and considerable in the case of ebony), they will fall with very different speeds. Hence, one can conclude without risk of error that the difference in speeds are directly caused, not by the different specific weights of bodies more or less diminished by that of the medium, but by the proportionally unequal ways in which different media exert their lifting effect on the bodies. What the diversification of the natural speeds of fall in a given medium reflects are the different values that the ratio W_{sm}/W_{sb} takes for each body in that medium.

(9) The main interest of this result lies, however, in the reasoning to which it leads. All the preceding argumentation is based on the observation that when the density of the medium diminishes, the differences of speeds diminish correlatively. It is, therefore, normal to ask oneself what would happen if the density of the medium totally disappeared. That is, if one considered, not a full medium insubstantial as it may be, but the void. No empty medium being at our disposal, only reasoning can guide us from now on. What does it teach us?

The ratio W_{sm}/W_{sb} obviously could not be constructed in a medium of null density because O/W_{sb} is meaningless. The lifting effect, in other words, would *eo ipso* disappear. But the differences in speeds in any concrete medium are precisely due to the existence of this lifting effect. In these conditions, what can one conclude except that the differences of speed (smaller when the density of the medium diminishes) would be completely cancelled in the void and that all bodies would consequently fall at the same speed? Galileo takes this step: "This seen, I say, I came to the opinion that if one were to remove entirely the resistance of the medium, all materials would descend with equal speed."[25] Thus, not only no direct relation exists between differences in specific weights and differences in speed: in a medium of specific weight zero, where the specific weight of bodies would become unimportant, all observable differences would disappear between the natural motions of bodies.[26]

Formally speaking, there is a remarkable similarity between this reasoning and that which above led to the conservation of uniform motion on a surface equidistant from the common center of the heavy bodies. In both cases, it is by progressively cancelling the factors responsible either for the deceleration and acceleration or for the difference of speeds that Galileo came to a revolutionary conclusion. The strategy is the same in each case: to introduce by a *transfer to the limit* a conclusion relative to an ideal, but also *physically significant*, situation – thanks to which an essential condition for geometrisation is set up. For to assume that in the void all bodies would fall with the same speed, amounts indeed to assuming that the natural immanent motive force acts in the same way in all bodies;[27] were it free from any obstacles, it would impart to every body an identical natural motion, assuring thereby the physical relevance of a unique geometrical model. Hence, without removing gravity, but by standardizing its action by the roundabout way of the void, Galileo succeeds in bringing to the fore the existence of a *general fact* of the fall of bodies which, beyond the various motions of the different bodies, constitutes the genuine reference of the theory.

GEOMETRISATION OF NATURALLY ACCELERATED MOTION

(10) A rightly admired definition opens the geometrisation of naturally accelerated motion. After having recalled his concern to achieve a successful explanation of the real motion of heavy bodies, and his belief that nature is in the habit of using "only those means which are most common, simple and easy", Galileo lays out his description:

Thus when I consider that a stone, falling from rest at some height, successively acquires new increments of speed, why should I not believe that those additions are made by the simplest and most evident rule? For if we look into this attentively, we can discover no simpler addition and increase than that which is added on always in the same way. We

easily understand that the closest affinity holds between time and motion, and thus equable and uniform motion is defined through uniformities of times and spaces; and indeed, we call movement equable when in equal times equal spaces are traversed. And by this same equality of parts of time, we can perceive the increase of swiftness to be made simply, conceiving mentally that this motion is uniformly and continually accelerated in the same way whenever, in any equal times, equal additions of swiftness are added on.[28]

This text offers first a striking summary of all the difficulties that Galileo had to overcome in order to find a consistent representation of naturally accelerated motion. Is the increase in speed a function of space or time? Is it continuous in the sense that it occurs in any time interval, however small? Is it uniform, repeating itself identically in each successive temporal interval, or is it subjected to variation, becoming, for instance, greater as the time of fall increases? To these questions which had long preoccupied Galileo, and which he will recall in the following pages,[29] the definition brings unequivocal answers, removing once and for all the uncertainties which, since the fourteenth century, had made sterile the reflection on the natural motion of heavy bodies. Moreover, no modern reader can remain insensitive to the way Galileo reaches the threshold of the concept of instantaneous speed. As no significant difference separates the acceleration in the actual sense of the term and this elementary increase in speed which, by repeating itself from instant to instant, ensures the increase of speed, Galileo's analysis leads, 'by hand as it were', to the formula $v = \gamma t$, where v stands for the speed of a body in free fall at any given moment in time. Geometrisation, it would seem, should easily follow from such a good start.

Now this highly penetrative definition – and this point cannot be overemphasized – does not play any role in the subsequent development of Galileo's argument. Should we be surprised by this? We have seen how traditional geometry, on which Galileo is still dependent, forbids the construction of the concept of average speed; *a fortiori* was it unable to provide an adequate expression for the concept of acceleration (or differential of speed). To form the ratio of a space by a time, be they the one and the other as small as one wants them to be, remains by definition meaningless. To make the most of his analysis, Galileo would have had to anticipate fifty years of mathematical development and to be already in possession of the calculus of Leibniz and Newton. Hence, there is no surprise that, unlike the definition of uniform motion which, with its four axioms, directly underlies the proofs of Theorems I and II, the definition of naturally accelerated motion does not occur in any of the theorems or in any of the constructions which will achieve the geometrisation. The conceptual analysis could only remain exterior to the mathematical analysis, incapable of supporting it or of being supported by it. Inspired as it may be, it was bound to remain a simple preliminary elucidation, and it is by totally different means that geometry will be introduced into the study of naturally accelerated motion.

(11) This first point being clarified, let us consider the Third Day of the *Dialogues* in its general organisation. Whoever makes the effort of reading closely the theorems and propositions will find that the demonstrative apparatus is principally based on two relations of proportionality established independently one from the other. The first is worded in these terms: "... if at the beginning of motion there are taken any two spaces whatever, run through in any [two] times, the times will be to each other as either of these two spaces is to the mean proportional space between the two given spaces."[30] As Galileo presents it, however, this proposition is a simple corollary to Theorem II – the time-squared law – which it renders operative. Through it, therefore, it is Theorem II that clearly appears as one of the two pillars of the theory.[31] As for the other relation on which Galileo relies, it is entitled Theorem III and is stated in these terms: "If the same moveable is carried from rest on an inclined plane, and also along a vertical of the same height, the times of the movements will be to one another as the lengths of he plane and of the vertical."[32] As these two relations (from Theorem IV onward) suffice for the expression as well as for the mathematical treatment of the problems relative to naturally accelerated motion, to understand the geometrisation of this motion clearly amounts to understanding how Theorems II and III are introduced. What then is the *way* followed by Galileo in this introduction? What particular *suppositions* is he led to make use of, either explicitly or implicitly? What *reasonings* does he use and what is their worth?

Such are the questions which, through a methodical examination of Theorems II and III, should help us to perceive in its historical singularity the steps by which Galileo, after geometrizing uniform motion, was also able to geometrize naturally accelerated motion.

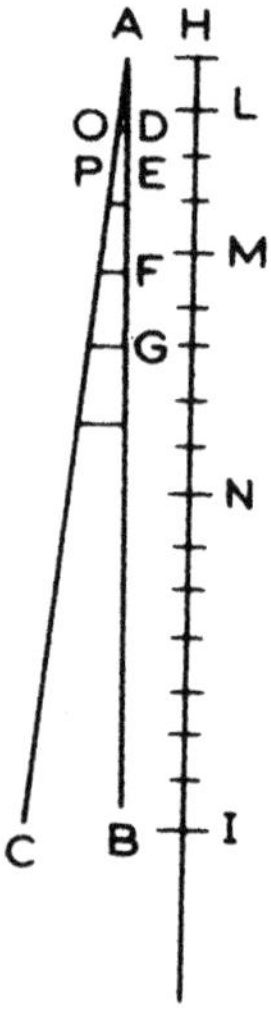

(12) To begin with, let us take Theorem II. Let us represent by *AB* a 'flux of time' beginning at *A*, and by *AD* and *DE* two equal intervals of this time. *HL*, on the line *HI*, will represent the distance a uniformly accelerated body travels during time *AD*, and *HM* that which it travels during time *AE*. Let us then draw the line *AC*, making with *AB* some angle, and from the points *D* and *E*, let us draw toward *AC* the parallel lines *DO* and *EP*: "*DO* will represent the maximum degree of speed acquired at instant *D* of time *AD*, and *PE* the maximum degree of speed acquired at instant *E* of time *AE*". Now it has just been demonstrated (Theorem I) that the spaces covered by two moving bodies are equal when one, starting at rest, moves in uniformly accelerated motion whereas the other, during the same interval of time, moves in uniform motion with a degree of speed which is half that of the greatest speed acquired by the first. It follows immediately, if we designate by V_1 the speed of uniform motion 'corresponding' to the accelerated motion along *HL* and by V_2, the speed of the uniform motion 'corresponding' to the motion accelerated along *HM*, that one can write: $V_1 = DO/2$ and $V_2 = EP/2$, that is, $V_2/V_1 = EP/DO = AE/AD$. But according to Theorem IV of uniform motion, the spaces covered

by moveables moving uniformly but with unequal speeds and in unequal times have the ratio compounded from the ratio of speeds and from the ratio of times – which will be expressed by taking t_1 for time AE, and t_2 for time AD:

$$\frac{\text{space } HM}{\text{space } HL} = \frac{v_2}{v_1} \times \frac{t_2}{t_1}.$$

It suffices, then, to replace v_2/v_1 by the equivalent ratio AE/AD to obtain the time-squared law:

$$\frac{\text{space } HM}{\text{space } HL} = \frac{AE}{AD} \times \frac{AE}{AD} = \frac{AE^2}{AD^2} = \frac{t_2^2}{t_1^2}.$$ [33]

A theorem of uniform motion and a very special "theorem" first put into place: such are the two foundations of Theorem II. Since the first causes no problem, it is on the second (Theorem I of naturally accelerated motion) that actually depends the possibility of formulating the first essential relation for the geometrisation of naturally accelerated motion. How and at what price?

(13) To settle the question, let us begin by reading, in a literal translation, the text and demonstration of this Theorem I.[34]

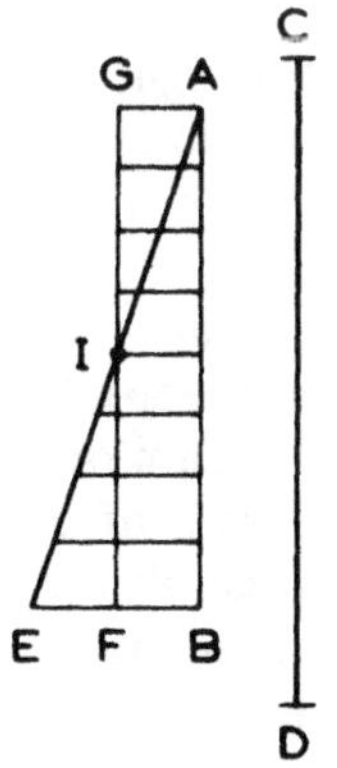

The time in which a certain space is traversed by a moveable in uniformly accelerated movement from rest is equal to the time in which the same space would be traversed by the same moveable carried in uniform motion whose degree of speed is one-half the maximum and final degree of speed of the previous, uniformly accelerated, motion.

Let line *AB* represent the time in which the space *CD* is traversed by a moveable in uniformly accelerated movement from rest at *C*. Let *EB*, drawn in any way upon *AB*, represent the maximum and final degree of speed increased in the instants of the time *AB*. All the lines reaching *AE* from single points of the line *AB* and drawn parallel to *BE* will represent the increasing degrees of speed after the instant *A*. Next, I bisect *BE* at *F*, and I draw *FG* and *AG* parallel to *BA* and *BF*; the parallelogram *AGFB* will [thus] be constructed, equal to the triangle *AEB*, its side *GF* bisecting *AE* at *I*.

Now if the parallels in triangle *AEB* are extended as far as *IG*, we shall have the aggregate of all parallels contained in the the quadrilateral equal to the aggregate of those included in triangle *AEB*, for those in triangle *IEF* are matched by those contained in triangle *GIA*, while those which are in the trapezium *AIFB* are common. Since each instant and all instants of time *AB* correspond to each point and all points of line *AB*, from which points the parallels drawn and included with triangle *AEB* represent increasing degrees of the increased speed, while the parallels contained within the parallelogram represent in the same way just as many degrees of speed not increased but equable, it appears that there are just as many momenta of speed consumed in the accelerated motion according to the increasing parallels of triangle *AEB*, as in the equable motion according to the parallels of the parallelogram *GB*. For the deficit of momenta in the first half of the accelerated motion (the momenta represented by the parallels in triangle *AGI* falling short) is made up by the momenta represented by the parallels of triangle *IEF*.

It is therefore evident that equal spaces will be run through in the same time by two moveables, of which one is moved with a motion uniformly accelerated from rest, and the other with equable motion having a momentum one-half the momentum of the maximum speed of the accelerated motion; which was [the proposition] intended.

Perhaps the best way to shed light on this 'demonstration' would be to first bring out its argument's organisation. Using obvious abbreviations, one can distinguish three successive moments in it:

① Aggregate of the // of the △ = aggregate of the // of the ▱.

② // of the △ = instantaneous degrees of the increasing speed of the uniformly accelerated motion between *A* and *B*, and // of the ▱ = instantaneous degrees of a uniform speed equal to half the greatest speed of the accelerated motion.

③ Aggregate of the instantaneous degrees of the increasing speed of the uniformly accelerated motion between *A* and *B* = aggregate of the instantaneous degrees of a uniform speed equal to half the greatest speed of the accelerated motion.

Moment 3 , which comes from 1 and 2 by simple substitution, does not raise any particular difficulty. In the first two moments are therefore concentrated the essentials to get to the heart of our problem: to determine the true foundation of Theorem II which, together with Theorem III, renders possible the geometrisation of naturally accelerated motion. What, then, do we learn from examining ① and ②?

①, as the reader will have already understood, relies on an assertion closely related to the so-called method of indivisibles which consists of considering that an aggregate of infinitely many lines can be made equivalent to a surface or, better, be geometrically specifiable as a surface, which allows *eo ipso* to ascribe to this aggregate the properties of the surface to which one deems possible to assimilate it. Such an idea is obviously totally outside the Euclidean system, and Galileo, more than anyone else, is aware of its difficulties. The fact remains that since Democritus, many authors, for whom rigor is not all that matters, never ceased to be sensitive to its heuristic value. In that respect, its recurrence in Galileo, the creator of a new science, cannot be mistaken for an accident.[35]

The most important part is actually located in ②. And first of all the recourse to a surprising representation of speed. Theorem II of uniform motion had shown us how Galileo, sliding from time to speed, had managed to include this latter in his geometrized reasoning. This subterfuge, however, was only conceivable because of the very particular nature of uniform motion. In naturally accelerated motion, where speed varies from instant to instant, one must forego concluding directly from elapsed time to space covered: the primordial fact becomes the continuous increase of the speed which it is essential to apprehend in itself. The representation of speed to which Theorem I resorts provides this means. Making speed into a quasi-physical magnitude, co-extensive to motion in its temporal development and characterized at every instant by a determined degree (or intensity), it allows Galileo, in the absence of an adequate mathematical tool, to follow as closely as possible its continuity and increase.[36] Moreover, combined with the method of indivisibles, it immediately leads to representing this increase under the form of an aggregate of infinitely many lines, identifiable as a determined geometrical figure in which is, so to speak, visualized the relative magnitude of the distance that a body, making with such a speed, is able to cover. What this assimilation of speed to an intensive magnitude brings to Galileo is thus precisely *the essential relay to go from the increase of speed to the space covered*. But, at once, the correspondence established in ② between the parallels of the triangle and the instantaneous degrees of the increasing speed of the naturally accelerated motion on the one hand, and the parallels of the parallelogram and the instantaneous degrees of a uniform speed on the other, takes its real sense. By this correspondence ③ becomes the equivalent of an integration – or, more precisely, leads to the same result as an integration – and allows to assert the proposition baptized 'Theorem I'.[37] Such is finally the procedure which, through this 'theorem' (true foundation of Theorem II), makes possible the formulation of the first of the two relations thanks to which the study of naturally accelerated motion within elementary geometry will become feasible. One can judge unfortunate – albeit historically explainable – the way Galileo takes. The Euclidean orthodoxy of Theorem II (or of its corollary) makes it, however, as we will see a little farther on, an exaggeration to accuse the Galilean construction of inconsistency.

(14) It is with a reasoning just as disconcerting from the mathematical point of view that Theorem III introduces the second relation necessary to the geometrisation of naturally accelerated motion. Let us attend to Galileo's own words:[38]

If the same moveable is carried from rest on an inclined plane, and also along a vertical of the same height, the times of the movements will be to one another as the lengths of the plane and of the vertical.

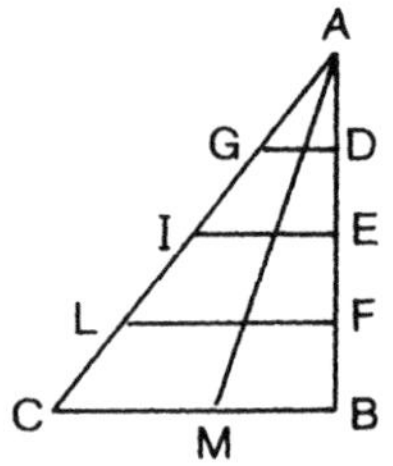

Let the inclined plane *AC* and the vertical *AB* each have the same altitude above the horizontal *CB*, that is, the line *BA*. I say that the time of descent along plane *AC* has, to the time of fall of the same moveable along the vertical *AB*, the same ratio that the length of plane *AC* has to the length of vertical *AB*. Assume any line *DG*, *EI*, and *FL* parallel to the horizontal *CB*; it follows from our postulate that the degrees of speed acquired by the moveable from the first beginning of motion, *A*, to the points *G* and *D*, are equal, since their approaches to the horizontal are equal; likewise, the speeds

at points *I* and *E* are the same, as are the speeds at *L* and *F*. Now, if not only these, but parallels from all points of the line *AB* are supposed drawn as far as line *AC*, the momenta or degrees of speed at both ends of each parallel are always matched with each other. Thus the two spaces *AC* and *AB* are traversed at the same degrees of speed. But it has been shown that if two spaces are traversed by a moveable which is carried at the same degrees of speed, then whatever ratio those spaces have, the times of motion have the same [ratio]. [*Theorem I of uniform motion*] Therefore the time of motion through *AC* is to the time through *AB* as the length of plane *AC* is to the length of vertical *AB*: which was to be demonstrated.

Evidently, Galileo's technique is the same as for Theorem II: to introduce the desired relation by relying on a theorem of uniform motion. In the case of Theorem II, the necessary 'reduction' was obtained by Theorem I. It depends here on a postulate, stated even before Theorem I and with some emphasis:[39]

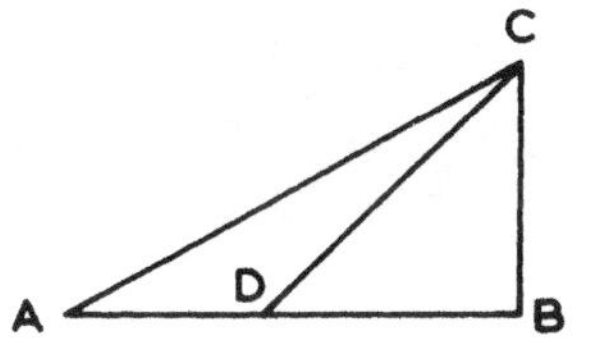

This definition established, the Author requires and takes as true one single assumption; that is:

> **I assume that the degrees of speed acquired by the same moveable over different inclinations of planes are equal whenever the heights of those planes are equal.**

He calls the 'height' of an inclined plane that vertical from the upper end of the plane which falls on the horizontal line extended through the lower end of the said inclined plane. For an understanding of this, take line *AB* parallel to the horison, upon which are the two inclined planes *CA* and *CD*; the vertical *CB*, falling to the horizontal *BA*, is called by the Author the height [or altitude, or elevation] of planes *CA* and *CD*. Here he assumes that the degrees of speed of the same moveable, descending along the inclined planes *CA* and *CD* to points *A* and *D*, are equal, because their height is the same *CB*; and the like is also to be understood of the degree of speed that the same body falling from the point *C* would have at *B*.

(15) Now the recourse to this principle poses an important problem. To understand it, consider the experimental illustration which Galileo gives and that in his eyes confers on this principle a certainty almost equal to that of a demonstrated proposition.

Given a pendulum *AB* whose bob is at *B*. If we bring the bob to *C* and let it go, it will swing back all the way to *D*, a point located at the same height as *C* above the horizon; similarly if, by means of nails fixed at points *E* and *F*, we block the motion of the part *AE* or *AF* of the pendulum, the bob, carried by its momentum (or *impeto*), will nevertheless rise up to *G* or *I*, *always located at the same height*, after following the arcs *BG* or *BI*.

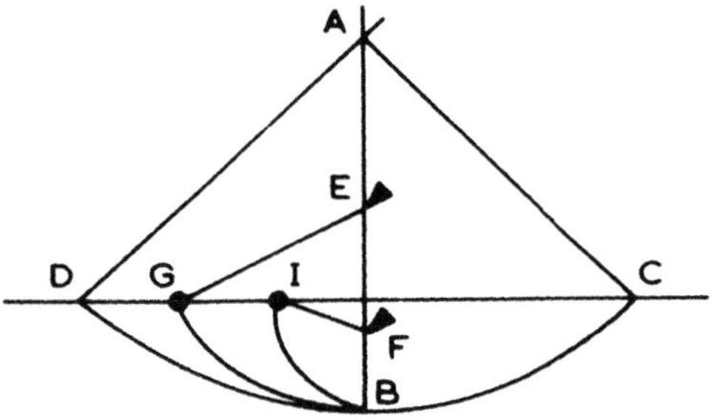

This experiment leaves no room for doubt as to the truth of our assumption, for the two arcs *CB* and *DB* being equal and similarly situated, the acquisition of momentum made by descent through the arc *CB* is the same as that made by descent through the arc *DB*; but the momentum acquired at *B* through arc *CB* is able to drive the same moveable back up through arc *BD*, whence also the momentum acquired in the descent *DB* is equal to that which drives the same moveable through the same arc from *B* to *D*. So that in general, every momentum acquired by descent through an arc equals one which can make the same moveable rise through that same arc; and all the momenta that make it rise through all the arcs *BD*, *BG*, and *BI* are equal, because they are created by the same momentum acquired throught the descent *CB*, as experiment shows. Hence all the momenta acquired through descents along arcs *DB*, *GB*, and *IB* are equal.[40]

Galileo's idea, as we see it, is to present his 'principle' as a necessary consequence of the fact that the *impeto* acquired by a moving body during a given descent from a given point always takes it back to the same height as that point, whatever the route followed. This repeats, in another form, the fundamental axiom of the "mechanical science" (or theory of simple machines) according to which a moving body is incapable of rising under the effect of gravity alone.[41] Now this is undoubtedly a dynamic assertion, hence immediately the following problem: when Galileo, in order to prove Theorem III, appeals to the principle that "from equal heights, equal speeds"; does he lay at the very foundation of his geometrisation of naturally accelerated motion typically dynamic considerations? Together with the theorems on uniform motion and the concepts subjacent to Theorem I, is not his successful application of geometry to the natural motion of heavy bodies resting on an intuition about gravity as a natural motive force? To this question, which is very important for anyone who wants to understand Galileo's geometrisation, one needs a straightforward answer.

(16) A significant first step will be to notice that this principle, whatever its dynamic origin, does in no way specify the *mode of the action* of gravity. This already excludes the possibility of an immediate rapport with the analysis of accelerated motion. A more thorough examination shows, in fact, that its role is totally different. Inasmuch as the general intuition that it brings into play is also the basis of statics and of the theory of simple machines, it gives to the new science which is then taking form the guarantee of a long established science.[42] This is not negligible, but certainly not enough to assert that Galileo is building, at least partially, his theory of accelerated motion on the basis of dynamic considerations. Furthermore, it suffices to look attentively to the demonstration of Theorem III to perceive, in its originality, the real role of the principle.

The situation is very different from that which prevailed in Theorem I. *AB* and *AC*, in particular, now represent *spaces* and not times, and this forbids (to avoid returning, in a way, to the idea that speed increases like space) the consideration of the aggregates of the successive instantaneous speeds of the two motions on *AB* and on *AC*. A new idea is thus essential to operate the necessary reduction to uniform motions. Starting from the assumed principle, Galileo begins by concluding that in each pair of points ⟨D, G⟩, ⟨*E, I*⟩, ⟨*F, L*⟩, etc., located at a same height above the horizon, a moving body descending along *AB* and a moving body descending along AC will always have the same speed. Now the trajectories of these two bodies are made from an infinity of such couples: consequently, it will never be possible, whichever pair is chosen, to ever notice the slightest difference in their respective speeds, and *this situation is exactly similar to that which we would encounter if our two bodies were moving uniformly at the same speed*. It is therefore quite legitimate, Galileo deems, to reduce these two uniformly accelerated motions to uniform motions, which immediately renders possible the recourse to Theorem I of uniform motion and the completion of the "demonstration". In this way, the equality of the speeds in each pair of points (a consequence of the principle), far from signifying a dynamic derivation of Theorem III has for its sole function to allow the assimilation of two naturally accelerated motions to uniform motions. Such a role is no more dynamic than the assimilation of an infinite aggregate of lines to a determined geometrical figure, on which the demonstration of Theorem I relied. What Galileo is looking for, in both cases, is the means of bringing back a situation which, by definition, varies from instant to instant, to a uniform one. Mathematically speaking – and Galileo is well aware of it – the procedure is anything but rigorous: its intuitive aspect and its aptitude to espouse the phenomenon considered in its internal continuity are nevertheless in his eyes sufficient assets to compensate for this lack or rigour.

(17) We believe that several conclusions can be drawn from this study of the geometrisation by Galileo of the natural motion of heavy bodies.

We started from the idea, suggested by a long acquaintance with Galileo's works, that only a previous or, at least, simultaneous reflection on the role of gravity in natural motion had enabled its geometrisation. Whether it is a matter of the conservation of uniform motion on an equidistant plane from the common center of heavy bodies or of the disappearance, in an empty medium, of the differences in speed, no geometric analysis of motion could take shape without this double progress. But the study of Theorems I and II of accelerated motion showed also that, in spite of this relation, the geometrical imposition does not directly depend on any dynamic consideration. By neutralizing twice the action of gravity, Galileo rendered natural motion thinkable in itself and not only as the always transitory effect of a force: the resulting study could, nevertheless, only be a kinematics. Several texts, in Galileo's last years, will use this kinematics to make the first steps towards a new dynamics, notably by a drawing closer together of the motive effect of gravity and the moment, or elementary quantity, of speed which, adding itself up instant to instant to the speed already acquired, assures the uniformaly accelerated character of natural motion.[43] The crude fact remains that the new dynamics was born from a kinematics itself inseparable from a reflection on the motive role of gravity.

(18) The second conclusion concerns the way in which the *Two New Sciences* succeeded in extending to naturally accelerated motion the methods of Euclidea geometry. The successive analyses of Theorems II and III have clearly revealed the presence of a same guiding idea: to anchor in the theory of

uniform motion the necessary and sufficient relations for this extension. To achieve this, Galileo must, however, pay a high price. On the one hand, he must often use confused and badly explained presuppositions, and on the other hand, he has to resort to anti-Euclidean methods. This being so, is not the discrepancy between his intentions and his actual realization momentous? Is not the geometrisation of accelerated motion in many respects more apparent than real? Or, which amounts to the same thing, is it not burdened with deficiencies which considerably reduce its scope?

We believe that with our analysis a relevant perspective is being created. For if one cannot see in the "demonstrations" of Theorems I and III models of mathematical rigour, one cannot make an argument from it to conclude the fragility of the theory as a whole. The number and the ingenuity of the theorems and propositions which came later – their perfect Euclidean orthodoxy, too – sufficiently prove, that such is not the case. *The sound perspective is rather to see in Theorems I and III postulates – the true postulates of the theory of accelerated motion – and in their demonstrations, simple ordered recapitulations of the main reasons for their acceptance* Such a reappreciation not only alters in no way the historical singularity of the Galilean work, but, by bringing out its real articulations, it allows a restoration of its indisputable mathematical consistency. The theory of naturally accelerated motion is not the simple continuation of the theory of uniform motion. Nonetheless, it grafts itself directly on this latter as soon as one adds, as so many further claims, the content of Theorems I and III. Seen this way, the Third Day of the *Two New Sciences* reveals a mathematical framework carefully and rigorously woven.

There would remain to examine how the Fourth Day, dedicated to the motion of projectiles, is related in its turn to the Third Day. Though it does not yet possess the deductive perfection brought later by Huyghens and Newton, this last part is developed much more clearly than the theory of accelerated motion. While only an analysis of the demonstrations of Theorems I and III allows us to pick out the true bases of the theory of accelerated motion, Galileo states explicitly the two principles by which the geometrisation of the motion of projectiles can follow from the preceding study: on the one hand, the principle of conservation of uniform motion[44] and, on the other, the principle of composition of motions. The form given to the principle of conservation and the particular fashion in which the principle of composition is presented would deserve by themselves long commentaries:[45] geometrisation is nonetheless assured with striking effectiveness and scope. The geometery of conics takes over from as well as reinforces Euclidean geometry proper, and Galileo devises a method allowing the determination of the *impeto* (or speed) of a projectile at each point on its path.[46] In this way, the traditional relations between mathematics and motion are totally modified. The Greek geometers, especially Archimedes, had long since shown how to construct curves by the composition of motions.[47] Galileo, as for him, puts geometrical theorems into service of motion and, from an auxiliary to geometry, he makes it the object of an applied geometrical theory. Nothing better than this inversion could symbolize the success of his enterprise.

University of Paris, Sorbonne

NOTES

[1] *On Motion*, in *On Motion and On Mechanics*, translated respectively I. E. Drabkin and Stillman Drake, University of Wisconsin Press, 1960; for a systematic analysis of the position elaborated by Galileo, one may consult our book *The Natural Philosophy of Galileo* (abridged *NPhG*), The MIT Press, 1974, Chapter 3.

[2] A negative value naturally means upward motion.

[3] *Two New Sciences*, translated by Stillman Drake, The University of Wisconsin Press, 1974, p. 77.

[4] The weaknesses of the aristotelian doctrine are shown in *On Motion*; cf. *NPhG* pp. 120 sq.

[5] This critic is taken up again in the *Two New Sciences*, pp. 70–71.

[6] Galileo never clearly distinguished sliding and rolling.

[7] In accordance with the expression used in the *Dialogue Concerning The Two Chief World Systems*, translated by Stillman Drake, University of California Press, 1967, p. 147.

[8] *On Mechanics*, p. 171.

[9] Such is the formulation given in the *Dialogue*, p. 149. The Fourth Day of the *Two New Sciences*, as one knows, will replace this spherical plane by a horizontal plane. A letter of Benedetto Castelli allows us to assert that Galileo had reached this conclusion at least as early as 1607; cf. *Opere di Galileo Galilei*, a cura di A. Favaro, Firenze, 1890–1909, Vol. X, p. 170.

[10] *Two New Sciences*, p. 149.

[11] *Ibid.*, p. 148.

[12] Galileo devotes several lines of commentary to this term in a *Note*. In a word, the term "any" is necessary to avoid considering as uniform motions in which distances still equal would be covered in times of a certain determined duration, but yet in which "the spaces completed in smaller parts of those same times, themselves equal, are not equal".

[13] *Ibid.*, p. 148.

[14] *Ibid.*

[15] *Ibid.*, pp. 149–50.

[16] *Ibid.*, p. 150.

[17] It is Descartes who first, by asserting the principle of a correspondence between

the number and the continuous magnitude, will lift this interdict and will open the way to a mathematical genesis of the physical concepts; cf. Pierre Boutroux, *Principes de l'analyse mathématique*, tome I, Paris, 1914, pp. 121 sq.

18 *Two New Sciences*, p. 148–49.

19 *Ibid.*, p. 149.

20 As we have alreday noted, Theorem I is practically taken from Archimedes, *On Spirals*, I. If Galileo makes the resort to definition 5 of Book V of the *Elements* more explicit, he brings nothing substantially new.

21 According to the *Elements*, VI, proposition 23; by definition, A/B is a ratio composed of two ratios if one has $A/B = A/C \times C/B$. Theorem IV, notably, will play a decisive role for the theory of accelerated motion.

22 Or inversely, by the excess of gravity of the medium on that of the bodies, in the case of upward motion.

23 I am reconstructing, by abridging it, the long analysis developed by Galileo in the First Day of the *Two New Sciences*, pp. 71 sq.

24 These numbers are from Galileo, *ibid.*, pp. 79–80.

25 *Ibid.*, p. 75. In the still unsure vocabulary of Galileo, 'resistance' simply signifies 'density'.

26 *Ibid.*, p. 86.

27 " . . . the internal heaviness [*interna gravita*] of different moveables has no part at all in diversifying their speeds . . . , " *ibid.*, p. 90.

28 *Two New Sciences*, III, p. 154.

29 *Ibid.*, pp. 155–61.

30 *Ibid.*, p. 170–71. If SX = $\sqrt{ST \times SV}$, then $Tst/Tsv = ST/SX$ or $Tsv/Tst = SV/SX$.

31 Theorem II is indeed never used directly . . . except for establishing this corollary.

32 *Ibid.*, p. 175.

33 *Ibid.*, pp. 166–67.

34 *Ibid.*, pp. 165–66.

35 Moreover, we come across it several times, for instance farther in the Third Day, proposition 23, Scolie, or already in the *Dialogue*, Second Day, pp. 227–28. Galileo discusses its difficulties – but also its plausibility – at length in the First Day of the *Two New Sciences*.

36 Besides, he is not at all its inventor: it dates back to the 14th century when, with its help, the philosophers from Oxford and Paris could revive in a remarkable way the traditional analysis of motion, and even got close on certain points to the threshold of great discoveries; cf. *NPhG* Chapter 2, pp. 62–83. One must obviously not conclude that Galileo would content himself with perfecting the works of the 14th centruy, *ibid.*, chapter 6, pp. 292–98.

37 If ③ can be taken as the equivalent of an integration, it is of course by a way which does not resemble at all what we call like that nowadays, and does not anticipate it either. The same procedure reappears in the Scolie of proposition 23: same assimilation of aggregates of lines – *via* the method of indivisibles – to determined geometrical figures, same postulate of the proportionality of the space covered to these aggregates geometrically specified.

38 *Two New Sciences*, pp. 175–76.

39 *Ibid.*, p. 162. The Author in question is that of the treatise that the interlocutors are supposed to read, i.e., Galileo himself.

40 *Ibid.*, pp. 163–64.
41 If the *impeto* acquired by a moving body after various descents, but of constant height, did not always take it back to this same height, nothing would prevent us to imagine devices which, by deflecting the moving body from an itinerary to another, would allow to make it go back up to a greater height; cf. *NPhG*, pp. 365–66.
42 *NPhG*, pp. 366–67.
43 *Ibid.*, pp. 345–59.
44 Let's not forget that this principle does not intervene at all in the geometrical constructions of the Third Day.
45 The principle of conservation holds on a horizontal surface, and not spherical anymore; as for the principle of composition, it is stated in answer to an objection of one of the interlocutors, after Theorem I which uses it.
46 For the reconstruction of this method, cf. *NPhG*, Appendix 6.
47 Which Galileo recalls, for instance, on the threshold of his definition of the naturally accelerated motion, *Two New Sciences*, p. 153.

WILLIAM R. SHEA

THE GALILEAN GEOMETRIZATION OF MOTION: SOME HISTORICAL CONSIDERATIONS

Maurice Clavelin has offered a brilliant reconstruction of the conceptual and technical aspect of Galileo's geometrical analysis of the motion of freely falling bodies. He has captured the gist of Galileo's method, and he has shown with remarkable clarity (I am tempted to say with typical French clarity) the importance of Book V of Euclid's *Elements*, which deals with the theory of proportions, on Galileo's interpretation of motion both in his earlier works and in his mature masterpiece, the *Discourses on Two New Sciences*. Clavelin underlines the significance of the change that was introduced by thinking of motion not as a process, as the Aristotelians did, but as a state, as has become familiar to students of physics since the seventeenth century. He rightly emphasizes the gradual discovery of the law of inertia and he shows how the ambiguities of the early formulations should not blind us to the suggestiveness they acquired for the generation immediately following Galileo. Clavelin also reminds us of some of the limitations of traditional geometry, for instance, that metrical formulae involving the ratios of unlike quantities were not entertained. This explains why we never find our general expression $v = s/t$ in Galileo for whom v, like s and t, is a basic and non-reducible magnitude. Finally, Clavelin distinguishes between the *momento del descendere* and the *momentum velocitatis* in his attempt to capture the *gravitational* and the *motive* role of weight. I shall have to qualify some of Clavelin's assertions about the relation between the dual function of weight, but let me say at the outset that I owe much to Clavelin who was the first historian of science, to my knowledge, to draw attention to the crucial role that the notion of *momento* plays at the junction between Galileo's kinematics and his dynamics.[1]

Clavelin writes as a philosopher of science who is in real sympathy with the historian of science. I shall offer a few comments from the viewpoint of the historian who is deeply appreciative of the light that philosophy can shed on the genesis and the methodology of science even when he has occasional misgivings about the very intensity of this illumination.

I am struck by a dual perspective, and hence a twofold requirement, in Clavelin's paper. On the one hand, he appears to favour a rigorously philosophical approach, as when he writes in the opening paragraph of a history of

William R. Shea (ed.), Nature Mathematized, 51–60.

science "more concerned with internal logic than with chronological details", and adds the warning that "in the following pages, one will not find discussions of unpublished documents, or chronological conjectures, or even a comparative study". On the other hand, Clavelin states, in the same paragraph, that he intends "to restore in their order and originality some of the initiatives that made Galileo the creator of an entirely new theory about nature". In a later passage, Clavelin even speaks of perceiving "in their historical singularity" the steps that Galileo took to geometrize naturally accelerated motion. This is the goal of the historian who recognizes that what actually happened has to be determined in the light of available evidence, that the history of ideas, scientific or otherwise, is not invariably dictated by sound logic, when soundness is determined with the advantage of hindsight, in this case, by the tenets of the Newtonian system.

I wonder whether Clavelin, in his attempt to spell out Galileo's reasoning, has not conferred upon it a greater coherence and more logical rigour then it will bear? The historical Galileo seems, to me, to fall rather short of the Cartesian ideal of clarity. Was it not Descartes himself who confided to Mersenne that although Galileo "philosophizes much better than the average", he "fails signally in continually making digressions and not stopping to explain fully the matter at hand, which shows that he has not proceeded in an orderly fashion but that he has merely sought the causes of a few particular incidents without considering the first causes in nature. Hence what he has built does not rest on any foundations".[2] My own pragmatic predilections make me see in Galileo's approach the fruit of a rich and varied experience that may lack the tidiness of the Cartesian method but is no less useful in discovering the truth about nature. But let us come to grips with a few instances of Galileo's method.

I. THE ROLE OF GRAVITY OR WEIGHT

Clavelin raises the following question: " . . . is it not the case that the success of the application of geometry to the natural motion of heavy bodies rests on the intuition of gravity as a natural motive force? This question, which is very important for anyone who wants to understand Galileo's geometrization, calls for a straightforward answer". Alas, a straightforward answer is precisely what the available evidence will not allow us to give. Clavelin himself admits that the principle upon which Galileo rests his case, namely that the final speed reached by a body falling along the perpendicular is equal to the final speed of a body sliding down the incline of an inclined plane, "does not specify the mode of action of gravity and hence precludes the possibility of a direct connection with the analysis of accelerated motion".

Galileo wrestled with this problem over a long period of time. In his *De Motu* written around 1592, Galileo affirms against Aristotle that all bodies are intrinsically heavy and that lightness is merely a relative property. Fire rises not because it has the quality of lightness but because it is less heavy than air. What looks like spontaneous and 'natural' ascent is really a process of 'violent' extrusion. In *De Motu* Galileo combines the hydrostatic model of Archimedes and the model of the lever from Aristotle *Mechanical Questions* to construct a new theory of motion based on the concept of *moment*. At this stage, Galileo was less concerned with speed as such and more with the ratio of the speeds of different bodies in the same medium, or of the same body in different media, or of different bodies in different media. In line with the anti-Aristotelian tradition he holds that the speed of natural motion is proportional to the difference between the density of the body and of the medium. Hence two bodies of the same material but of the same size would have the same speed of natural motion. For instance, the speeds v_a and v_b of bodies a and b, falling in media m_1 and m_2 of density d_a and d_b will be: $v_a/v_b = d_a - d_{m_1}/d_b - d_{m_2}$. In this perspective, motion in a void is no longer impossible as it was for Aristotle who made the speed of natural motion downward proportional to the density of the medium. But Galileo still maintained in *De Motu* that the speed of falling bodies depended on weight, not, however, on the total weight, but on the difference between the weights of units volumes of the body and of the medium through which the motion took place. On this view, one would not expect bodies to be accelerated but to fall at a uniform velocity. Galileo met this difficulty by stating that the initial acceleration of bodies was accidental and came to an end as soon as the body reached the downward speed appropriate to the medium. This constant downward speed is not to be confused with the terminal velocity due to the increased resistance of the medium, as Galileo in later years conceived it. It would be arrived at even in a void and is ascribed to the gradual disappearance of the 'lightness' initially impressed in the body when it was raised.[3] For all its ingenuity, this solution undermined the very thrust of Galileo's argument since it reintroduced 'lightness' as a real, and not merely relative quality. Galileo recognized the difficulty and it was his inability to deal with it effectively that explains why the *De Motu* remained incomplete and was never edited for publication.

Later, during his eighteen-year stay in Padua, Galileo found that all bodies, regardless of their weight, fall at the same rate. This discovery appeared to rob weight or gravity of the motive function that he had assigned to it. Galileo never abandoned, however, his initial conviction that gravity is the essential cause of motion. Unable to account for the role of gravity, he skilfully skirts around the subject in the *Discourses* where he avoids speaking explicitly of gravity as the cause of acceleration and limits himself to the properties (*sintomi*) of motion. It is this apparently positivistic stance that led many to consider Galileo as the father of modern science which, from their point of view, replaces the futile quest for causes by the search for regularities that can be translated into the objective language of mathematics. The historical Galileo, however, only gave up the investigation of the cause of acceleration because he was unable to identity it. In the *Diaglogue on the two Chief World Systems* and in the *Discourses on Two New Sciences*, acceleration is clearly acknowledged as a law of nature:

a heavy body has from nature an intrinsic principle of moving towards the common center of heavy objects (that is, of our terrestrial globe) with a continually accelerated movement, and always equally accelerated.[4]

In his postils to the *Esercitazioni Filosofiche* of Antonio Rocco Galileo even speaks of innate speed:

I found an axiom not to be called in doubt by anyone, and I supposed that any heavy falling body had the degrees of speed of his motion fixed and limited by nature . . . the innate speeds of all falling bodies must be similar and equal in a vacuum.[5]

In order to affirm that all bodies were equally accelerated Galileo assigned to gravity the function of opening a path through a resisting medium. This is the position held by Themistius in his paraphrase of Aristotle's *Physics*, a theory that Galileo had mentioned and rejected in his *De Motu* along with its consequence that all bodies falling in a vacuum move with the same speed, since in a vacuum there is no resistance to overcome and, hence, no role to be played by gravity.

It is also of considerable interest to note that Galileo, between 1604 and 1609, arrived at the correct law that if the same body, starting from rest, falls along an inclined plane and also along a vertical of the same height, the times of descent will be to each other as the lengths of the inclined plane and the vertical. He derived this law, however, from the erroneous principle that "the moments of speed increase continuously as the perpendicular distance to the horizontal line".[6]

It is not easy to reconstruct the stages of Galileo's reasoning on this matter, and it may well be that having discovered (perhaps in the light of actual experiments along inclined planes) that, for *equal vertical* heights, the speed of a falling body along the incline and along the perpendicular are identical, he drew the mistaken inference that acceleration was proportional to the vertical distance at all times. If this were the case, I would have to conclude that Clavelin makes a far too sweeping generalization when he writes: "It is in the wake of time that speed enters, as if a perfect reciprocity existed between the two, as if what holds for time holds *eo ipso* for speed".

Once Galileo discovered the correct relation between time and acceleration (which we express in our notation as $t^2 = 2s/a$, he had to vindicate his assumption that the speeds were equal when the vertical distances were equal. In the *Discourses*, he appeals to the experimental proof quoted by Clavelin, of arresting the motion of a pendulum by driving nails inside the surface of the arc of oscillation. In 1639, however, three years after the publication of the *Discourses*, Galileo indicated to his young amanuensis, Vincenzo Viviani, a new way of dealing with the problem. Let *AB* be an inclined plane whose height above the level *BC* is *AC*. Galileo reasons as follows:

> the *impeto* of a body falling along the vertical *AC* is to the *impeto* of the same body along the inclined plane *AB* as *AB* is to *AC*. On the incline *AB*, lay off *AD* a third proportional to *AB* and *AC*; then the *impeto* along *AC* is to that along *AB* (i.e., along *AD*) as *AC* is to *AD*. Therefore the body will traverse the space *AD*, along the incline *AB*, in the same time which it would take to cover the perpendicular distance *AC* (since the *momenti* are in the same ratio as these distances), also the degree of speed at *C* is to the degree of speed at *D* as *AC* to *AD*. But, according to the definition of accelerated motion, the degree of speed at *B* is to the degree of speed of the same body at *D* as the time required to traverse *AB* to the time required for *AD*; and, according to the last corollary of the second proposition, the time of passing through *AD* the same ratio as *AC* (a mean proportional between *AB* and *AD*) to *AD*. Accordingly, the degrees at *B* and *C* each bear to the degree in *D* the same ratio, namely that of *AC* to *AD*; hence they are equal. This is the theorem that I set out to prove.[7]

The demonstration is clear and to the point. Galileo shows the relationship that obtains between *momento* and acceleration for the *same* falling body. Within these limits, the demonstration is unassaillable. But can it be geometrized to include the case of different falling bodies? Galileo makes the generalization, but the step is unwarranted and Galileo does not seem to have been satisfied with it.

The question of acceleration is also discussed in the Third Day of the *Discourses* when the deceleration of a body projected at right angle to the surface of the earth is analysed. Sagredo proposes the explanation put forward by Galileo in *De Motu*. A body projected upward gradually loses the impressed *virtù* or *impeto* until gravity gains the upper hand and the fall begins, but slowly at first because a large portion of the *virtù impressa* still remains in the body. As it continues to decrease, the body falls faster and faster. Hence the explanation of the continuous acceleration of the body.

This theory is discarded by Salviati as unsatisfactory. Yet it is not only an ingenious but a quantifiable solution to the problem of the identical acceleration of all bodies in free fall. All that is needed, at the conceptual level, is to postulate that the ratio of the *virtù impressa* to gravity has a fixed proportion. This was indeed suggested by Giovan-Battista Baliani in his *De motu gravium* published in 1646 where he makes the following conjecture:

> perhaps gravity is an active agent while matter, or if you prefer the material body, is the passive one. Hence heavy bodies fall according to the ratio of gravity to matter, and where they freely fall without impediments in a straight line they move at the same speed because where there is more gravity there is also more matter or material gravity (*quia ubi plus est gravitatis plus pariter sit materiae, seu materialis gravitatis*).[8]

A year ealier, in a letter to Marin Mersenne, one of Galileo's last two disciples, Evangelista Torricelli, had expressed the same theory:

> It is not surprising that metal, stone, wood, etc. fall at the same speed in free fall or along the same plane since each has the same momentum, that is a moving virtue equal to itself (*cum singula aequale momentum, hoc est aequalem sibi ipsis virtutem moventem habeant*.)[9]

We have seen that Galileo entertained two solutions to the problem of acceleration. The first recognized gravity or weight as a causal factor but reduced acceleration to an accident, a transitory phenomenon, ascribed to the gradual wearing away of the "impressed virtue". This is the position in *De Motu*. In the *Dialogue* and the *Discourses*, acceleration is said to be natural (hence, ultimately, a law of creation) and no causal influence is assigned to gravity. There is, however, a third solution, namely the one proposed by Baliani and Torricelli, and it is of considerable interest to note that Galileo adopted it or, at least, entertained it after the publication of the *Discourses*. Our evidence is a fragment in the handwriting of Vincenzo Viviani (Galileo was blind from 1639) in which he compares the action of gravity on a freely falling body to that of the wind on an accelerating sailboat and makes much the same point as Baliani and Torricelli.[10]

II. INSTANTANEOUS VELOCITY AND MEAN VELOCITY

Clavelin shows in his paper how Galileo arrived at the "threshold of the concept of instantaneous velocity" and he expresses surprise that the definition of accelerated motion that Galileo offers at the beginning of the section on "Naturally Accelerated Motion" in the Third Day of the *Discourses* should not be used in the subsequent discussion.

Actually, Galileo uses the phrase *momentum velocitatis* in two different senses. In the first sense, *momentum velocitatis* indicates the instantaneous velocity arrived at after a fall lasting a certain period of time, as in the concluding paragraph of theorem I of the section on "Naturally Accelerated Motion":

> Hence it is clear that equal spaces will be traversed in equal times by two bodies, one of which, starting from rest, moves with a uniform acceleration, while the momentum of the other, moving with uniform speed, is one-half the maximum of the moment of speed of the accelerated motion (*momenti maximi velocitatis accelerati motus*).[11]

In the second sense, *momentum velocitatis* or *celeritatis* refers to the equal increments of speed that corresponds to the succession of instants in time, as in the following definition:

> A motion is said to be uniformly accelerated when starting from rest, it acquires, during equal time-intervals, equal moments of speed.[12]

Momentum velocitatis is more commonly used in the first sense, and it is the one that most readers of Galileo will remember from the triangle that Galileo used to illustrate accelerated motion where the lines parallel to the base stand for different instantaneous velocities.

The ambiguity, and the occasional confusion, between the two ways in which the expression is used, arises, I believe, for two reasons. The first is the difficulty Galileo experiences in formulating clearly a definition that involves the ratio of s/t in a mathematical notation that precludes such a direct mode of representation. The second stems from the fact that the relation of instantaneous speed to time posed a twofold problem to Galileo: (a) how to measure the speed at a given instant, and (b) how to determine the constant increment of speed at each instant. *Momentum velocitatis* is used in both instances.

This ambiguity leads me to raise another question about Galileo's procedure. As Clavelin notes, Galileo considers "quite legitimate . . . to reduce two uniformly accelerated motions to uniform motions", and hence to use theorem I in which Galileo states both that the speed can be divided into an infinite number of *momenta* (corresponding to the speeds reached at each successive instant of time from the initial state of rest) and that these *momenta seu gradus* refer to uniform, not accelerated motion. That the *momenta* refer to uniform motion is a postulate (for how else could Galileo reduce constantly accelerated motion to uniform motion) but it is not presented as such. One gets the impression that Galileo is denying that acceleration actually occurs at the level of the infinitesimal, but he never states outright that all the *momenta velocitatis* are uniform. Why this silence? I believe the answer is not difficult to find. The postulate may be necessary to reduce a uniformly accelerated motion to a corresponding uniform motion but it clashes with Galileo's own definition of naturally accelerated motion where he stresses that no body can reach a *determinato grado* or *momento di velocità* without passing through an infinite number of lesser degrees of speed. Since the acceleration is never interrupted, uniform motion is impossible, even for an instant.

Galileo is skilful but occasionally devious. Let me mention another instance. In the corollary to theorem I, he reduces the progressively greater speeds during the time of free fall to a series of equal *momenta* which he indicates by the base of equal rectangles. He states that each of these uniform *momenta* (which operate during an instant of time) correspond to the same traversed distance. What Galileo does not point out is that the rectangles stand both for the infinite number of the *momenta velocitatis* and the spaces traversed by these *momenta*. In other words, it would appear that Galileo does not distinguish between the sum of infinitely numerous *momenta velocitatis* (accelerated or uniform) and the distances traversed (constantly increasing or always equal). Yet, Galileo was not unaware of the difficulty since in the figure accompanying theorm I he represents the path of descent by an independent line to the right of the main figure.

Why, we can ask, doesn't Galileo state that the areas produced by the *momenta velocitatis* give the ratio of the distances traversed in the same interval of time?

Several reasons appear plausible:

(1) Galileo could have been embarrassed by a procedure that arrived at a finite size by adding an infinite number of parts whose sizes were not quantitatively determined. Hence he was wary of representing the space traversed by the sum of the infinitely numerous degrees of speed corresponding to an equally infinite number of instants of time.

(2) Or, again, Galileo might have felt reluctant to represent space – a linear distance in common experience – by a surface.

(3) A third conjecture is rendered possible by a curious fact. Some thirty-five pages after theorem I, in the scholium to Problem IX, Galileo uses a figure that is identical with the top half of the main figure accompanying

theorem I but without the separate line *CD* that stood for the distances. He recalls that if the body continued to move uniformly at the speed reached at *B*, it would cover twice the distance covered during the naturally accelerated motion from *A* to *B*:

Accordingly the distance traversed with these degrees during the same interval of time will be twice that traversed with the degrees of speed represented by the triangle *ABC*. But along a horizontal plane the motion is uniform since there is here no cause of acceleration or retardation. Therefore, we conclude that the space *CD* traversed during a time-interval equal to *AC* is twice the distance *AC*.[13]

In his fascinating analysis of Galileo's discussion of accelerated motion Paolo Galluzzi has suggested that it is only here that Galileo could clearly state that the area of the rectangle *CD* represents the distance traversed.[14] In theorems I and II he wished to demonstrate that the same distance is traversed in equal times by a body in free fall (naturally accelerated) and a body moving at a uniform speed equal to half the final speed reached by the body in free fall. Had Galileo explicitly declared that the sum of the accelerated and uniform *momenta* (namely the areas of the triangle and the rectangle) represented the distances traversed, he would have met with the objection that he was begging the question by supposing what he had to prove, since, in the diagram, the triangle and half the rectangle are equal *by construction*.

The text that I have just quoted is the one on which Galluzzi rests his case. The telling passage reads: " . . . we conclude that the space *CD* traversed during a time-interval equal to *AC* is twice the distance *AC*." So runs the standard edition edited by Antonio Favaro. The original edition of 1638, however, reads *CT* for *CD*, namely it assumes that Galileo is referring to the previous diagram printed alongside Theorem I. If this is the correct reading, then Galileo is *not using an area to represent a distance traversed*. It would indeed be incongruous for an Euclidean mathematician to do this. This is why the seemingly superfluous line, *CT*, was introduced to indicate the distance traversed. A closer scrutiny of Galileo's writings, both published and unpublished, may shed additional light on this problem, but, for the time being, we are denied anything approaching Cartesian clarity on the matter.

McGill University

NOTES

[1] Maurice Clavelin, La philosophie naturelle de Galileo, Armand Colin, Paris, 1968.
[2] Letter to Mersenne, 11 October 1638, in René Descartes, *Oeuvres*, Charles Adam and Paul Tannery (ed.), 12 vols. and index, 1897–1913, Vol. II, p. 380.
[3] Galileo Galilei, *Opere* A. Favaro (ed.), 20 vols. Barbèra, Florence, 1890–1909, Vol. I. *De Motu* is partly translated with introduction and notes by J. E. Drabkin in Galileo Galilei, *On Motion and On Mechanics*. University of Wisconsin Press, Madison, 1960.
[4] Galileo, *Opere*, Vol. VIII, p. 118.
[5] *Ibid.*, Vol. VII, p. 731 and pp. 742–743.
[6] Galilean mss., Biblioteca Nazionale, Florence, Vol. 72, F. 179*v*, reproduced in *Opere*, Vol. VIII, p. 389.
[7] Galileo, *Opere*, Vol. VIII, p. 218.
[8] Giovanbattista Baliani, *De motu gravium*, Genoa, 1646, quoted in Raffaello Caverni, *Storia del metodo sperimentale in Italia*, 6 vols. Florence: Stabilimento G. Civelli, 1891–1900. Reprint, Johnson Reprint Corporation, New York, 1972, Vol. IV, p. 278.
[9] Evangelista Torricelli to Marin Mersenne, 15 January 1645 in *Opere dei Discepoli di Galileo Carteggio*, Vol. I (1642–1648), P. Galluzzi and M. Torrini (eds.), Giunti Barbèra, Florence, 1975, p. 182.
[10] Galileo, *Opere*, Vol. VIII, pp. 441–442. The most recent discussion of the problem of acceleration is Paolo Galluzzi's remarkable *Momento Studi Galileiani*. Edizioni dell'Ateneo e Bizzarri, Roma, 1979.
[11] *Ibid.*, p. 209.
[12] *Ibid.*, p. 198.
[13] *Ibid.*, p. 243.
[14] Galluzzi, *Momento*, pp. 370–371.

ASHOT GRIGORIAN

MEASURE, PROPORTION AND MATHEMATICAL STRUCTURE OF GALLEO'S MECHANICS

The subject of measure and proportion as an object of scientific analysis in Galileo's works is closely connected with the broadest and most general problems of the history of the 17th-century science. When did the classical science arise? What concepts and notions designated its origin? In this respect the concepts of uniform motion of a body left to its own resources and of uniformly accelerated motion of a freely falling body, on the one hand, and methods of experimental and quantitative mathematical investigation of nature, on the other, are usually regarded as the most characteristic. The synthesis of the concepts mentioned is the most typical of the 17th century. These components of the new science existed earlier, but the synthesis changed the character of both. The concepts of uniform and uniform-difform motion were used by the Paris nominalists. In the 17th century and, first in the *Discorsi* (and partly in the *Dialogo* already) they became quantitative characteristics of motion and made the quantitative mathematical approach possible in principle. Bacon stressed the decisive role of experiment, successfully used by Gilbert later in the 15th and 16th centuries. But only with Galileo did it become a quantitative experiment answering not only the questions "Is it so?", and "What is the cause of this or that phenomenon?" but most often the question of "Is it equal or unequal?", and "Equal to what?", "How much is this?" With Galileo mathematics had not yet become scientific tool but had all the necessary prerequisites.

We are going to discuss some features of the mathematical structure of Galileo's mechanics in connection with the fundamental problem of the history of the origin of classical science.

Measurement, and consequently, the notion of measure was not connected with the most important experiments of Galileo as well as with his basic ideas. For Aristotle and his followers, to define motion was to reply to the question: Is a body in its natural place (in case of 'natural' motion) or had it been pushed forward (in case of 'forced' motion)? Galileo defines motion by its speed which characterized for Aristotle not motion itself but the form of a moving body and the medium in which it was moving. That is why the measurement of velocity is of fundamental importance to Galileo in the study of motion. In his *Dialogo* he speaks mostly about absolute velocity, about

William R. Shea (ed.), Nature Mathematized, 61–65.

uniform motion of bodies participating in the daily motion of the Earth. The velocity of uniform motion becomes an object of this analysis. The question here is the proportionality of space and time. This concept of direct (and inverse) proportionality was familiar to Galileo's predecessors. But Galileo introduced a principally new thing as compared with all his forerunners; in fact Galileo was the first to introduce quite distinctly (and opposing the new notion to the traditional definitions) the concept of the proportionality for all values of time. In the *Discorsi* there is a definition of uniform motion (distances covered in any equal periods of time are equal) and the following addition to it: "Visam est addere veteri difinitioni (quae simpliciter appelat motum aequabilem dum temporibus aequalibus transiuntur spatia) particulam quibuscunque, hoc est omnibus temporibus aequalibus: fieri enim potest, ut temporibus aliquibus aequalibus mobile pertransiat spatia aequalia, dum tamen spacia transacta in partibus eorundem temporum minoribus, licet aequalibus aequalia non sint" (G. G., VIII, 191).

A more general concept of accelerated motion is anticipated; in such a definition of uniform motion. One can say, to use modern terms, that the instantaneous constant speed is introduced in connection with the possibility of changing velocity.

As soon as a certain change of velocity comes into play, a uniform velocity becomes the object of changes, the object undergoing changes, i.e. the object of analysis. This signifies a great advance: before Galileo the object of search for active causes had been the change of co-ordinates. With Galileo, velocity becames such an object, that is to say, proportions of time and space.

Galileo does not seek a physical cause to the change of velocity, but neither does he dismiss the concept of such a cause for uniform motion, as it was done in the *Dialogo*. He discovers the law determining the change of velocity and the change of direction. The velocity of a freely falling body is in proportion to the time, the distance to the square of time. This square-law immediately enlarges the number of physically valid abstract formulas of proportionality. Before Galileo such were the direct and inverse proportionalities only. Neither Galileo nor Newton nor the mechanists of the 18th century introduced the concept of acceleration: it appeared as late as the beginning of the 19th century. The velocity is the object of changes for Galileo. The law of such a change under uniformly acclerated motion is the law of a change which has not yet gained the form of constancy of a certain proportion. The proportionality of force and acceleration became the fundamental thesis in post-Newtonian classical mechanics. This constancy of proportion or, in other words, constancy of mass seemed to be an absolute truth until Einstein.

Here an unalterable proportion becomes an unalterable measure – a dynamic variable.

The ratio of the covered space and passed time has already become an independent quantity – constant (as in the *Dialogo*) or variable (as in the *Discorsi*). One can say that velocity has become with Galileo already measure (before Galileo only space and time but not their proportion were measured), and acceleration still remained proportion and had not yet become an independent local predicate characterizing a body. That is why Galileo spoke of an unalterable proportion in the case of uniform motion but did not speak of it in the case of uniformly accelerated motion. We discover two basic ideas of proportionality in Galileo's mechanics:

(1) A moving body left to its own resources covers portions of space which are proportional to time.

(2) A body undergoing 'natural' motion, i.e. freely, covers portions of space which are proportional to the square of time.

The first idea is characterized with converting proportion into measure: the proportion of space and time becomes an independent variable – an object of causally explained alterations. The second idea had not yet acquired this character: the proportion of space and square of time had not so far been converted into a certain measure – acceleration. Galileo does not examine the alterations of this acceleration. That is why acceleration as a subject of changes does not appear in Galileo's mechanics.

Later on, different kinds of functional dependence lost their specific physical sense and were used in an abstract way. But with Galileo, the mathematical categories figure as definitions of motion. In this concrete form we can clearly see the connection between measure and proportion. Preservation of a certain quantity of a certain measure (of velocity – explained in the *Dialogo*, of acceleration – implied in the *Discorsi*) is formulated as an immediate result of preservation of proportionality expressing a physical law.

A physical law is expressed in a one-to-one correspondence of two sets: the set of positions of a body and the set of instants. Since motion is continuous, we have one-to-one correspondence of the two indefinite sets. The finite trajectory of a particle becomes an infinite set of points and the finite time of motion becomes an infinite set of instants. Laws of mechanics give us a one-to-one correspondence between the elements of these sets.

Obscure as they may appear, these internal mathematical and logical peculiarities of Galileo's dynamics enabled him to express some very profound ideas about infinite sets.

Galileo devoted many pages in the *Discorsi* to infinity (infinity as a result of an infinite division of a finite quantity). One of the most important ideas here is the assertion of the impossibility to apply concepts discovered in the study of finite sets to infinite ones.

Simplicio pays attention in the *Discorsi* to a paradoxical feature of segments composed of an infinite number of points: one infinite set can be bigger than another. Salviati's answer is: "Queste son quelle difficoltá che derivano dal discorrer che noi facciamo, col nostro intelletto finito intorno a gl'infiniti, dandogli quelli attributi che noi diamo alle cose finite a terminate . . . " (G.G., VIII, 77–78).

Galileo did not know that when the notions 'bigger' and 'smaller' are generalized and the notion of 'Mächtigkeit' is introduced, the problem of comparison of infinite numbers can be solved not only in the negative (impossibility to apply the logic of finite numbers) but also in the positive. But he saw the beginning of the way leading later in the works of Cantor to the positive solution. Salviati gives an extremely interesting example of such an idea. The set of squares is smaller than the set of all numbers (not all the numbers are squares), but is equal to the set of roots (each square has its root), and the set of roots is equal to the set of all numbers (each number can be a root). Here it can be seen that Galileo does not give a positive solution of the problem but sees the initial way to such a solution. He passes from vain searches for an actual infinity as a calculated innumerable set, which had been fairly frequent before Galileo, to the comparison of sets according to the one-to-one correspondence of their elements ("each square has its root", "each number can be a root").

Coming back to Galileo's dynamics we discover physical prototypes of a similar trend. A physical variable is measure, which had not broken away with proportion. Measure itself became an independent object of analysis (velocity) or, *explicit*, stays within limits of proportion (acceleration). But in all the cases one deals with the comparison of sets: instants and positions, instants and velocities.

The ultimate physical meaning of Galileo's ideas about infinity is the fact that they preserve the 'birthmarks' of their origin and had not so far become so abstract as they did later, in the explanation of one more characteristic point. We find out in Galileo's ideas some very vague conjecture about infinitesimal quantity as a variable – about "un terzo medio termine, che à il rispondere ad ogni segnato numero". In other words, it is a quantity which can be equal to any part of the whole, to its infinitesimal part as well.

Proportion is an early pseudonym to the later simple functional dependence, expressing the principal absolute trustworthiness of scientific statements about the position of a particle in this or that point at the given moment. The trustworthiness is the basis of classical notions about nature. With Galileo mathematics is not only an instrument of science but also a criterion of its trustworthiness.

In the wide-known lines of the *Dialogo* devoted to the extensive and intensive trustworthiness of scientific knowledge Galileo says: "L'intellecto umano ne intende alcune cosi perfettamente, e ne ha cosi assoluta certezza, quanto se n'abbia l'istessa natura; e tali sono le scienze mathematiche pure, cioè la geometria e l'aritmetica" (G.G., VII, 128–129). Mathematical cognition, Galileo proceeds, "arriva a comprendere la necessità, sopra la quale non par che possa esser sicurezza maggiore".

Necessity is a physical law; it is expressed in the form of constancy of some proportions and measures while other ones are being altered. Thus, the concepts of Galileo mentioned above gave way, to some extent, to the idea of a univocal determination of all the processes in nature.

Institute for History of Science and Technology, Moscow

NOTES

[1] Galileo Galilei, *Opere*, Ed. naz., Vol. 1-XX, Firenze, 1890–1904, Vol. VIII, p. 191.
[2] Galileo Galilei, *Opere*, Vol. VIII, pp. 77–78.
[3] Galileo Galilei, *Opere*, Vol. VII, pp. 128–129.

PART II

J. E. McGUIRE

SPACE, GEOMETRICAL OBJECTS AND INFINITY: NEWTON AND DESCARTES ON EXTENSION

INTRODUCTION

In an early treatise, now entitled *De Gravitatione et aequipondio fluidorum* (c. 1668), Newton attacks Descartes' doctrine of material extension. The occasion for the attack occurs in a discussion of the nature and existence of space, in which Newton argues that the extension of space is uncreated, eternal, immutable, and endowed with an autonomous geometrical structure.[1] Not surprisingly, the proposition on which Newton concentrates is his positive claim that "Space everywhere extends to infinity. [*Spatium in infinitum usque omnifariam extenditur*]" (*De Grav*. p. 101).

In order to establish this claim against Descartes' position Newton advances five principal theses: (1) that the extension of space is in fact infinite and not merely 'indefinitely large'; (2) that the infinity of space is an object of the understanding but not of the imagination; (3) that we can have a positive conception of the infinity of extension, though language is compelled to express it in a negative mode; (4) that infinity in itself is not a perfection, but merely 'completes' those perfections or imperfections which are intrinsic to natures; and, lastly, (5) that geometrical argument can provide a positive basis for grasping the actual infinity of spatial extension.

It is clear that Newton's theses are philosophical in character. Moreover, they have an intrinsic interest which goes beyond a mere rejection of Cartesian extension. In the main body of my discussion I shall critically analyze Newton's views on infinitely extended space, and also compare them with the relevant doctrines of Descartes' philosophy. In what follows, my analysis is restricted to Newton's early views on extended space (c. 1664–1668), and no attempt is made to consider the evolution of his thought. It is fair to say, however, that the general account of space presented in *De Gravitatione* sets the framework for Newton's later thought on the subject. Furthermore, the epistemic and metaphysical questions that I shall examine, comprise an intricate web of concepts, presuppositions and levels of thought; consequently I shall probe similar sorts of questions in various but related contexts.

A brief statement on the methodology of the present study is in order. It is recognized nowadays that Newton's *De Gravitatione* reveals a systematic

William R. Shea (ed.), Nature Mathematized, 69–112.

grasp of Descartes' natural philosophy, as well as an understanding of the metaphysical principles on which it is based. An earlier source, a unpublished pocket book, entitled *Questiones quædem philosophicæ* (c. 1664), further testifies to Newton's careful reading of the Frenchman's *Meditations* and the *Objections and Replies.*[2] This is important, as it provides textual evidence that Newton was familiar with Descartes' epistemology, and especially with the latter's approach to the representative character of knowledge. When I consider Newton's use of the distinction between imagining and understanding, in the context of conceiving extension, the fact that he knew Cartesian epistemology will become important. But at the outset, this fact allows me to put Newton and Descartes into dialogue. Not only does the device help to dramatise their individual response to various conceptions of extension and infinity, but it helps in understanding the subtle ways in which Newton and Descartes redefine elements common to Ancient, Scholastic, and Renaissance thought concerning problems of extension and the infinite.

It is not possible, of course, to give a full treatment of the doctrines of infinity, space, and extension which the 17th century inherited from the past, though I shall have something to say about Aristotle. Moreover, I shall also consider ancient atomism as a source for Newton's views on the infinity of extension. For the most part, however, I can only refer to those thinkers whose doctrines usefully highlight the views held by Newton and Descartes. In this regard, I shall attempt to show the Newton's early views on how we conceive of infinite extension rest on an epistemological basis entirely different from the one later proposed by John Locke in his *Essay*. In making this suggestion I intend nothing anachronistic; rather, my argument will be that Newton's claim to understand *that* extension is infinite, cannot be accounted for by the sort of epistemology which Locke will later attempt to justify in Book II of the *Essay.*[3] Indeed, Newton's views emerge from an entirely different tradition. Furthermore, I shall suggest that Newton's epistemic approach to extension, *per se*, has significant affinities with Descartes' treatment of extension (*qua* the essence of matter) as found in the *Meditations* and elsewhere in his writings.

In an earlier study I have pursued the question of Henry More's influence on Newton's doctrine of extended space.[4] I conclude that it is minimal in the period from 1664 to 1668, on the ground that Newton's conception of how space relates to the being and existence of God is subtly different from More's Neo-Platonic conception. Moreover, unlike More, Newton does not conceive infinity as being constitutive of divine essence, but as second level attribute which characterizes each of God's perfections (see Section 4).

Consequently, though it is true that Newton's commitment to an ontology of natures is Platonic in character, I shall suggest that the primary influence in this respect is Descartes' *Meditations*, rather than the eclecticism of Renaissance Neo-Platonism, of which we find little evidence in *De Gravitatione.*

1. SPACE IN INFINITELY EXTENDED

I shall begin by simply stating Newton's position before justifying my interpretation. The *Questions* provides early evidence of Newton's commitment to the actual infinity of extension. Under the heading 'Of Quantity' he says:

> If extension is indefinite only in greatness and not infinite, then a point is but indefinitely little; and yet we cannot comprehend anything less. To say that extension is but indefinite (I mean all the extension which exists and not so much only as we can fancy) because we cannot perceive its limits, is as much to say God is but indefinitely perfect because we cannot apprehend his whole perfection (folio 131^{r}).

There is no doubt that Descartes is the target here: it is he who uses the term 'indefinite' when claiming not to know whether extension is truly infinite. In fact, many entries in the *Questiones* testify to Newton's careful reading of the Frenchman's views on extension from both the *Meditations* and *Principles of Philosophy* (folio 83^{r-v}, folio 92^{r-v}, folio 93^{v}). Newton's main point is straightforward enough: the scope of what is imaginable (i.e., what we can 'fancy') provides no grounds for the conclusion that in reality extension is "indefinite only in greatness and not infinite", any more than our inability to comprehend the fullness of God's perfections in any way affects the reality of divine being. While Newton gives no indication here of how we know the infinity of extension, he clearly rejects what he takes to be a negative thesis, i.e., that if things are epistemologically indeterminate, this provides grounds for claiming that they are indeterminate in reality.

A similar anti-Cartesian argument is made in *De Gravitatione*:

> Nor is it an objection that space is said to be indefinite in relation to us, that is, that we are simply ignorant of its limits and do not positively know that there are none (Part I, Art. 27 [Descartes' *Principia*]). This is because, although we are ignorant beings, God at least understands that there are no limits, not merely indefinitely, but certainly and positively, and because though we negatively imagine it [space] to transcend [*transcendere*] all limits, yet we positively and most certainly understand that it does so (p. 102).

Here Newton rejects what he understands to be the Cartesian view, namely, that since we can only determine negatively that extension lacks limits, this merely provides grounds for claiming that it is "indefinitely great". Moreover, the distinction between what is properly understood by the intellect and what can be grasped by the imagination is made explicit. In Newton's view, knowledge of the infinity of spatial extension falls under the scope of the understanding, and he seems to imply that Descartes conceives extension to be indefinite because he illicitly restricts our conception of it to what can be determined by the imagination.

There are a number of interrelated issues here which need to be disentangled and considered in their own right. Let's ask first: what conception of the infinite extension of space does Newton have? He begins his account of space's ontology by defining extension as "the uniform and unlimited stretching out [*distensio*] of space in length, breadth, and depth"; (*De Grav*. p. 100) for "space everywhere extends to infinity" and "we cannot imagine [*imaginari*] any limit anywhere without at the same [*simul*] time understanding [*intelligamus*] that there is space beyond it" (*De Grav*. p. 101). Thus our conception of space is not restricted to negating in imagination any boundary that can be assigned to extension; rather, the understanding grasps *that* extension transcends all limits "not merely indefinitely, but certainly and positively" (*De Grav*. p. 102). It is clear, then, that Newton holds that the understanding not only grasps *that* space is infinite, but also implies that it has a direct and positive conception of the infinity of extension, even though it can never be sure that it has an adequate comprhension of all that is contained in that nature. So while it is impossible that the mind can imagine the infinity of extension, Newton contends that "we can understand it. We can imagine a greater extension, and then a greater one, but we understand [*intelligamus*] that there exists a greater extension than any we can imagine. And here, incidently, the faculty of understanding is clearly distinguished from imagination" (*De Grav*. p. 101).

Newton's reasoning turns on two commitments which he draws on explicitly in the section from which these passages are taken. The first is the epistemological principle that knowing how something lacks with respect to a kind presupposes a positive conception of the genus in question. I shall return to this principle presently. In the second place, Newton holds that any real being "either has limits or not and so is either finite or infinite" (*De Grav*. p. 102). Accordingly, whether or not we are in a position to justify making true statements about extension lacking limits, or able to imagine an obtaining state-of-affairs of that sort, it is the case the extension is in fact either limited or not. And if it is not limited, it is true that it is infinite, and true to say that we can understand *that* it is. When combined, these principles allow Newton to hold that the imagination's capacity for successively negating the assigned limits of extension presupposes that the understanding grasps that extension is an infinite nature. So though it is indefinite how much extension can be imagined, this in itself indicates no ground for denying that extension is infinite in the sense of being bound by no limits.

We are now in a position to reconstruct the conception of infinitely extended space which is implicit in Newton's thought. Though the imagination can't grasp it, the intellect understands that space's extension is in fact infinite in the sense that is is not intrinsically self-delimiting: it is therefore impossible that anything can exist beyond it since it lacks boundaries. This conception is clearly to be distinguished from the fact that beyond the boundies of a finite being it is possible that something other than itself exist. And since Newton holds the non-Aristotelian principle that that which is unlimited is fully actual, he at once links actually of spatial infinity with the completeness of being. So, instead of the Aristotelian notion that "The infinite is not where there is nothing further, but where there is always something further" (*Physics* III, 6. 207a 1–2), Newton holds that the infinity of space's extension is a completed reality beyond which there can be nothing. The Newtonian position is consequently far removed from Aristotle's claim that the infinite is that in which always more can be taken, such that each taking is non-identical with anything taken before (6. 207a 5–10).

It is evident that Newton's conception of the magnitude of spatial infinity is 'Epicurean' in spirit. As Newton's contemporaries understood the atomistic tradition, an infinite quantity is that whose existence cannot be limited by anything separate from, or other than, itself. In a work which Newton probably read, Thomas Stanley's *History of Philosophy*, Stanley puts the Epicurean claim as follows:

– that which is finite hath a bound, that which hath a bound, is seen from some other thing, or may be seen from out of an interval beyond, or without it. But the Universe is not seen out of any other things beyond it; for there is no interval or space which it containeth not within itself.[5]

Of such an entity it can be said that it lacks boundaries, and anything that lacks boundaries has no limit; hence it is impossible that anything other than itself exist beyond it; therefore it is unlimited or infinite. The reasoning here is similar to my reconstruction of Newton's position, and it exemplifies a pattern of thought to be found in Epicurus' *Letter to Herodotus* as preserved in Diogenes Laertius, Book X, an edition of which Newton owned and used.

"The universe is boundless. For that which is bounded has an extreme point: and the extreme point is seen against something else. So that as it has no extreme point, it has no limit; and as it has no limit it must be boundless and not bounded".[6] As well as these sources, Newton was familiar with Charleton's *Physiologia Epicuro-Gassendo-Charletoniana*, and probably (in this early period) with Lucretius' *De Rerum Natura*, both of which conceive infinite quantity in a similar way.[7] Although these sources argue that the universe is boundless, it is easy to see that Newton could adopt the arguments to the claim that the extension of space is infinite.

We can get a better grasp of Newton's conception of the infinity of spatial extension if we compare it with the leading doctrines concerning quantity common in the period. According to the Aristotelian tradition, things are quantitative just in case they are finite, possess determinate boundaries, and are conceived to have parts which can be increased or indefinitely. Given this conception of quantity, the positive existence of an infinitely extended space is contradictory to the notion of quantity itself, since it supposes the actual existence of a quantity than which there could be none greter. But this is a claim which is inconsistent with the view that quantity is essentially capable of increase or decrease. It is clear that Newton must deny that the infinity of extension presupposes quantity thus understood, i.e., that infinity of extension is to be understood in terms of the potentiality of endlessly dividing or increasing a determinite quantity such that at any given time it cannot consist of an actual infinity of parts. And he must deny this, because he holds that the extension of space is an unlimited expanse, at once containing an infinite number of simultaneously actualized parts. In other words, it is a completed totality, beyond which it is impossible that anything separate from itself exist.

But can Newton accept the view, common to thinkers like Cusanus, Bruno, and Leibniz, that though the notion of a completed infinity is incompatible with the notion of divisible quantity, there are grounds for speaking of actual infinity? This is the view that when we truly conceive of a positive infinity, we grasp that it pertains to natures which cannot fall under the category of determinate quantity.[8] For given that completed infinity – conceived as a mode dependent on quantity – is a self-contradictory notion, it provides a *reductio ad absurdum* of the view that the notion of a positive infinity pertains to that which is capable of parts. That is to say, according to the tradition which Leibniz represents, the infinite pertains to unities which lack composition, and whose being or essence are one, eternal and indivisible.[9] I shall have more to say about this conception presently.

It is clear that Newton cannot accept the tradition that opposes completed infinity and quantity, since the infinite extension of space, though not a quantity in the Aristotelian sense, is nevertheless quantitative in character: it is a completed and actual totality. Furthermore, Newton holds in *De Gravitatione* that space is an indivisible quantity in the sense that it comprises of an infinite number of indivisible units which cannot in any way be actually changed or altered (*De Grav*. p. 103). To the notion that space is an indivisible quantity Newton relates his theory that infinity is a second-level character which characterizes the intrinsic reality of the extended nature of space itself (see Section 4). Thus the grounds on which Newton is willing to legitimize talk of an actual infinity are different from Leibniz's. They are involved in the conception that space is a differentiated being beyond which nothing other than itself can possibly exist, a reality which (as we shall see) embodies in its nature a link which connects its indivisibility and infinity. And in Newton's view, spatial extension is infinite in the straightforward sense that it is without limit.

Early in 1665 Newton defined the notion of an 'infinitely Greate' quantum in a little treatise entitled 'quantity or Muchness'.[10] Euclidean in origin, the treatise deals, among other topics, with proportions, ratios and incommensurables. On folio 10^r Newton writes: "That quantum is infinitely Greate or greater y^n any finite quantity wch is not increased or diminished by being joyned to a finite quantum". It is clear that the phrase beginning "wch is not – ", refers to the term 'quantum'. Accordingly, Newton is attempting to capture the notion that there is a 'something' which can be expressed in absolute terms in a geometrical context. But what does he mean to convey by the term 'quantum'? We get a clue at the beginning of the treatise. "The subject of Quantity or Muchnesse, *as it is such*, may bee called a Quantum" (folio 9^r. My italics). If we don't talk of quantity in general, *as such*, but always of this quantity or that quantity, when answering the question how much? Newton's 'quantum', the subject of "Quantity or Muchnesse", is not an appropriate answer, i.e., it is not a quantity. To speak of a thing as a quantity is to refer to it's length or to its area; whereas, to say that something is a quantum is to say that it is quantifiable in some respect or other. Accordingly, when Newton says that a quantum is infinitely great, he is not saying that there is a *quantity* which is non-comparatively or intrinsically great. Rather, he is claiming that the 'measure' of an infinite quantum is in no way altered in comparison with the 'measure' of a finite quantum. Indeed, the point of that comparison is to attempt a definition of an infinite quantum which denies that it is capable of addition and subtraction in the manner of a finite quantum.

It is clear that Newton's definition is couched in negative terms. Nevertheless, it seems to me that he has in mind a Platonic distinction between something's being a nature and something's having a nature, where the nature defines the sort or class, and something is finite in a class if it lacks in an appropriate way what an infinite member of that class has. Accordingly, when Newton says that there is an infinitely great quantum, he is presupposing an infinite nature which defines the category of the quantitative, a nature presupposed by our understanding of the quantitative character of individual things, just in case they have determinate embodiments of the nature. A full justification of this claim must await Section 5, in which I consider infinite extension and its geometrical determinates; at this point let it suffice that the interpretation is at lest consistent with Newton's text.

2. IMAGINING AND UNDERSTANDING

So much for a statement of Newton's general position; it is now time to justify the interpretation. Let's begin with the epistemic distinction between imagining and understanding which figures prominently in both Descartes' and Newton's conceptions of extension. Interestingly enough, there is much that they both agree on, though the ways in which they relate this distinction to questions of infinity and extension differ. Admittedly, in what follows I shall have to reconstruct Newton's reasoning on the basis of the evidence provided by the *Questions* and *De Gravitatione*. Fortunately, the *Questiones* show that Newton, as well as knowing Descartes' scientific works, was familiar with the epistemology of the *Meditations*. He made notes on, and listed extensive page references to the 1656 Amsterdam edition of Descartes' writings. His references pertain to topics such as necessary being, the idea of conceiving infinity and of grasping an infinite nature, the distinction between existence and essence, the mind-body distinction, the conception of a true and immutable nature, and the distinctions between objective, formal, and eminent reality, etc.[11] Besides these and many other references to Descartes' epistemology, Newton wrote an essay on Descartes' theory of objective and formal reality as it is presented in *Meditations Three, Five*, and *Six*, and gave his own version of Descartes' approach to the ontological argument (folio 83^{r}). Furthermore, many of Newton's page references pertain to the *Objections and Replies* to the *Meditations*, in which Descartes employs his distinctions between infinity and the indefinite, between understanding and comprehending, and between imagining and understanding. In fact, Newton's page references show his familiarity with Descartes' replies to *Objections One,*

Two, and *Four*, replies which contain important statements of Descartes' epistemological principles (folio 83^r). Though Descartes and Newton draw on common epistemological traditions, it's my belief that Newton was influenced by Descartes' approach to the ontology of natures, as well as by his use of the distinction between imagining and understanding the nature of extension. After all, it's part of Newton's general strategy in *De Gravitatione* to argue that, by genuinely espousing the same principles as Descartes, conclusions can be established which differ from Descartes'.

Both thinkers draw a distinction between the 'faculties' of imagination and understanding, together with the objects appropriate to these ways of conceiving. Though hedged with many qualifications by different thinkers, the distinction is essentially this: the imagination represents the sensuously based features of a given object, but the understanding is capable of producing a non-sensuous representation of an object' essential nature. On this conception, the mind has the capacity of knowing one and the same object in different ways, and, correspondingly, there are levels of knowing appropriate to each way of conceiving. Moreover, we can be said to have confused as well as distinct representations of things.

Accordingly, the epistemological framework presupposed by Newton and Descartes is this. The mind in understanding or imagining 'grasps' or 'apprehends' natures in so far as they are embodied in the reality of things, and in so far as that reality is present in sensuous experience. Furthermore, like Descartes, Newton accepts the representative character of knowledge. In the epistemological tradition followed by both thinkers, what the mind is said to be immediately aware of is the content of its constructs or ideas, where to conceive of an idea is not necessarily to imagine or to picture. On the contrary, abstract reasoning concerning the properties of ideas and their relations, involves the mind's conceiving ideas as objects of mental vision, in much the way in which things are said to be perceived as objects of sight, but with this difference: when the understanding 'perceives' something, it *sees* through the inessential to the essential nature in question.

There is a further distinction that Descartes explicitly accepts, which provides an ontology for his use of the distinction between imagining and understanding. This is an act/content conception of thought, which is found in Medieval writers and in the works of the Scholastic commentators. A mental act exhibits the mind's power of representing, while the content is what the representation is about. Thus, the content of any mental act has something of which it is about; whether it is an act of imagining or understanding, in each case there 'exists' an intentional object *in* the act.

In other words, the object has intentional inexistence, the mode of being an item has when it is thought of *qua* thought; in Descartes' language, the "objective reality" of the idea. Both representings and representings *qua* what they represent (i.e., their contents) are modifications of mind. Furthermore, if the representative content of an act is *true*, it can be said to exist not only *in* the act, but also to possess a ground which exists independently in the world. In other words, everything that is represented by a *true* represented as existing in the world *is* in the world.

While there is not sufficient evidence to show that Newton accepts (as does Descartes) a fully developed act/content ontology of mind, he certainly holds that there is a correspondence between the 'objective' content of true ideas *qua* thought of, and their extra-mental ground in the natures of existing things (*Questiones*, Folio 83^{r}). In other words, he accepts the 'mentalization' of inner representations of outer reality. Given these epistemological principles, 'grasping' or 'apprehending' describe the mind's attempt to conceive an independent world through the presumed accuracy of the representative content of its ideas. Thus we see again that there are levels of knowing in as much as representative contents can refer to the sensing, imagining or understanding of a given object through one and the same type of experience.

While Descartes accepts the distinction between conceiving through the imagination and conceiving through the understanding, he denies that the mind is endowed with distinct faculties. His is a dynamic conception of the mind's capacities, in which affirming, denying, willing, conceiving, understanding, and imagining are all *modi cogitandi*, or cognitive powers appropriate to one and the same indivisible mind.[12] Thus each of these activities is an appropriate modification of a mind endowed with the innate capacity to think, but not a distinct power belonging to a separate cognitive faculty.

Descartes' use of the distinction between imagining and understanding occurs at two important places in his *Meditations*: the argument concerning the essential nature of a piece of wax; and the argument concerning the manner in which the mind apprehends figures such as the chiliagon. In both cases, Descartes argues that through imagining alone, the mind attains an incomplete and confused representative content of these respective natures. Moreover, ideas of imagination presuppose changing sensory representations of particular things, the images or 'pictures' of which the mind produces, while the mind's power of understanding forms a purely conceptual representation, in the content of which the thing's essential nature is grasped. Of the wax, for example, Descartes says that the nature of its extension, flexibility and changeability is not captured by the mind's imaginative capacity to picture what is now round as becoming square or triangular, etc. On the contrary, " – I conceive it (the wax) capable of undergoing an infinity [*innumerabilium*] of similar changes, and I cannot compass this infinity in my imagination. Consequently, this conception that I have of the wax is not achieved by the faculty of imagination" (AT. VII. *Meditation* II, 31). Rather, the essential dispositional nature of the wax is conceived by the power of the understanding; for it grasps distinctly and clearly that the wax is an extended, flexible and changeable nature; in contrast, that fact is conceived confusedly in terms of the sensuously-based and occurrent representations, the contents of which are present in the imagination.

It would be wrong to suppose that Descartes denies that the wax is perceived by the senses. The mind grasps the wax through a notion formed by the understanding, a notion that is originally confused by various sensory ideas and images which we mistakenly suppose are a part of what the wax is. However, once the understanding apprehends that the wax is something purely extended, flexible, and changeable, the conception its nature is clear and distinct. "But what is here important to notice is that perception is not a vision, a touch, nor an imagination, and has never been that, even though it formerly appeared so; but is solely an inspection by the mind, which can be imperfect and confused as it was formerly, or clear and distinct as it is at present, as I attend more or less to the things of which it [the wax] is composed." (AT. VII. *Meditation* 11, 31). Here we have an epistemological argument which has metaphysical consequences, at a stage in the dialectic prior to Descartes' being in a position to show that extended things exist extra-mentally.

To a large extent, the Cartesian understanding of the distinction between the imagination and the understanding is one which Newton accepts, though not necessarily Descartes' complete conception of the act/content ontology of mind. Newton's statements to the effect that we can successively imagine greater and greater extension, and yet can "negatively imagine it [extension] to transcend all limits" (*De Grav*. p. 102) in order to achieve a positive representation of its actual infinity, are clear evidence of this. Moreover, I shall argue that Newton's conception of the extension of space is in no way based on a doctrine of innate ideas, or on a view of the mind's capacity to form *a priori* constructs which are determinative of experience.

But what is it that the understanding grasps when it represents the infinite extension of space? Newton's answer is surprising: the understanding apprehends through its representative content that extension is an "eternal, infinite, [and] uncreated" nature which is revealed by the power of the mind alone (*De*

Grav. p. 111). It is the understanding that has an "exceptionally clear idea of extension" (*De Grav*. p. 99) based as it is on a nature which is "real and intelligible" (*De Grav*. p. 110) in and of itself. Newton's conception is Platonic in spirit. He is claiming that the mind has a non-sensuous prepresentation of an eternal nature, in this case, infinite extension, a nature which in itself is apprehended by the mind prior to any of its specific determinations. As we shall see, however, for Newton this does not mean that sensory experience plays no role in the mind's ability to grasp extension's nature.

Newton's epistemic approach to the infinity of extension, involving as it does the notion of apprehending extension *qua* extension, can usefully be compared with Aristotle's and Locke's; Though the infinite is not among *aistheta* for Aristotle (*Metaphysics*, K. 10. 1066b 25), he nevertheless considers problems of infinity in the context of studying objects of *aísthēsis*. In Aristotle's view, the core notion of the infinite is "that which cannot be gone through (*Physics*, 111. 6. 204a 15, *Met*. K. 10. 1066a 35sqq.), as when it is said paradigmaticallly of a series of numbers, say the positive intergers, that they are an endless progression. Consequently, he explicates his conception of infinity on the notion of number, or 'the countable', as it pertains to quantity, i.e., to continued quantity construed as the repetition of a determinate unit (*Phys*. 111. 6. 204a 17–19, *Met*. K. 10. 1066b 5–10).

Aristotle is emphatic that the infinite is not a thing which exists sparately; in fact, it is simply a category mistake to think of it as being in any sense 'a thing or object'. It is this conception which motivates his appeal to the 'mode of being' which 'a day' or 'the games' can be said to have (*Phys*. 111. 6. 206a 20–25). We can say of 'a day' that it is occuring and will occur, and of 'the games' that they are going on and will go on. Of 'Socrates' or 'a horse' this cannot be said; they are individuals which, *qua* individuals, exist as a whole at anytime. But this cannot be said of 'a day' or 'the games', as their 'mode of being' is successive in character. Aristotle is here mobilizing a distinction between completed and uncompleted 'things'. 'A day', 'the games' or 'a walk to somewhere' fall into the second category. Of each it can be said that it is *now* occurring or is *now* going on, as well as that it will occur or will go on. Being successive in character, they possess a characteristic *pêras* which indicates that they are processes which can be completed, i.e., we can *at* any time be walking to Thebes, though at a given time the process is incomplete. However, of individuals like 'Socrates' or 'a horse', *qua* individuals, it cannot be said that they are occurring or are going on.

Despite Aristotle's claim that 'the infinite' is unlike a thing, but like 'the games' is successive in character, we must recognize an important disanalogy between 'the infinite' and 'the games'. As Aristotle understands the *ápeiron*, it of course lacks any characteristic *pèras*; as such, it is unlimited and hence *uncompletable*. We can say of 'the games' that they are going on *at a time*, as there is a characteristic *pèras* applicable to that type of phenomenon. However, it is self-contradictory to say that the infinite is going on *at a time*; that would be to complete the uncompletable. So while we can be playing a game at a time, we can't be said to be infinitely dividing *at* a time, since an infinite operation cannot ever be complete and actual. Furthermore, we can say of 'the games' that they take place in a stretch of time, but temporal notions are inapplicable to the way in which the infinite is successive. It is not a process that takes time; rather it is successive in the way in which a numerical series is successive. In short, for Aristotle, the intelligible root of the *ápeiron* is that it is a well-formed potentiality, the features of which can be understood by means of numerical series and asymptotic procedures.

Though there is no definitive text for this interpretation, I think it captures the spirit and thrust of Aristotle's reasoning. His works are everywhere filled with talk of 'number' and 'measure'. For example, when considering infinites in Bk 111 of the *Physics*, Aristotle embarks on an extensive discussion of various rules of division which pertain to a given finite quantity (6. 206b 5–30; 207b 5–25), rules in the formulation of which the notions of number and measure play a prominent part. But also involved in his discussion, at least implicitly, is the notion that a determinate algorithm or formulae can be given which captures the mind's capacity to go on in accordance with a rule. As we can articulate an algorithm for generating the positive integers, so we can articulate an algorithm for any determinate rule of division which pertains to quantity. In both cases, in defining the algorithm, appeal is made to an appropriate unit, as well as to a principle of iteration. Thus, to paraphrase Euclid, we can say that given two unequal quantities, from the greater we can subtract more than its half, and from the remainder more than its half, such that a quantity smaller than a given smaller quantity is always reached (*Elements* X. prop I). Here a rule of division *ad infinitum* is stated without appeal to the notion of infinity: all that is mentioned explictly is an appropriate unit and a rule of iteration. It is to procedures of this sort which Aristotle appeals when he says that a finite quantity is infinitely divisible potentially, a claim which does not entail there ever being a stage at which the quantity is divided into an infinity of parts.

It would be a mistake to assume, however, that Aristotle's aim is merely to provide techniques which make the potential infinity of quantity conceivable. He must and does go beyond a conceivability claim: he intends that his account of potential infinity pertains to the quantity of things as they exist in nature. For Aristotle there is a significant correlation between objects, considered as quantities which are geometrically conceivable, and the way objects are. Space permits me to give only a brief and programmatic sketch of what I take to be the core of Aristotle's position.

What must be understood are the doctrines of *aphairesis* and intellgible matter (*noete hýlē Met*. Z. 1036a 10; H. 6. 1045a 33–35) as they relate to Aristotle's claim that the genus is matter (*Met.* Δ. 28. 1024b 36–b4, *Post. Analytics*. A. 10. 76a 31–36, 76b 11–16). When Aristotle says of a genus that it is the matter, he means that the genus term refers to what underlies the differentiae; "for that to which the differentia – belongs is the substratum, which we call matter" (*Met.* Δ. 28. 1024b 9–10). It is important to realize, however, that Aristotle's doctrine of *aphairesis*, the notion that the mind 'separates off' or 'takes away' certain features of a sensible object, in order to attend to others, involves the procedure of taking a genus together with its differences. Thus, in order to show that a circle is a plane figure we must take the relevant differentiae in the right order. What becomes clear is that a plane – the continuous and quantitative in two dimensions – is the matter and a figure the form, a circle in this case being a determinate sort of plane figure established by taking the right differentiae. Accordingly, when we consider a sensible object under the genus of quantity, what we grasp is not the thing's quantity as such, but a hierarchy of differentiae which move the mind from an initially vague grasp of the object to a progressively more refined determination of it; at all times, however, its intelligible matter, the continuous and quantitative in three dimensions, is presupposed. This is not to say that a thing's intelligible matter can exist devoid of all differentiation; rather, it is to claim that intelligible matter underlies its differentiae. For example, the solid is in the genus of solid figures. And given that each figure is such and such a solid, that which underlies (the reference of the genus term) never exists separately or apart: it is always one or other of some determinate sort of figure. Thus, according to Aristotle, mathematical objects are never separate from their substances in any ontological sense. Moreover, they can only be thought of as 'this – suchs' in the category of quantity. So while we cannot conceive an object as just extended, but only as an extended 'this – such' with a certain shape, it is possible nevertheless to understand its underlying extendedness rationally. I shall return to Aristotle's views in Section 5, where I consider Newton's treatment of infinite extension and its relation to the geometricals.

But how does Aristotle's doctrine of the potential infinite relate to taking matter as genus in the category of quantity? After all, in so far as he construes infinity by an appeal to the way in which a numerical series is generated algorithmically, doesn't Aristotle detach the apparatus for handling infinity from its applications? Here it is important to appreciate the role that measure plays when a genus and difference analysis is given. As an analysis becomes progressively refined, the claim that an object *is* a determinate length, can be substituted for the notion that it merely *has* a length. In other words, together with a consistent rule of operation, we can apply a unit of measure to an object.

But we can go further. A quantity can be marked off in successive intervals which can be halved, doubled, or otherwise considered in accordance with a rule of iteration. Book III, Chapter 6, of the *Physics* provides an important discussion of this. Aristotle says that in a finite quantity "the infinite by addition comes about in a way inverse" to the infinite of division. "For in proportion as we see division going on, in the same proportion, we see addition being made to what is already marked off" (6. 206b 4–10). Aristotle buttresses his claim by drawing a distinction between proportional and aliquot parts in the context in which the divisibility of a finite quantity is being considered. An aliquot part is an integral fraction of a whole number which, when multiplied by itself, reconstitutes the number, *viz*, 2 is an aliquot part of 10 but not 3. So an aliquot part of a finite quantity is an equal part of the whole which, when appropriately added to itself, reconstitutes the whole. It is clear that a division by aliquot parts exhausts any finite quantity and so fails to capture the *ápeiron* of divisibility and addition (6. 206b 10–15).

But a rule of division by proportional parts does capture the *ápeiron* of divisibility, as well as its inverse additivity (6. 206b 5–10). Aristotle's reasoning can be illustrated by an example. If a third of a finite quantity is taken, a third of what remains, a third again of the next remainder, and so on, each subtraction proceeds endlessly without exhausting the quantity: so too does the corresponding addition of each part taken. Here is a rule of divisibility which subtracts the same proportional part from what remains of a quantity after each division, as successive remainders converge assymtotically through smaller and smaller intervals. There can be no doubt that Aristotle's approach to infinity is algorithmic in character. Moreover, it involves the claim that the mind exercises its capacity to conceptualize iterative procedures in accordance with rules. Consequently, it need not be asked whether infinity is a genuine possibility, that is, whether or not it can remain unactualized through time: Aristotle's *ápeiron* is not a 'thing', nor is it a process which takes time or exists at a time. Rather, it is a well-formed potentiality, the complete nature of which is captured by the mind's capacity to formulate rules which generate iterative procedures, the individual *lógos* of which is established through a genus and differnce analysis in the category of quantity.

Now Locke's account of infinity in the *Essay* has similarities with Aristotle's approach, though ultimately it lacks the latter's sophistication. For Locke, the notions 'finite' and 'infinite' are simply "modes of quantity", that is, characteristics of things which have parts, and are capable of increase and decrease by the "addition or subtraction of any least part" (Bk. II. Chapter 17.1). Moreover, essential to Locke's program is showing how the *idea* of infinity can be constructed on the basis of ideas of sense and reflection, together with the simple and distinct modes of number (*Essay*, Bk III, 16.4). For example, given that we have sensory-based ideas of finite bits of extension, the mind can go on doubling or adding those ideas for as often as it pleases, a procedure that in fact applies to anything that has proper parts which can be increased by addition. And since there is no reason to stop these mental progressions, we are led to the *idea* of the *infinity* of space (Bk. II. 17.3).

As for Aristotle, so for Locke, number is basic to the notion of infinity. "But of all other ideas, it is number – which I think furnishes us with the clearest and most distinct idea of infinity we are capable of". For the idea of infinity "is nothing but the infinity of number applied to determine parts of which we have – the distinct ideas" (Bk. II. 17.9 & 10). If the infinite for Aristotle is something which cannot be gone through, for Locke it is a process which requires continued addition a process "which affords no prospect of stop or boundary" (Bk. II. 17.9).

There are, however, some important differences between the two thinkers. Locke speaks as though the infinite were a process, like walking, which takes time. As we have seen, this is precisely what Aristotle denies: to that extent he has freed himself from metaphors associated with ongoing processes. But metaphors continue to entrap Locke's account of the infinite. To be sure, Aristotle is committed to holding that calculating according to a rule is a mental procedure which takes time: but the rule itself which immediately defines the iterative procedure does not involve time. It is the characteristics of the latter which Aristotle's account of the *ápeiron* relies on.

In the second place, Locke's notion of a positive idea is that which pertains to countable units established through sensory experience; it follows that there can be no positive idea of infinity, whether of number or whatever. " – we have no more a clear idea of infinite parts in matter than we have a clear idea of an infinite number by being able still to add new numbers to any assigned number we have". In neither case can the mind grasp an *idea* of an actual infinity, "they both being only in a power still of increasing the number, be it already as great as it will". However, regarding "what remains to be added (wherein consists the infinity) we have but an obscure, imperfect and confused idea – " (Bk. II. 29.16). Consequently, since number is the foundation of the notion of infinity, and as we cannot have a positive idea of an infinite number, so we cannot have a positive idea of an actual infinity of numberable things. Though he would agree with Locke that it is self-contradictory to suppose that the infinite is *actually* completable, there is nothing in his position which prevents Aristotle from holding that we can have a positive conception of infinity as understood algorithmically. Aristotle's conception of the infinite is fundamentally not imagistic, nor does it depend on a genetic epistemology of ideas such as we find in Locke.

Newton's *Questiones* provides evidence that he was concerned to understand infinite processes in terms of the properties of numerical series. Under the influence of John Wallis' *Arithmetica Infinitorum,*[13] Newton explores series in which, for instance, a fraction is said to become arbitrarily large as the denominator x approaches zero (folio 90^r). In this early manuscript, we witness Newton's attempt to formulate the idea that the 'limit' of an infinite sequence can be evaluated mathematically without the notion of an ultimate or infinite ratio. Though Newton never writes Wallis' symbol for infinity ∞, when he says that a/x becomes arbitrarily large as x approaches zero, in effect (put anachronistically) he is saying that as the variable x approaches zero the numberical value of the corresponding variable a/x increases indefinitely, i.e., $a/o = \infty$ (*Questiones*, folio 90^r). Thus it is clear that Newton is willing to model a notion of 'potential' infinity on the properties of numerical progressions. However, with regard to the quantity of objects that exist extramentally, Newton is an Epicurean actualist. He holds that "nothing can be divided into more parts than it can possibly be constituted of" (*Questiones*. folio 89^r). Accordingly, if a finite quantity is said to be infinitely divisible, this states a genuine possibility; so to say that the quantity is infinitely divisible is to say that it *can* be so divided. But this is contradictory, as a finite quantity cannot be constituted of infinite parts (folio 89^r). It is also clear that Newton does not confuse infinite extension, regarded as an eternally complete form of being, with the idea of the infinity of quantity construed as "an endless growing idea" (*Essay*, Bk II. 17.7). In Newton's view, "all the extension that is" is actually infinite (folio 90^r). After all, the infinity of spatial extension is coeternal with divine existence (*De Grav*. p. 103); as such it is a nature of which the mind can have a positive and not merely a negative conception.

Since Locke strongly opposes a doctrine of infinite space which is reminiscent of Newton's, it will be useful to contrast Locke to Newton. Locke's account differs from Newton's in two important ways. First, Locke holds that we have an idea of the *infinity* of space, but denies that we can have a *positive* idea of an actually infinite space: " – I think it is not an insignificant subtilty if I say that we are carefully to distinguish between the idea of the infinity of space and the idea of space infinite" (Bk. II. 17.7). The former idea results from the mind's power of continuing without end a process of multiplication. The latter idea is impossible. It involves having in mind the idea of a completed *and* endless progression. On Locke's account this is a contradiction in terms (Bk. II, 17.7).

Newton, however, affirms what Locke denies. He claims that we do have a positive idea of an actually infinite extension, and not merely an idea of the *infinity* of extended space (*De Grav*. pp. 101–102). While for Locke the infinity of space is merely an indefinite extension of the finite (the latter the sole basis of whatever positive idea we have of it) for Newton the infinite extension of space is positive and real, the conception of which is presupposed by any notion of finite extension we may have. As Newton puts the point in the context of discussing Descartes' doctrine of the indefinite, "the limit or boundary is the restriction or negation of a greater reality or existence [in respect of] limited beings – " (*De Grav*. p. 102). So rather than the infinite being the negation of the finite, for Newton the finite is itself the negation of the infinite. In this conception, he is probably following Descartes (AT. VII, *Meditation* III. 45), and both betray their Neo-Platonic heritage. Though Newton and Descartes make the limit a negative concept and the infinite a positive one, their position involves more than play with words in a game in which Locke merely reverses the order. As we shall see in Section 3, both Newton and Descartes treat the claim that something is without limits as having its epistemic basis in the understanding's capacity to apprehend that fact non-discursively, without the sort of appeal that Locke makes to endless mental progressions. Furthermore, Locke's account of the boundlessness of extension is concerned to show how that idea is constructed by the iterative powers of the mind; whereas, for Newton, to say that we can understand that extension is boundless, is to say how it is with extension *per se*.

Newton's position differs from Locke's in another important way. If the infinite is a mode of quantity, and if quantity is essentially capable of increase and decrease, there is no actually infinite quantity. This would be a quantity incapable of increase or decrease, and hence contradictory to the notion of quantity itself. On Locke's acccount of how we get the idea of infinity, there could be no idea of an actually infinite extension: this would be an idea of a quantity than which it is impossible that there be any greater. It is clear, then, that there is an intrinsic connection between Newton's claim that extension is an infinite and eternal nature, of which the understanding can form a positive conception, and his denial that the idea of infinite space is based on the mind's indefinitely repeating sensory based ideas of the finite. In Newton's view, it is the understanding that grasps the unchanging and essential nature of extension through its non-sensuous representations, and not the imagination which merely repeats sensory orginated ideas of the finite.

I conclude that Newton's epistemology, based as it is on the distinction between the understanding and the imagination, is close to Descartes', especially on the question of how the mind grasps the notion of the infinite. After all, it is Descartes who claims, in a statement which expresses a view similar to Newton's, "that the notion I have of the infinite is in me before that of the finite because, by the mere fact that I conceive being or that which is, without thinking whether it is finite or infinite, what I conceive is infinite being; but, in order that I could conceive finite being, it is necessary that I remove something from this general notion of being which consequently must procede" (AT. V. 352). While it is true that Descartes (writing to Clerselier) is commenting on the problem of apprehending being *qua* being, and not talking about extension, as such, Newton will argue that the same epistemic principle applies to extension, i.e., that the finite is the negation of the infinite. So the question that must be asked is this: given that Descartes and Newton have similar epistemological views, why does Descartes hold, but not Newton, that we grasp extension as 'indefinite' in character?

3. DESCARTES ON NEGATION AND THE 'INDEFINITE'

I shall begin by considering the principles which lie behind Descartes' claim that extension is to be regarded as 'indefinite'. As regards the limits of extension, the divisibility of a body, and the number of the stars, Descartes tells us in his *Principia Philosophiae* (Part I, XXVII) that:

> – these things we call indefinite rather than infinite in order to reserve to God alone the name of infinite, because in him alone in every way we apprehend [*agnoscimus*] no limits, and because we positively understand [*intelligimus*] that he has none; with regard to other things, we do not in the same way positively understand them to be in every part unlimited, but negatively admit that their limits, if they have them, cannot be discovered by us (At. VIII 15).

Descartes' distinction between the indefinite and the infinite is this: a thing is indefinite just in case there are no reasons by which to show that it has limits, i.e., that it is epistemically possible that it is boundless. On the other hand, a thing that is infinite just in case it can be shown that is impossible that it is bounded in any respect whatsoever. According to Descartes, it is in the nature of God alone that it is possible to ground a demonstration of this sort (*Meditation* IV. At. VII 64–69). This is Descartes' official version of the distinction between the negative and the positive infinite. What can be said of the extension of the Universe, for example, can be expressed negatively, as it involves a successive negation of all limits that can be assigned to it. And though we have "no arguments to prove, and not even being able to conceive, that the world has limits", nevertheless, we are only entitled to call it indefinitely large. (Letter to Chanut, 1647, AT. V. 50). But were it possible to show that the extension of the Universe is in fact unbounded, there would be positive warrant to say that it is infinite.[14]

Descartes' position rests on a particular view regarding the relationship between what is conceivable and what is real. He repeatedly distinguishes between that which the finite mind is capable of grasping – such that what is conceivable is the ground of that which can be asserted an denied – and reality itself as it exists independently of conception. So though the mind is not the measure of reality, it is the measure of what can be consistently asserted and denied of it:

> I consider only what I can conceive and what I cannot conceive, and I take great pains that my judgement should accord with my understanding. And so I boldly assert that God can do everything which I conceive to be possible, but I am not so bold as to deny that he cannot do whatsoever conflicts with my understanding – I merely say that it involves a contradiction [in conception]. (To More, 1649. AT. V. 269).

Though Descartes claims that the mind is the arbiter of the conceivable, he is certainly not committed to holding that everything encompassed in our ideas exists in reality, any more than it is true that God "cannot do what we cannot comprehend" (AT. III. 545). So given that there are separate reasons to think that our clear and distinct conceptions are not deceived, "whatever we perceive clearly is true, and so it exists, if we perceive that its existence is possible" (To Mersenne, 1642. AT. III. 545).

So what is conceivable is possible, but what is not conceivable is not necessarily impossible: if our beliefs involve clear, distinct and consistent (noncontradictory) ideas, and if there are separate grounds for believing that God is not deceditful, we can truly assert that there is a correspondence between what is said and what is in fact the case. In this sense Descartes holds that there is an intrinsic connection between conceiving and that which is conceived. However, while it is true that in conceiving something clearly and distinctly, what the mind conceives is in the nature of the thing, it does not follow on Descartes' view that its attentive operations necessarily exhaust the real, or that the mind is in a position to know that it possesses complete and adequate knowledge of things.[15]

It is clear that Descartes does not draw unwarranted ontological conclusions about what actually is from his epistemological account of the conceivable. When he says that from the mind's point of view extension is regarded as indefinite, Descartes is saying that that characterization is consistent with extension as we conceive it, but not that the mind necessarily exhausts everything that pertains to the nature of extension itself. In this viewpoint, there is nothing which compells Descartes to conclude that extension is in fact ontologically indeterminate. So though there may be properties of extension of which the mind lacks ideas, there is nothing in extension which is inconsistent with conceiving it as indefinite. In this restricted sense the mind is the measure of reality.

Given these epistemological commitments, it is not surprising that Descartes is noncommittal on whether or not extension is infinite. Moreover, he leaves open the possibility that extension may in fact be infinite. Referring again to the extension of the world, and the number of parts into which matter is divisible, Descartes writes to Henry More in 1649 that, " – I do not know whether or not they are straightforwardly infinite" (AT. V. 274). It is precisely because he thinks that the infinity of extension is *conceivable* that Descartes makes this statement. "It conflicts with my conception to attribute any boundary to the world; and I have no measure of what I should affirm or deny except my own conception". Descartes goes on to point out, however, that this is also consistent with holding that the mind cannot demonstrate whether or not the world's extension is in fact bounded or unbounded (AT. V. 342).

In sum, though it is epistemically possible that extension is infinite, in Descartes' view it is not epistemically possible to show that this is a fact. On th other hand, he sometimes allows that extension could be finite (AT. V. 274). This, however, cannot be a genuine possibility for Descartes, as it violates his fundamental doctrine that wherever there is extension there is material substance. So if Descartes were seriously to posit a finite and bounded world, this doctrine would be contravened. A bounded world can only be demarcated by something other than itself, by something that is not a material substance, i.e., a void or an immaterial space; both sorts of entities, however, are impossible on the principles of Descartes' philosophy (At. V. 269 & 401). It is true that he tells Chanut in 1647 that the extended world may have limits "which are known to God, though incomprehensible to me: and that is why I do not say decisively that it is infinite" (AT. V. 51–52). Later in 1649, the same possibility is mentioned to Burman, namely, that God may have "a conception and understanding of fixed limits in the world, and, in number, quantity, etc. – and so for him these things may be finite". (At. V. 167). It is difficult, however, not to read these statements as pious gestures to divine power and wisdom, as well as characteristic affirmations that the finite mind is incapable of comprehending God's intentions regarding creation.

Recall, that in the *Questiones* passage (cited in Section I), Newton seems to accuse Descartes of illicitly moving from a claim about the epistemic indeterminacy of the limits of extension, to the claim that extension is indeterminate in reality. In *De Gravitatione*., Newton returns to this issue. He tells us that Descartes should be corrected by the grammerians", since "the indefinite is never said of that which actually is, but always pertains to a future possibility – ". In Newton's view, the term 'indefinite' is to be reserved for unrealized yet possible forms of being, and is therefore inappropriately said "of that which actually is". Of that which is, it must be possible to say that it "either has limits [*terminos*] or not, and so is either finite or infinite" (*De Grav*. p. 102). As early as the *Questiones*, Newton makes it clear into which category extension falls: "This indefinite, that is undetermined, how great a sphere may be made, how great a number may be reckoned, how far matter is divisible, how much time or extension we can fancy; but all the extension that is [and] Eternity – are infinite" (folio 90^{r}).

It is clear that Newton is challenging Descartes to make a genuine ontological claim. But how are we to interpret his accusation that with respect to extension Descartes applies the principle that "that which actually is, is not to be defined [*non est definiendum*]? (*De Grav*. p. 102). On one interpretation, we might say that Newton is accusing Descartes of holding that extension is indefinable just because he holds that it *is* ontologically indeterminate. On this view, the basis of Newton's criticism is ontological, i.e., extension can't be indeterminate in nature, as that conception is contradictory to what is, whether or not we are able to define it. In order to sustain this view, however, it is necessary to suppose that Newton fails to understand Descartes' epistemological conception of how the conceivable relates to that which is.

On the other hand, as John North observes, Newton may be accusing Descartes of equivocating on the notion of the 'indefinite'.[16] I shall make a related observation. The Greek *horismós* and the Latin *definio*, both convey the notion that to define is to set bounds or limits. Accordingly, in defining something we limit or confine it to a kind or species: we say what it is. But the root meaning of these traditional notions also involves reference to the fact that in reality things are said to have measurable boundaries or limits. So if we say that the extension of the world is defined, this means that we know something about it, i.e., we know that it has limits in reality. On the other hand, if we say that the world is 'indefinite', this can mean *either* that we are not in a position to know whether it is finite or infinite, *or* that it is not defined, i.e., that its limits do not exist. On this reading of Newton's criticism, he is accusing Descartes of equivocating between the notion of epistemological indeterminacy, and the view that extension is 'indefinite' in the sense that it is without limits, i.e., an extended *ápeiron*. On the second view, Descartes can reasonably be taken as saying that extension is infinite. It is clear from my general discussion of Descartes' epistemology that he is able to separate these two sorts of claims. But it also seems likely that he is attempting to run both together under the rubric of the 'indefinite', much to Newton's consternation.

Descartes' ambiguous use of the 'indefinite' allows him to develop another plausible pattern of reasoning. Since we cannot demonstrate that the world is limitless, all that is required is a negative conception of it. To every boundary assigned to the world, "or the totality of corporeal substance", another can be imagined, such that no upper bound can be assumed in the procedure (AT. VIII. *Principia*. Part II. XXI. 52). But this provides a concept of negative infinity, or, in Descartes language, the 'indefinite'. However, one "cannot infer because limitation involves in itself [*contineat*] the negation of infinity, that the negation of limitation involves knowledge of infinity". On the contrary, that which "makes the infinite different from the finite is something real and positive" (AT. III. 427). But it is only God or "the greatest being" that can be called infinite in this sense. Without doubt Descartes has a twofold conception of infinity: a negative idea of the indefiniteness of material extension, in virtue of the imagined negation of each boundary assigned to it, and a positive conception of the infinite reality of divine being.

4. NATURES, PERFECTION AND INFINITY

Is it the case that Descartes must restrict the notion of positive infinity to the nature of divine essence? Consider first the views of Descartes and Newton on whether infinity can be properly understood when expressed through the vehicle of the grammatically negative. According to Descartes, the mind cannot understand the infinite by "the negation of limitation". His reason is straightforward. If what makes the infinite different from the finite is positively real, "limitation, by which the finite differs from the infinite, is a non-being or a negation of being [*est non ens sive negatio entis*]; and that which is not, cannot lead us to a knowledge of that which is" (AT. III. 427). On the contrary, the act of negating a thing's nature presupposes positive knowledge of the thing itself. So though infinity can be characterized in terms of negating negations, it does not follow that the positive nature of infinity is known through this procedure, but only the nature of what is said to be 'indefinite'.

Newton's comment on this kind of reasoning is as follows:

> Should it be – said that we do not understand what an infinite being is (*quid sit ens infinitum*), save by negating the limits of a finite being, and that this is a negative and faulty conception, I deny this. For the limit or boundary is the restriction or negation of a greater reality or existence [in respect to] the limited beings, and the less we conceive any being whatsoever to be constrained by limits, the more we understand that something applies to it, that is, the more positively we conceive it. And thus by negating all limits, the conception becomes most positive. End [*finis*] is a word negative as to sense, and thus infinity, as it is the negation of a negation (that is, of ends), will be a word most positive according to our sense and conception, though it seems grammatically negative (*De Grav*. pp. 101–102).

There is much here that Descartes can agree with. Both thinkers hold that the procedure whereby the mind negates the negation of a nature in expressing the infinite, depends on a prior and unrestricted conception of that nature. If a restricted conception of a kind defined by a nature presupposes a cognitively prior apprehension of the kind itself, in Newton's view "negating all limits" in our idea of it, is a legitimate way of expressing that prior conception. Thus in conceiving the infinite, "though we negatively imagine it to transcend all limits, yet we positively and most certainly understand [*positive et certissime intelligimus*] that it does so" (*De Grav*. p. 102). Though our language is grammatically negative, it does not follow that what we understand is devoid of positive ground. Descartes agrees on this point, saying that it is "common usage – to conceive of a thing bound by no limits in order to understand the infinite" (At. III. 427). Furthermore, Newton as much as Descartes, is committed to holding that the negation of finite being does not in itself provide a positive conception of infinity. This merely posits a lack of limits in a thing in accordance with our mode of conceiving it, but it cannot change the finite reality from which the mind begins. What is characteristic of Descartes' position, however, is the conviction that the positive infinity of divine being cannot adequately be expressed in the language of the 'not-finite', as divine reality cannot be conceived in terms of what it is not. Descartes admits, as does Newton, that we are nevertheless forced to use the grammatically negative in expressing it.

In the first version of this paper I wrote: "In Descartes' view, extension contradicts the nature of infinity, because it is an imperfection, and infinity cannot be admitted where there is not perfection". As it happens, what I intended to state is elliptically and misleadingly expressed. In the first place, I ought to have claimed explicitly that "extension contradicts the *positive* nature of infinity" where infinity is conceived to be constitutive of divine essence. Unfortunately, the claim is at best contextual implied, and North was understandably led to suppose that I was claiming that the idea of infinite extension is *per se* self-contradictory for Descartes. In the second place, I argued that according to Descartes, extension is an imperfection in comparison with other realities. This, too, was disputed by North in his original reply.

However, nothing in my position then or now commits me to the view that extension is self-contradictory for Descartes. After all, basic to my interpretation is the claim that Descartes is willing to conceive of extension as limitless, and consequently infinite in that sense. Moreover, it is because he thinks that he lacks positive reasons for showing that it is impossible for extension to be limited, that Descartes remains epistemically non-committal regarding its nature. Furthermore, if the argument thus far is correct, Descartes is committed to the principle that grasping the finite of a kind presupposes positive knowledge of the infinite of the kind (AT. VII. 45 *Meditation III*) a principle that applies equally to extension.

There are a number of passages in Descartes' writings in which he suggests that extension is without limits, and in which these commitments are apparent. To More he says: "It conflicts with my conception, or, what is the same, I think it involves a contradiction, that the world should be finite or bounded; because I cannot but conceive a space beyond whatever bounds you assign to the world" (AT. V. 341). Similar arguments are put to Chanut and Burman (AT. V. 51 and AT. V. 167), and in each case Descartes seems to admit the possibility that the extended universe is infinite. The structure of his argument is this. For any boundary assigned to the extension of the universe, it is always possible to conceive of an area beyond that boundary which must be included in the universe. But for every boundary assigned to accomodate an additional area, the same procedure applies, and so on. It is thus impossible to talk of a bounded universe, or of a universe that has the greatest extension. This is no doubt what Descartes has in mind when he tells Burman that "we can perhaps say that the world is infinite, and the same for number, etc."; for "when the indefinite is multiplied again and again – it is identical with infinity" (AT. V.). This reasoning seems to legitimize the term 'infinity' as an appropriate characterization of extension. Again, it is clear that Descartes sees nothing self-contradictory in the notion that extension is infinite. Moreover, his argument is reminiscent of Newton's claim that though we negatively imagine successive limits to extension, we nevertheless understand *that* there is extension beyond any limit (*De Grav*. p. 101).

Thus far it seems that Descartes ought to accept the view that God and extension, *qua* extension, are both infinite according to their kind. This is precisely the view that Newton argues for. He says:

> But I see what Descartes feared, namely, that if he should consider space infinite, it would perhaps constitute God because of the perfection of infinity. But by no means, for infinity is not perfection except when it is attributed to perfections. Infinity of intellect, power, happiness, and so forth, is the height of perfection; but infinity of ignorance, impotence, wretchedness, and so on, is the height of imperfection; and infinity of extension is so far perfect as that which is extended (*De Grav*. pp. 102–103).

Here Newton claims that infinity is a character which can be predicated of different things. Moreover, it is a second-level attribute; that is to say, it is a way of characterizing the mode of instantiation of first-level properties in so far as they allow something to be the highest exemplification of its kind. Given these claims, it is clear that Newton construes infinity as an extrinsic attribute which can be said of many things. Thus infinity *in itself* is neither a perfection nor an imperfection, nor is it defining of anything's nature. As Newton says, "infinity is not a perfection except when it is attributed to perfections" (*De Grav*. p. 102).

Newton's reasoning turns on the notion that things exemplify the natures which specify the kind or class to which they belong. Furthermore, he holds that things can be conceived as either perfect or imperfect in accordance with the nature that they exemplify.[17] This allows him to conceive of things as constituting levels of reality in which a difference in level is defined in terms of the specific nature of the kind to which the things belongs. In accordance with this scheme of predication, a thing is said to be infinite just in case it manifests in an exemplary manner the nature of its kind, whether the reality in question is taken to be a perfection or an imperfection.

Thus infinity is not a real or first-level attribute which defines the nature which a thing exemplifies, nor is it identical with either the concept of perfection or imperfection. Newton's formal theory of infinity therefore allows the existence of different kinds of infinity, and is consistent with distinguishing divine infinity from the infinite extension of space. Consequently, Newton argues that God is infinite because it can be said of him alone that he is infinitely wise, infinitely powerful, etc., where to say that God's attributes are infinite is to say that he possesses in an exemplary manner attributes which other agents own deficiently. Accordingly, "not everything eternal and infinite will be God, nor will God be prevented from the eternal and infinite exercise of his omnipotence in the creation and governing of things, by the imperfect nature of created things". In contrast to God, space by "reason of its eternity and infinity" will not be "wise nor powerful nor alive, but will merely be shown to be increased in duration and magnitude". The fact that God and space each possess their attributes infinitely, does not imply that their respective natures are identical, for "no thing is by eternity and infinity made better or of a more perfect nature".[18]

On the face of it, Newton's position on infinity is compatible with Descartes' pronouncements, and indeed may in part have been derived from the latter's philosophy. Why then does Descartes characterize extension as 'indefinite', rather than infinite? Is this merely a pious sop to the theologians? Such an explanation is not sufficient. In the first place, as we have already seen, he holds that extension may be infinite, though we are unable to justify the truth of the statement. And we cannot do so, because we lack positive reasons for showing that extension is necessarily infinite, and are able to claim only that its boundlessness is an epistemic possibility. Accordingly, Descartes' position comes to this: though we can consistently conceive that extension is infinite, we are not in a position to show that it is.

What is at issue can be brought out by considering Descartes' attitude towards the ontological proof for God's existence. On the basis of that proof, Descartes claims that the intellect grasps God's perfections, and sees that he is a necessary being who exists eternally and infinitely (AT. VII. 66–67. *Meditation V*). Moreover, the act of apprehension is intuitive and non-discursive; in the instant it grasps divine nature, the intellect sees that the mind's objective idea of the determinate, true and immutable nature of divinity is directly grounded in God's formal and persistent reality, the essence and existence of which is necessarily and inseparably connected (AT. VII. 65–66. *Meditation V*). The *cogito* argument has the same formal structure. The intellect moves directly and non-discursively from the objective reality that it is a thinking thing, to the truth that, at the very moment it is thinking, it is impossible that it think without being something that actually exists (AT. VII. 25. *Meditation II*).

It is on this basis, the existence of self and the existence of God, that Descartes grounds the possibility of systematic and discursive knowledge. In his view, the ontological argument shows that divine nature possesses all perfections in a limitless manner, and consequently that God's substantial presence embodies the absoluteness of infinity. He makes this claim emphatically:

> By an infinite substance, I understand substance which has actually infinite and immense and true perfections. This is not an accident superadded to the notion of substance, but the very essence of the substance taken absolutely and bounded by no defects; these defects, in respect of substances, are accidents, but infinity or infinitude is not (AT. V. 252).

Infinity, truly conceived, is intimately connected with possessing perfection; and since all true perfections are possessed by divinity, infinity is constitutive of its essential nature. Moreover, the possession of absolutely all perfection implies supremacy for Descartes. Therefore, only one infinite substance can exist, as it is contradictory to suppose that more than one supreme being exists.

It is clear, then, that Descartes thinks the ontological argument provides positive reasons for holding that God's perfections are absolutely without limit, and therefore that he is truly an infinite being. This does not mean that infinity itself is comprehensible to the human mind, nevertheless "in so far as clearly and distinctly to understand a thing to be such that no limit can be found it in, is to understand clearly that it is infinite" (AT. VII. 112). Notice that Descartes distinguishes comprehending x from understanding or knowing x. To comprehend something is to have adequate knowledge of it, i.e., to know all the properties that belong to it. Moreover, it is possible that we may in fact achieve adequate knowledge of something, while it remain possible that the mind not know that its knowledge is adequate. However, while it is possible to know a finite nature adequately, though impossible to be certain of it, it is impossible that the mind possess adequate knowledge of the infinite itself, since "incomprehensibility itself is contained in the formal concept (*in ratione formali*) of the infinite" (AT. VII. 368). While it is impossible that the finite mind comprehend the infinite, as that entails grasping at once and entirely all that is involved in the notion, it does not follow that we cannot understand the infinite. As we understand the notion of a triangle completely and distinctly, when we conceive of a figure bounded by three lines in a plane, "so it is enough to understand a thing bounded by no limits in order to have a true and complete idea of the whole of infinity (*ut vera et integra idea totius infiniti*)" (AT. VII. 368). Again, we see a basic epistemological principle at work: though it is the case that what is conceived clearly and distinctly by the mind is true, the mind's distinct conception does not necessarily exhaust the reality of what is thus understood.

Descrates is clearly distinguishing infinity, as formally understood from things which are infinite (AT. VII. 113). While we can understand infinity *per se* as that which lacks limits, "the thing itself, which is infinite, is indeed positively understood, though not adequately, that is, we do not comprehend the whole of that which is intelligible in it" (AT. VII. 113). So infinity, abstractly and formally considered, must be understood negatively as a conception that pertains to anything without limits. However, when it is said of something that it is infinite, this is positively understood, in so far as it can be shown that a thing's nature cannot possibly have any limitations whatsoever, and even though the reality that is grasped must be expressed in negative language. It is only of divine nature that we can establish positive grounds, by means of the ontological argument, for claiming that its perfections are limitless in all respects. Unfortunately, Descartes' distinction between understanding *that* something is infinite, and the impossibility of comprehending its infinity, raises a difficulty in the context of the ontological argument. When the intellect non-discursively apprehends the nature of divinity, and grasps the fact that it is a true and immutable nature, the locus of all true perfections, and the ground of its own existence, this is surely to comprehend the infinitude of divine nature, and not merely to understand *that* God's nature is infinite. Indeed, the ontological argument demands that the intellect positively grasps God's nature. How else are we to make sense of Descartes' claim that we have positive reasons which show that divine nature is without limits? But how is this to be reconciled with Descartes' claim that the finite mind cannot grasp all that is involved in the nature of infinitude? In other words, how can Descartes reconcile the view that infinity itself is incomprehensible, with the claim that the mind can have "a true and *complete* idea of the *whole* of infinity" (AT. VII. 368, my italics).

But is there anything about the ***nature*** of extension that leads Descartes to characterize it as 'indefinite'? It is clear that it does not embody infinity *qua* essence, the conception that Descartes thinks is shown of divine nature by the ontological argument. Moreover, it is manifestly not self-contradictory to hold that extension may be infinite in the sense that it lacks an end point.

It is well to recognize, however, that (a) extension, *qua* created reality, depends continuously on God's creative power, and (b) that it is the principal attribute of a created nature, i.e., corporeal substance.

Let us consider (b) first. In so far as we think that being extended is both necessary and sufficient for something's being of material kind, we think of a particular this or that which is shaped or bounded. On the other hand, one and the same extension can be considered as space; but this is just a way of regarding the disposition of, or the relation among, a set of shaped, bounded, and extended quantities (AT. VIII. 45–46. *Principia*, II, articles 9–17). In reality, there is no distinction between the three dimensions that comprise corporeal substance and those which comprise the basis of the idea of space; that is, that which is extended and extension are the same. But this is just to deny the existence of an absolute, vacuum, if to be extended is to be material (AT. VIII. 46). So when Descartes speaks of the generic unity of spatial extension, he is not referring to anything that can exist independently of matter, but only to extension, *qua* extension, as that which is abstracted by the power of the understanding. So considered, extension is no longer an object of the imagination, that is of any particular configuration of extension, and can therefore be conceived as lacking limits; i.e., as infinite. But so considered, extension does not exist in *rerum natura*, since this is tantamount to considering the existence of an extension not connected inseparably with material reality. Given Descartes' principles, this is contradictory; "for wherever extension is, there, of necessity, is body also" (AT. V. 219). Consequently, there is nothing in the principles of Descartes' philosophy which compells him to posit the infinity of extension, even though the mind can imagine extension beyond any boundary taken.

In his reply to the *Second Objections*, Descartes tells us that "it is self-evident that it is greater perfection to be undivided than to be divided" (AT. VII. 138). While this claim is made in the context of discussing divine attributes, it is clearly a general principle that Descartes adheres to. It must be understood, however, in relation to Descartes' Platonic conception of unity. In *Meditation Five*, Descartes connects the indivisible perfection of the non-extended with unity and simplicity (AT. VII. 66–69). Divine being is an absolute unity because its nature is simple and all-inclusive. It is simple because God is prior to all differentation: in him being is everywhere one and the same. He is an absolute unity because he possess all perfections completely and inclusively: there is nothing *ab alio* on which God depends (AT. *Meditation* V). These characteristics of unity and simplicity are foreign to the nature of extension. Not only can it be differentiated conceptually, but its parts can be separated one from another. This means that the unity of any extension can be conceived as contituted by its parts of which the number and disposition can change. Since Descartes holds that any aggregation of extension is divisible, it follows that extension is an aggregate of aggregates of aggregates, without end. Accordingly, its unity is merely notional and relative to the possibility of a changing plurality of parts. It must be admitted that the implications of Descartes' position on extension do not show that it cannot be infinite. After all, granted that the unity of extension is supervenient on its parts, nothing is said as to its infinity or otherwise. I suspect, however, that Descartes, while not equivocating on the concept of unity, choose to restrict talk of positive infinity to that whose being and unity are one, i.e., to the nature and perfection of divinity as shown in the ontological argument.

I turn now to (a), Descartes' claim that extension *qua* created reality, is dependent continuously on divine power. Descartes certainly holds that any form of dependent being cannot be infinite, since "independence, conceived distinctly, involves in itself infinity" (AT. III. 191). It is clear that this statement is prompted by the nature of divine being, and equally clear that Descartes makes a categorial distinction between the independent existence of necessary being and the radical contingency of created being. Anything, he seems to say, that does not embody in its nature the power of existing beyond a given moment is not intrinsically infinite. Clearly, no reality but God can satisfy this conception. If this is a fair reconstruction of Descartes' reasoning, it seems that he has excluded everything from infinity in that sense except God. And we can well imagine, that in this frame of mind Descartes would think it of little consequence to consider whether or not extension is infinite. Suffice it to say that it is indefinite, on the grounds that the mind cannot show that it is bounded, or show that there are limits to its divisibility. Moreover, though he vacillates on whether extension is finite or infinite, Descartes' reasoning surely presupposes the traditional distinctions of being (*divisio entis*) into *necessarium* and *contingens, creans* and *creatum*, and *infinitum* and *finitum*. According to these division, infinity is intrinsic to divine perfection, and finitness is a necessary defect of created entities. So conceived, an infinite but created entity falls under a self-contradictory notion, and is thus an impossible existent which God cannot create. This does not impose any limitation on God's power as the impossible is not an intrinsic limit.

Descartes' position on the ontology of created extension is therefore different from Newton's. For the Frenchman, extension is an eternal archetype which exists eminently in the divine mind, and is that in accordance with which divine power brings created extension into being. For Newton the extension of space is coeval with the omnipresent existence of God.

It exists because God exists. It is little wonder, then, that Newton conceives extension as an eternal, uncreated and infinite nature, and as part of the defining essence of space itself.[19] And why he, but not Descartes, is explicit in holding that the parts of infinite extension exist at once and altogether: in the language of the schools Newtonian extension is *simul et semel.*

5. GEOMETRY AND INFINITY

Unlike Descartes, Newton holds that we can understand the infinity of extension through geometry. Why does Newton think this, and what relationship does he suppose exists between the geometry of infinity and the nature of space itself? Two important sources provide answers to these questions: the treatise that we have been considering, *De Gravitatione*; and a document of 1692 which I have entitled *Locus et Tempus.*[20]

Immediately after his claims in *De Gravitatione* that space "extends to infinity", and that everywhere it contains inherent, volumetric figures of all sizes and kinds, Newton says that "you may have in truth an example of something infinite [*Verum ut infiniti specimen aliquod habeatis*]": he asks us to consider any triangle under the following conditions:

> Imagine (*fingite*) any triangle whose base and one side are at rest and the other side so turns about the contiguous end of its base in the plane of the triangle that the triangle gradually opens at the vertex (*in vertice gradatim apperiatur*): and meanwhile animadvert to the point where the two sides meet, if they are produced that far: it is manifest that all these points are found on the straight line along which the fixed side lies, and that they become perpetually more distant as the moving side turns further until the two sides become parallel and can no longer meet anywhere. Now I ask: What was the distance of the last point where the sides meet? It was certainly greater than any assignable distance, or rather none of the points was the last (*vel potius nullum e punctis fuit ultimum*), and so the straight line in which all those meeting points lie is in fact greater than finite (*est actu plusquam finita*) (*De Grav.* p. 101).

Newton's argument involves some interesting intuitions which can be illustrated by considering the following triangle:

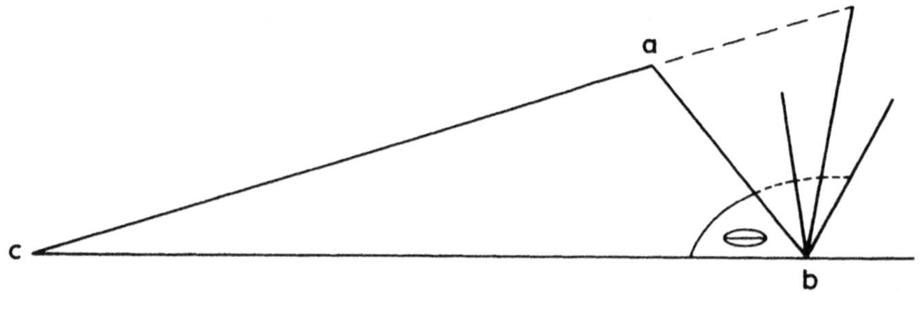

Point *a* is the intersection of sides *ac* and *ab*, and sides *ca* and *cb* are fixed. Now allow angle ⊖ to increase, so as to open at the vertex *b*. Also allow sides *ac* and *ab* to adjust so that *ab* and *ac* intersect as *ac* is produced. As the angle ⊖ opens out, sides *ac* and *ab* approach parallel, so that the array of intersection points transversed before parallelism is infinite. Thus in the geometry we can *understand* that a distance which has no upper bound is infinite.

What are we to make of this argument? Notice first what Newton is not maintaining. He is not saying that the two parallel lines meet *at* infinity; rather, he holds that as one line is rotated towards the position of parallelism with the other, there is *no* last point where they meet. What is infinite is not the distance of the last point, but the array of intersection points traversed before parallelism is reached: it is the line *ac* itself "in which *all* those meeting points lie" that is infinite.

Newton takes himself to have established the following: (a) that the line exists as a clear and distinct object of the understanding; (b) that all the intersection points traversed, lie along it; and (c) that the totality of intersection points is infinite. Can he maintain these claims? I think that the geometers of his period would give Newton the existence of the line as an object of understanding. After all, the property that his argument attributes to it, is consistent with its nature, namely, that since there is no last intersection point on the line traversed, it is infinite. From a modern set-theoretical perspective, however, we would demand a separate proof, one which shows that there is a line which satisfies the claim that an infinite totality of intersection points lies along it. In other words, we would want to know whether it is possible for a line satisfying this condition to exist. In Newton's case, this would mean a proof that any denial that the totality of intersection points is infinite, is open to a *reductio ad absurdum* argument.

But what response would Descartes make to Newton's argument? He would probably construe it from a constructivist perspective. For example, he could say that though Newton's argument shows that we cannot determine that there is a last point on the line traversed, this in itself does not warrant the claim that the line is infinite, but rather the claim that it is greater than any assigned number of intersection points traversed. Now a construal of this sort is certainly consistent with Descartes' conception of the nature of quantity. It does not, however, capture the intuition on which Newton's argument rests, which is simply that there is no last point on the line where the sides meet. And if it is impossible that there be a last point, this is tantamount to the claim that the line is infinite. For the length of the line is greater than any finite length, and so we can understand an infinite length or distance. In fact, Descartes' claim that we can speak of an intersection point beyond any assigned, rests on the Newtonian intuition that the line is infinite.

But Newton has another triangle argument. This argument claims that the infinite line is not only understandable, but that it actually exists:

Nor can anyone say that this is infinite only in imagination, and not in fact (*non actu infinitam esse*); for if a triangle is actually drawn, its sides are always in fact directed towards some common point, where both would meet if produced, (*ubi concurrerent ambo si modo producerentur*) and therefore there is always a point where the produced sides would actually meet, although it may be imagined (*fingatur*) to be beyond the limits of the corporeal universe. And so the line traced by all these points will be real (*erit actualis*), though it extends beyond all distance (*De Grav.* p. 101).

This is a strange argument, for it rests on a mis-stated conterfactual. To say that if the sides of the triangle are in fact drawn, they are always directed to some common point where they *would* meet if produced, makes no sense. If the sides make a closed figure – in this case a triangle – they must *actually* meet. Newton must mean that 'if two non-parallel line segments are drawn in a plane, they are always in fact directed towards some common point, where both would meet if produced'. The argument now makes sense, but we are given no reason for believing that the common point exists. To say "it may be imagined to be beyond the limits of the corporeal universe" does not help. Of course we can *imagine* the point to be wherever we please, but this is no argument to show that it actually *is* anywhere.

Some twenty-four years after writing *De Gravitatione*, Newton returned to his claim that infinity is understandable through geometry in the document entitled *Locus et Tempus*. As I have argued, this document is in all likelihood part of a revision of the *Principia* which engaged Newton during the early 1690's. In it he presents arguments that are related to the Scholium on space and time of the *Principia*, and others that anticipate the theological views put forward first in the General Scholium to the second edition of *Principia* (1713).

Once again Newton is concerned to show that infinities are understandable, especially the notion of an infinite distance. While admitting that infinities are difficult to conceive of, Newton nevertheless maintains that "there are many things concerning numbers and magnitudes which to men not learned in mathematics will appear paradoxical, and yet are entirely true" (*Locus et Tempus*, p. 119). He then proceeds to give examples of infinities, which fall into four categories: (a) the properties of infinity as determined by the geometry of conic sections; (b) the generation of infinite loci by moving points and lines; (c) the divisibility of lines to infinity; and lastly, (d) the mutual approach of things converging to limit at infinity.

It is an example of the kind of infinity mentioned in (a) which concerns us here. Newton characterizes it in general terms as follows: "that a point by moving according to a given law in a finite time, can go off to infinity, and conversely can return to us from infinity in the same time". The example he takes is this: "that if any straight line is revolved round a fixed point with uniform motion, and meanwhile cuts another straight line fixed in position, the point of intersection in each revolution will twice go to infinity and twice return from infinity" (*Locus et Tempus*, p. 119).

This example is geometrically isomorphic with the opening vertex example of *De Gravitatione*, and the logical structure of the argument is the same. As the moving line (fixed at one end) sweeps along the side of the given line, there is a moment such that the point of intersection of the two lines is infinitely far out. This can be understood as before. The moving line generates a sequence of points until it becomes parallel to the given line, the points being the points of intersection of the two lines determined by this motion. The length of the given line is greater than any finite length; hence we can understand an infinite length or distance. Once again the geometry enables us to express the intuition that a distance which has no upper bound is infinite. In this example, however, Newton conceives that the rotating line completes a total revolution around its fixed end. In this case there is an asymmetry in the motion of the rotating line at opposite ends of the line given in position. Thus at one end of the given line, the array of intersection points traversed *before* the rotating line changes direction is infinite; at the opposite end, the array of intersection points traversed *after* the rotating line changes direction is infinite. So given that the points of intersection generated by the rotating line "in each revolution will twice go to infinity and twice return from infinity", the length of the given line in either direction is greater than any finite length; once again we can understand an infinite distance. In this example, Newton has dispensed with the triangle as a device for generating a succession of intersection points, and speaks directly of proceeding to greater and greater distances along the line. There can be no question here that he is trying to make the notion 'actually more than finite' understandable by seeing it as a function of more readily intuited concepts exemplified by the triangle, i.e., other lines, angles, intersection, parallelism, etc. As a matter of fact, many of the same intuitive notions are here appealed to, only now we simply have a line sweeping out 360° as it intersects a fixed line!

It is clear from these arguments that Newton holds not only that we understand an infinite length or distance, but also (as against Descartes) that what we understand is positively conceived. This immediately prompts the following question: what is the ground of conceivability for the geometrically expressed and positive idea of infinite distance which we have? In other words, what is the relationship in Newton's view between conceiving of an infinite distance and the basis in reality for that conception? This question is especially pressing, since Newton's attempt to show that the 'infinite' triangle exists not only in the imagination, but also in actuality, does not succeed. The question at hand comes to this: is there a correspondence between the geometry of infinite distance and Newton's conception of the nature of infinitely extended space?

One response open to Newton is one evident in both *De Gravitatione* and the document *Locus et Tempus*. Mathematics provides us with a sure foundation in terms of which to conceive the properties of the infinite. On the traditional assumption that geometry is the science of continuous magnitude, anything that can be said about magnitude consistently in geometry is at least consistent with the nature of spatial extension. So if we can speak of infinite distance in geometry, equally we can speak of infinite distance in space itself. This, of course, does not show that space is in fact infinitely extended, but it shows that we can coherently conceive of space as infinite.

There is no doubt that at a certain level this is Newton's view. From *De Gravitatione*, however, a more metaphysical approach to the question of how geometry and space correspond can be reconstructed. It involves three commitments: (a) the claim that space actually exists; (b) the claim that space possesses an infinitude of inherent, actual, and geometrical figures which are everywhere present in its infinite expanse; and (c) the implicit thesis that the objects of geometry exist in the reality of space itself.

I shall not examine the first commitment. Suffice it to say that Newton advances theological arguments by which he purports to show that space is actually infinite just because it is a consequence of God's real and omnipresent existence.[21] Given that this claim can be made out, namely, that space is actually infinite, what is the import of commitment (b)? After arguing that space contains a three-dimensional network of figures of all kinds which are everywhere fused together at their common boundaries Newton states:

And hence there are everywhere all kinds of figures, everywhere spheres, cubes, triangles, straight lines, everywhere circular, elliptical, parabolical and all other kinds of figures, and those of all shapes and sizes, even though they are not delineated by sight (*non ad*

> *visum delineatae*). For the material delineation of any figure (*Nam materialis delineatio figurae alicujus*) is not a new production of that figure as regards space, but only a corporeal representation it of, so that what was first insensible in space now appears to the senses to exist (*ut jam sensibus appareat esse quae prius fuit insensibilis in spatio*). For thus we believe all those spaces to be spherical through which any sphere ever passes, being progressively moved from moment to moment, even though a sensible trace of the sphere no longer remains there. We firmly believe that the space was spherical before the sphere occupied it. (*imo spatium credimus prius fuisse sphaericum quam sphaera occupabat*), so that it could contain the sphere; and hence as there are everywhere spaces that can adequately contain any material sphere, it is clear that space is everywhere spherical. And so of other figures (*De Grav.* p. 100).

A number of claims are made in this passage. In the first place, Newton outlines the ontological conditions that obtain in space so that God can be said to bring matter into being there. Briefly put, space is so structured that God, in virtue of his creative power, can 'pick out' and condition various of space's inherent figures so that they in turn can cause the mind, via sensory experience, to form representations of material phenomena. In other words, Newton is putting forward a theory to show how God can be said to create material phenomena, given that space's inherent figures are omnipresent 'place-holders' in which active powers of the sort requisite to cause the mind's inner representations of material phenomena can come to be.

This is not the place to examine the details of Newton's theory regarding the creation of matter, nor to consider whether the ontology of matter of *De Gravitatione* is ultimately coherent. But what is important here is that we get a firm grasp on his views concerning the structure of space. The figures in space actually inhere *in* space everywhere and eternally, and are immutably present in the sense that they cannot be separated from space itself. Thus Newton is not claiming that space is the potentiality of figures in the sence that bodies of different sizes and shapes can occupy it either successively or at the same time, but is rather claiming that figures of all sorts and sizes are *actually* omnipresent in its infinite expanse. Accordingly, we can conceive of space as having structure because it in fact has an immutable structure.

Again we see that Newton's conception of extension is very different from Descartes'. In Descartes' view extension is inextricably connected with materical reality. Cartesian *Res extensa* does not contain actual parts, but it can be conceived of as having parts because we can actually divide it, or lay down an algorithm for dividing it in theory. But in Newton's view extension is not inextricably connected with material reality; on the contrary, it exists eternally. Moreover, it is *eo ipsa* a condition presupposed by the very existence of matter itself. While it is not necessary to consider here whether Newton begs the question against Descartes in denying the latter's claim that extension is necessarily connected with the corporeal (Newton in fact considers the problem in *De Gravitatione*), it is abundantly clear that his conception of how spatial extension is structured differs fundamentally from Descartes' theory of material extension with respect to which space is defined derivatively.

In Newton's view the presence of matter presupposes ontologically the infinitude of spatial extension; whereas, in Descartes' case, space is merely a mode of material extension. But if space is a form of real being, as Newton claims, it must possess a differentiated structure which the mind can grasp. Otherwise it is in danger of slipping into the realm of indeterminacy and non-being, of which everything or nothing can be said. Moreover, in order to be a knowable reality space's structure must be intelligible. In this regard, there is no other candidate for Newton but the framework of Eucidean geometry.

As we might expect, however, Newton does not hold that Euclidean geometry merely affords a set of descriptions consistent with the nature of space. The constraints which delimit the scope of conceptual well-foundedness are to be drawn more determinately than that. Accordingly, in Newton's view, for anything that can be said consistently of space, there must exist a real and corresponding ground in the nature of space itself. In other words, in conceiving the quantitative nature of anything, Newton accepts a form of the real distinction, which holds that the mind can differentiate something quantitative just in case there are direct counterparts in the thing itself which allow it to be conceived thus and so (*Questions*, folio 88^{v}). The contrast to Aristotle's view of how the mind differentiates quantity could not be more marked. In answer to question (c) Newton is forthright. Where do the objects of geometry exist? They exist in the omnipresent expanse of infinite space. Accordingly, they are not *entia rationis*, or ideal entities whose *esse* is simply to exist in the mind alone. Rather, they exist, extra-mentally and independently of human conception, in the reality of space itself; whereas they exist in the mind derivatively in so far as it has the capacity to apprehend them.

But what connection is there between these sorts of ontological claims and Newton's argument that an infinite distance is a distinct object of the understanding? Do Newton's geometrical arguments presuppose this ontology? If so, is he not presupposing what he takes the geometry to demonstrate, i.e., that we can understand an infinite distance. Furthermore, what role does experience play in relating the geometrically established conception of infinite distance and Newton's claim that "there are everywhere – straight lines" in the expanse of space itself.

Newton's geometrical arguments for establishing an understanding of infinite length stand alone, and, in that context presuppose nothing beyond the geometrical framework which he employs. However, if we ask on what basis it is that the mind forms geometrical concepts, we are asking a different question. The empiricist would answer that geometrical objects are idealizations of shapes and lengths abstracted from sensory particulars by the power of the mind; or, put more crudely, that they are simply ideas arising from observations of recurrent features of experience.

This, of course, is not Newton's answer. Given that he accepts the epistemic principle that knowing that something is imperfect of a kind presupposes a conception of the kind's perfection, Newton claims that the understanding grasps the nature of extension *through*, but not in abstraction *from*, the deliverances of sensory experience. Put more precisely, the understanding, not the senses or the imagination, apprehends by means of its representative content the general nature of extension as it is variously and imperfectly exemplified by 'materially delineated' bits of extension, i.e., it apprehends that the extension of space is in itself infinite. Thus the relationship between the understanding's distinct idea of infinite distance and the infinity of distance in space itself is this. When the mind attends to its idea of infinite distance in the context of a geometrical argument, that idea of itself is not *what* we know; rather, it is the vehicle by *which* the thought of infinite distance takes place. In other words, the content, as such, of the mind's conception of infinite distance is not what the thought is *about*; it is about the infinity of distance in space itself, though this, of course, involves the mind mobilizing or exercising an ability to think a thought with that content. Furtherfore, the likeness between our geometrically expressed idea *of* infinite distance and infinite distance itself is not one of portrait to original, but rather that of the resemblance between effect and cause as mediated via active dispositions present in the objects of sensory experience. It cannot be otherwise, if for Newton our understanding of the infinity of distance is an affair of the intellect, and not of pictures or images in the imagination as it is for Locke. In this sense, then, the understanding possess a non-sensuous representation of infinite distance which has its ultimate ground in the real but uncreated nature of extension itself. Epistemological alternatives are thus rejected by Newton, i.e., that the mind possesses a special mode of intuition capable of recognizing directly the objective essence of space, or, alternatively, that it has an innate idea of space itself.

As I have reconstructed Newton's reasoning there appears to be a kinship with Descartes' arguments concerning the nature of the wax. Someone might object: Descartes' arguments are about the essence of a particular piece of wax, which in its entirety cannot be an object of experience. Moreover, Descartes throughout insists that extension is that which is imaginable and "has parts outside other parts of a determinate size and shape – " (AT. V. 268). So although the understanding alone grasps that the essential nature of wax is extension, at any given time the wax has some determinate shape or other which falls under the scope of the imagination, while in the case of the chiligon the imagination, though it cannot conceive it all at once, can at least attend to its sides successively. In no sense, however, can the infinite extension of space, as such, fall under the scope of the imagination.

These objections are well made, and they indicate a fundamental diffence in how Descartes and Newton conceive of the relationship between knowing the nature of extension and that nature itself. The difference can be put precisely: Newton, but not Descartes, is committed to the principle that the nature of extension, considered in itself, is ontologically prior to all finite determinations consistent with its nature. Given that Newtonian space is in fact an infinitely extended expanse, at once containing an infinitude of inherent and extended figures, in perceiving these figures through their 'material delineation' the mind comes to grasp the ontological priority of extension's infinity to that of its inherent determinations.

Reconstructed, Newton's position is this. By means of geometrical arguments the understanding can distinctly and positively conceive that space is infinitely extended; but since conceiving *that* involves conceiving *of* (and conversely), when the mind conceives *that* extension is infinite it conceives *of* extension. And consistent with the Euclidian conception *of* infinitely extended space is Newton's claim *that* it is differentiated by means of geometrical figures of various sorts and sizes. But the unity of infinite extension, which can be grasped distinctly by the understanding, is prior to the various figures which inhere in it. Thus, according to Newton there is a parallel between the ontological priority of infinite entension itself and the fact that understanding grasps it distinctly, a view consistent with claiming that confused sensory ideas of finite extensions are prior in the order of conceiving to those which are more distinct.

Newton's position can be compared with Aristotle's as there are similarities and differences. Moreover, Newton himself thinks that his position has affinities with the Aristotelians (*De Grav*. p. 106–107). For Aristotle, intelligible matter, the continuous and quantitative in three dimensions, is that which underlies all differentiation in terms of the genera of quantity and their specific differences. Furthermore, intelligible matter, the subject of all determinate figures, is always figured in some determinate way, and can never be said to exist apart from quantified individuals. Newton expressly states that there is an analogy between Aristotle's notion and his own doctrine of infinite extension, on the ground that the latter is the intelligible ground and subject of an infinitude of determinate figures (*De Grav*. p. 106–107). But here the similarity ends. According to Newton's conception, extension and its determinate figures can and do exist separately from the active powers of matter with which God endows certain figures in space. Moreover, though infinite extension underlies the 'material delineation' of space's inherent figures, Newton nowhere provides a detailed account of how the mind moves, in genus and species terms, from the perceptible material delineation' of space's figures to a grasp of the intelligible reality (extension) which underlies them. As I noted earlier in Section 2, Newton holds that natures are prior to natured things, and thus that extension is prior to the extended figures that inhere in it. But he gives no clues (as does Aristotle), as to how, for example, extension itself is prior to a particular triangle and how they are related by means of a genera/species patterning. Accordingly, I have tried to reconstruct it in a manner consistent with Newton's text.

What Descartes cannot accept is Newton's claim that infinite extension itself (conceived as existing all together and at once) is prior to the actual determinations of its inherent figures in the sense that their being depends on its existence. For Descartes, the metaphysical conclusion to be derived from his epistemic 'appraisal' of the wax is that extension in itself is a changeable quantity which is capable of increase or decrease in either thought or reality. So considered, it is an imperfect, dependent and created nature which is divisible potentially to infinity. And since Descartes conceives of material extension (generically) as a homogeneous, undifferentiated plenum, which is everywhere indistinguishably the same, its unity is that of a whole to its potential parts, a conception very different from Newton's view that the unity and infinity of extended space is prior to its inherent figures, a claim compatible with the conception that finite cases of extension are cognitively prior in the order of knowing.

There would appear, however, to be a difficult in Newton's position. If extension is an uncreated and infinite being which exists necessarily (*De Grav*. p. 105) is its infinitude not qualified by the finiteness of its inherent figures? In response, Newton can mobilize his conception of infinity as a second-level characterization of existing natures. Since extension and its actually indivisible figures are mutually compatible natures, one can be characterized as infinite in magnitude, the other as infinite in number. In other words, given Newton's account of how infinity is attributable, the infinity of space's extension is compossible with the infinitude of its inherent 'parts'. Although there are deep problems concerning the unity of extended space (i.e., Newton denies that it is a substance), which go beyond the scope of this paper, in terms of its gross structure space is a compositional unity in which infinity and the immutability and non-separability of its inherent figures are linked.

CONCLUSION

To say the least, the issues which I have been considering are complexly interrelated. I have attempted to show that they can be dealt with, effectively, only if we willingly engage in conceptual analysis which is sensitive to the intellectual and metaphysical framework of the thinkers under consideration. More is involved, of course, then merely exposing the shortcomings in the arguments of a Newton or a Descartes. To be sure, this must be done. But it cannot be done without displaying the coherence and implications of their reasoning in relation to the assumptions and principles which each thinker accepts or denies. And this means getting hold of the traditions that they assume, desperate though it seems.

It is probably the case that today we will reject any representative theory of knowledge which is based on an unanalyzed notion of mental stuff; probable, too, that we will deny that the mind 'intuits' objective natures; and it is more than likely that we will discard as misguided the metaphysical views of Newton and Descartes on infinities and geometry. However, if we take seriously the possibility of doing disciplined conceptual history we have got to enter the 'dynamics' that drive the thinking of men such as Newton and Descartes. This will enable us to see why they argue as they do, allow us to isolate the motivations for their arguments, and provide an understanding of the tacit and explicit implications which they draw from their intellectual commitments.

Newton and Descartes alike view knowledge as the mind's confrontation with an outer reality via its inner representations. For both, the 'grounding' of the known is not in terms of what is abstracted from the given of experience, but in terms of the understanding's power to apprehend the outer world *through* the medium of sensory experience.

I wish to thank the NSF (Grant 5–36061) for generously supporting the research for this paper.

University of Pittsburgh

NOTES

[1] A. Rupert Hall and Maria Boas Hall, *Unpublished Scientific Papers of Isaac Newton*, Cambridge University Press, 1962, *De Gravitatione*, pp. 99, 100–101, and 110. Subsequent page references to *De Grav.* will be in the text.
[2] U. L. C. MS. Add. 3996. folio 83^{r-v}. Subsequent references to the *Questiones* will be in the text. Professor Martin Tanny and I are completing a critical edition of *Questiones* which will be published by the Cambridge University Press. In all passages quoted from the *Questiones* spelling has been modernized.
[3] John Locke, *An Essay Concerning Human Understanding*, London, Everyman's Library, 1972, Vol. I, Bk. II, Chapters XVI and XVII. Subsequent references are to this edition will be in the text.
[4] J. E. McGuire, 'Existence, Actuality and Necessity: Newton on Space and Time', *Annals of Science* 35 (1978), 463–509.
[5] London, 1687, Part XIII, Chapter 11, p. 858.
[6] Diogenes Laertius, *De Vitis Dogmatis et Apophthegmatis – Libri X*, London, 1664. Epicurus Lib. X, p. 276. This page is dog-eared by Newton, as indeed are many pages in the edition having to do with the infinity of the world.
[7] London, 1654, Bk. I, Chapter 11, Section 11, 9–15. There are many late 16th century and early 17th century copies of Lucretius' poem in the Trinity College Library. There is no evidence, however, that Newton used them. Nevertheless, it seems improbable that he did not know of Lucretius.
[8] Nicolas de Cusa, *Opera Omnia*. Heidelberg Academy Edition, ed. Ernest Hoffmann & Raymund Klibansky, Lipsig, 1932, Vol. I, *De Docta Ignorantia*, Chapter XVII, p. 33. Since nothing determinate can be infinite, Cusanus never affirms the infinity of the world, as Descartes mistakenly claims he does (AT. V. 50, To Chanut, 6, June 1647). D. W. Singer, *Giordano Bruno, His Life and Thought*, New York: Abelard-Schuman, 1950. *On the Infinite Universe and the Worlds*. First Dialogue, p. 257. Gottfried Liebniz, *New Essays Concerning Human Understanding*, the Open Court Company, Chicago, 1916, the critique of Bk. II of Locke's *Essay*, Chapter XVII, pp. 161–164.
[9] *Ibid*., Chapter XV, pp. 158–159 and Chapter XVII, pp. 169–174. Leibniz's approach to infinity must be understood in the context of his claim that ostensibly extended phenomena are not true unities in the manner of indivisibles whose unity and being are one.
[10] U. L. C. MS. Add. 3995. folios 9^{r-v}–18^{r}.
[11] Folio 83^{r-v}. Both folios are copiously filled with citations from, and page references to, Descartes' *Meditations* and the *Objections and Replies*. In every case Newton's citations and page references are to the *Opera Philosophica*, 3rd edition, Amsterdam, 1656, which includes Descartes' complete works. The exact citations and references to this edition leave no doubt that Newton had read Descartes' philosophical writings carefully.

[12] *Oeuvres de Descartes*, Charles Adam and Paul Tannery (eds.), Libraire Philosophique J. Urin, Paris, 1973. *Meditation II*, pp. 28–29, Vol. VII. In the future, all references will be in the text as (AT. VII. 28–29).

[13] John Wallis, *Operum Mathematicorum Pars Altera*, Oxford, 1656, *Arithmetica Infinitorum in pars altera*, propositions 34 and 40.

[14] For a useful discussion of Descartes' distinction between the 'indefinite' and the infinite, see John Cottingham (ed. and transl., *Descartes' Conversations with Burman*. Clarendon Press, Oxford, 1976), Commentary, p. 101.

[15] *Ibid*., See Cottingham's discussion of the notion of adequate knowledge in Descartes, p. 65.

[16] In his reply to my paper at the Montreal meeting of the Union of the History and Philosophy of Science, pp. 18–19.

[17] J. E. McGuire, 'Newton on Place, Time, and God: An Unpublished Source', *British Journal for the History of Science, II*, (1978), 114–129, p. 119.

[18] *Ibid*., pp. 119 and 121.

[19] *Op. Cit*., note 4. See Section Three entitled 'Existence and natures', pp. 474–487.

[20] *Op. Cit*., note 17.

[21] *Ibid.*

J. D. NORTH

FINITE AND OTHERWISE. ARISTOTLE AND SOME SEVENTEENTH CENTURY VIEWS.

It is unwise to discuss so pervasive a cluster of concepts from 17th century natural philosophy as space, matter, extension, and infinity, without referring to Aristotelian and scholastic doctrine. One may argue far into the night over the question of whether Newton's words reveal a dependance on Burthogge, rather than Herveus Natalis, and whether Descartes at a particular place was using Anselm or John Damascene; but one thing we must surely accept is that the *language* used was that of the Aristotelian inheritance. Of course it was changed in many ways; but just as when trees are turned into telegraph poles, the grain of the wood remaining visible, so with the language of Aristotle. Since we are particularly concerned with doctrines of the infinite, I will give the merest outline of those parts of Aristotle which seem to me to have a bearing on the 17th century discussion. I will then place rather more emphasis than Ted McGuire is inclined to do on the common element in the thought of Descartes, Locke, and Newton. In the time available I shall not be able to make more than passing reference to scholastic discussions of the same themes, although I am sure that to do so would show that, whether he liked it or not, 17th century man was a lineal descendent of the scholastics. As it happened, Georg Cantor gave the palm to the seventeenth century for its arguments against the actual infinite – arguments which he judged to be more cogent than Aristotle's, but arguments which he believed he could refute. He commended Locke, Descartes, Spinoza, and Leibniz, while suggesting Hobbes and Berkeley as additional reading.

ARISTOTLE ON THE INFINITE

Aristotle opens his discussion of the infinite with a sentence that could hardly have been bettered as a source of questions for debate:

> The science of nature is concerned with spatial magnitudes and motion and time, each of these being necessarily either infinite or finite – even though some things dealt with by the science (e.g. a quality or a point) are not, and should perhaps not be, put necessarily under either heading.[1]

After a brief survey of the views of the Pythagoreans, Anaxagoras, the

William R. Shea (ed.), Nature Mathematized, 113–148.

atomists, and Plato (of whom more anon), Aristotle sets out his own position. He distinguishes (i) the infinite in respect of addition, and (ii) the infinite in respect of division – number being infinite in the first sense, space in the second, and time in both.[2] Even more fundamental, however, is Aristotle's (perfectly natural but also perfectly dangerous) insistence that he takes the word "infinite" in the sense "*that which cannot be gone through*". It might be of the nature of a thing that it is "incapable of being gone through" (and "this is the sense in which the voice is invisible" – an example repeated at *Met.* 1066a. 35–); the process of traversing might be possible, but have no termination; the traversal might present difficulties; or it might not be *actually* begun or ended. There are many echoes of these distinctions in the middle ages. Aquinas and Ockham accepted them, more or less – to name only two influential writers.[3]

Here, then, is one statement amounting to a definition of what Aristotle meant by "infinite". Even though the *Physica* is a work chiefly devoted to the study of the objects of sense, Aristotle clearly wanted the word to take its primary meaning from mathematics – where the integers provide a paradigm of a sequence which "cannot be gone through". In *Metaphysica* (K. 10), he shows that he has seen, and studiously left alone, two concepts beloved of later theologians. His infinite is not a "separate, independent thing", nor is it "the indivisible". The first of these would reduce to the second – or so he appears to think – and he is simply "not examining this sort of infinite, but the infinite as untraversable". For Aristotle, the infinite must be "of a certain quantity", and it cannot, in consequence, be indivisible.[4] It is emphatically not The One.

Another concept that Aristotle seems to be carefully avoiding is *space*, at least in the sense of χώρα. He has much to say about *place*, and *spatial magnitude*, and the *void*, however. Place is the limit of the container, the first (i.e. working outwards) unmoved body that bounds the thing with whose place we are concerned.[5] Everything in the universe is in place, but not the universe itself.[6] Clearly what he says about *body* has implications for place – and thus (since for Aristotle there can be no infinite body) it is misleading to say, as is occasionally said, that "Aristotle supposed space to be infinite in extent". Again, on the question of the void, it is too simplistic by far to say only that Aristotle "rejected the notion of void". Anticipating a little, in order to put the problem of the void to one side, I should like to draw attention to an extremely interesting analogy between the void and the infinite, as drawn by Aristotle in the *Metaphysica*. There is, he explains, no actual void (whether separate from bodies, occupied by bodies, or in the interstices in bodies), just as there is no actual infinite; but just as a line is infinitely divisible, so we can always imagine a body of lesser density than any specified density. Matter, for Aristotle, is a continuum, but there is no lower limit to the density of matter.[7] I imagine that the typical 19th century exponent of the physics of the continuum would have said much the same sort of thing.

Aristotle's discussion of the infinite is not one of his most systematic, but running through it all are two important conceptual distinctions. The first, between "infinite in respect of addition" and "infinite in respect of division" – between infinite and infinitesimal, as some would later have said – has already been mentioned. The second is the notorious distinction between *potential* and *actual* infinite. Since he is especially concerned with the concepts of number, space, time, and body, in this connection, we can see that such a question as "Is there an actually infinite body with respect to addition?" is one of sixteen questions that Aristotle deemed important (four topics with two twofold distinctions). Needless to say, there are other topics of interest than the four named; and all the questions of existence can be turned into questions of consistency – "Is the notion of an infinite body inconsistent?", and so on. In fact Aristotle attaches great importance to this last sort of question, as we shall see.

Consider first the *Physica*, where, although making passing reference to such general mathematical questions as "whether the infinite can be present in mathematical objects and things which are intelligible and do not have extension", he is mainly concerned with objects of sense. Is there an infinitely great body? Aristotle argues at length that there is not. His arguments take many forms, and are often unsatisfactory to the extent that they depend on his theory of the natural places of the four elements and the "balance" of contraries. He makes several strong philosophical points, even so. If, for instance, we mean by "body" something that is "bounded by a surface", then the notion of "unlimited body" is inconsistent.[8] On the other hand, to suppose that the infinite does not exist in any sense leads to difficulties: (1) there will be a beginning and an end of time; (2) a magnitude will not be divisible into magnitudes; (3) number will not be infinite. "Clearly there is a sense in which the infinite exists and another in which it does not".[9] It would be easy, but misleading, to say that for Aristotle the existent infinite is potential and the illegitimate infinite the actual (or better "completed"). He will certainly allow potential infinity to number, space, and time (with respect to addition) and to space and time (with respect to division). They are potentially infinite in the sense that they may grow beyond any given bound.

There were tensions here, of course, in the Christian commentaries, for obvious reasons: Christians and many Muslims supposed the world to have been created at a specific time, and to be destined to end at another definite time. But there were tensions too in Aristotle's own thought, for his universe was, after all, limited in spatial extent. Could there not be *something* potentially infinite beyond the bounds of the finite universe? The scholastic, no less than the Christian philosopher of later centuries, found problems here – problems of the heavens beyond the stars, problems of the implied limitation to God's power in saying that his creation is only finite, and so forth – but they were problems of a different sort from those of Aristotle, who does not seem to me to have come completely to terms with his potential/actual distinction. I will indicate briefly some of the difficulties.

"In respect of addition", he writes," there cannot be an infinite which even potentially exceeds every assignable magnitude unless it has the attribute of being actually infinite, as the physicists hold to be true of the body which is outside the world, whose essential nature [they hold to be] air or something of the kind".[10] We have seen why he rejects infinite body; but surely the quotation just given concerns more than the case of infinite body. Suppose we accept his various arguments for the finiteness of the universe, and his claim that no sensible magnitude is infinite.[11] In view of the quotation, why, however, will he not grant actual infinity ("with respect to addition") to *time*, to take an example of something to which he is prepared to grant potential infinity? It has potential infinity, we are told, in that it may grow beyond any given bound; but it is said to be no real or actual infinity in the sense of an infinite given whole, since *its parts do not coexist*. I think it is clear from this remark that Aristotle is putting a great deal of weight on an analogy – in the words of the scholastic, his "actually infinite" is "the infinite of the heap". We might be inclined to associate a number with each point of time, but the number is not a "magnitude" in the sense used in the quotation. This is not at all strange; but it does seem strange to us to refuse to call an *interval* of time a "whole", and therefore presumably to deny that it is a magnitude. Aristotle, like every other physicist of significance, was prepared to think of time intervals as something that could halved, doubled, and so on, whether or not the parts of time could be said to coexist.

One of the most tempting interpretations of Aristotle's "potentially infinite" is as simply "conceivably infinite". But what does this mean? It would be plainly trivial to say that it means "conceivably potentially infinite". The danger in the most obvious alternative is that it be treated as a visualisable thing, a conceptual object in a crude sense. There are remarks by Aristotle with a bearing on this. The geometers are said to neither need nor use the infinite; and yet – adds Aristotle – they postulate "that the finite straight line may be produced as far as they wish".[12] Aristotle's concession to the geometers has suggested to some that he accepted a potential infinite in the sense of a special sort of separately existing thing. The only "thing" it can be is the possibility of a certain sort of mental activity which is in principle unending. (Note how he speaks of mathematicians' thoughts as an actuality.)[13] In an unsophisticated sense of the word "actual", it is Aristotle's main purpose to put an end to talk of an actual infinite, and at its very heart his analysis rests on an analysis of meanings. "A quantity is infinite if it is such that we can always take a part outside what has already been taken", and yet "whole" and "complete" are such that there is nothing more to be taken (at least in the respect considered as making the whole a whole). An actual infinite is a contradiction in terms.

Without modifying such definitions, no "completed infinite" is possible, and the most interesting parts of Ted McGuire's paper seem to me to be where he is concerned with 17th century efforts at revising definitions. In the end, the revisions seem to have been more apparent than real. Before turning to them, however, let me mention a close analysis of many Aristotelian texts made by Jaako Hintikka, who asked why Aristotle, who certainly seems to have firmly believed that no genuine possibility can remain unactualised through an infinity of time ("the principle of plenitude", as Lovejoy called it), gave infinity potential but not actual existence.[14] His conclusion, if I may summarize it in an excessively brief way, is that the principle of plenitude *is* satisfied, in the sense that "the endless coming-to-be of further and further divisions is 'ensured' (Aristotle's verb is *apodidômi*) by the potential existence of the activity of diving".[15]

There are other ways of sidestepping Aristotle's conclusions, as I have indicated, by the expedient of modifying his definitions, but it is worth drawing attention to the stipulation that we must be able to "take a part *outside what has already been taken*". To those who, with Riemannian hindsight, would throw at him the modern notion of finite but unbounded, his answer is waiting in the middle of our *Physica* III. 6, where he considers the case of an endless ring. With such a thing "it is always possible to take a part which is outside a given part"; but – despite the similarity with the unacceptable notion of a "completed infinite" – the cases are not fully alike, for "it is necessary that the next part which is taken should never be the same". He looks for endless enumeration in the supposed counterexample, and does not find it. His ring is indeed finite and unbounded, and had he been more concerned with the property of boundedness, he would no doubt have said as much.

As Aristotle's doctrine of the infinite was handed on to the 17th century, it was modified, disjointed, and mixed with an extraordinary variety of extraneous ideas. There was the Platonic identification of the infinite with the indeterminate and formless – which could hardly therefore be identified with a Christian's God. We find a line of thought coming from Plotinus, who takes the Platonic doctrine (in connection with matter, which could be supposed evil with impunity, since it could not answer back), and adds a theory of infinite Mind, complete, self-sufficient, and all-powerful. Whenever God (or some mind-like entity somehow involved in the world) entered the discussion, it became doubly complicated, for not only were there extra classifications to be made (one might even say "hierarchical" ones, since God must always come out better than his creation), and problems of *relating* the different sorts of entity (God, the world, space, time, and so on), but there was also need for a revision of the very *concepts* of the infinite. *Number* tended to take a secondary place, elbowed out by the ineffable – whether it be essence, the unquantifiable act, reason, power, goodness, or whatever. All the mystery of the numerical infinite, for every eluding the human grasp, became transferred to the divine infinite – but in logical structure the two subjects were very discrepant, however hard the scholastics tried to run them together. The ever present danger was of course that of pantheism – with the world sharing in the divine infinity. This is found in forms as different as Giordano Bruno's and Spinoza's. (And neither of these seems to me to have handled the notion of the infinite in a conspicuously consistent way.)

DESCARTES AND THE INDEFINITE

One of the mysteries of the history of concepts of the infinite is why Descartes's distinction between the positively infinite and the indefinite has been adjudged – by his contemporaries no less than by latter-day historians of philosophy – so momentous. Some have gone the other way, and have seen it – but without good reason – as a reflection of Aristotle's distinction between the actual and potential infinites. Of course Descartes places God in a central position in his philosophy, and restricts the positively infinite to God; and Aristotle could never have been persuaded to do anything comparable with his "actual infinite", a contradiction in terms as far as the corresponding infinite by addition" is concerned. But an even more important difference is that where Aristotle has what one might call a *constructive* theory of the potentially infinite, Descartes inclines to the phrase "I do not know". To quote yet again his much quoted remark to Henry More:

God alone do I understand to be positively infinite: of other things, such as the extension of the world, the number of parts into which matter is divisible, and so forth, I confess I do not know whether or not they are straightforwardly infinite; I know only that I can recognize no end [*finis*] to them, and so – in respect of myself – I call them indefinite.[16]

It is easy to see in this passage from a letter of 1649 (which can be matched by many similar passages in his writings) a protestation of orthodoxy. Was the infinity of the physical world theologically acceptable to his contemporaries? Christina of Sweden, through Chanut, expressed her fears, but Descartes was able to cite Nicolas of Cusa, a Cardinal, who had got away with the idea.[17] Whether he understood Cusanus aright is immaterial, for – as we see – he confesses to being *ignorant* of whether the extension of the world is absolutely infinite. In fact he is here following the Cardinal rather closely, for just as Descartes distinguished infinite from indefinite, so Cusanus distinguished *infinitum* from *interminatum*. For him the world was simply *interminatum*, without termination, lacking bounds, undetermined – and therefore, according to him, subject only to partial and conjectural knowledge.[18] Resisting the temptation here to go back from the fifteenth century to the fourteenth in search of the genealogy of the concept *interminatum* (we might have looked at Ockham's *illimitatum*, but we might continue and come perilously close to Aristotle himself)[19] we ought at least to note that Nicolas of Cusa seems to have a tripartite classification in mind: "though the world is not infinite, yet it cannot be conceived as finite . . . ".[20] It must, he imagined, be the third thing, interminate. His proto-Hegelian synthesis of opposites coinciding in a transcendent absolute, all grasped through a "learned ignorance" which rises above discursive rational thought, is not Cartesian philosophy, but it does seem possible that Cusanus' conceptual distinction might have had some sort of influence on Descartes. (In a letter to Descartes, Burman credited him, Descartes, with having been the first to apply the infinite/indefinite distinction. Descartes does not appear to have offered any comment.) As I have already said, what seems to me puzzling is why the distinction should have seemed so world-shattering. I think some of its appeal must be put down to its appropriateness to a philosophy of doubt – which I will touch on shortly. Descartes' readers were certainly often in doubt as to its meaning.[21]

FEARS OF THE CHARGE OF IMPIETY

If Descartes was indeed worried by the thought that his utterances on to subject of infinity might lead to an accusation of impiety, he was not alone in his fear. The religious temper might vary as between different countries and different generations, but the eternity problem was felt keenly in most of them. Sir Walter Raleigh, for example, was not a dangerous intellectual heretic, and yet at his trial, in 1603, when his enemies made capital out of his atheism, Justice Popham addressed him in these terms: "Let no devil persuade you there is no eternity in heaven; if you think thus, you shall find eternity in Hell fire".[22] In one manuscript version of the speech, the word "devil" is replaced by "Hariot nor any such doctor". As for the more learned, and by inference more dangerous, Thomas Harriot, the story got around – to be repeated in Aubrey's *Lives*, that he rejected the creation of the world on the grounds that *ex nihilo nihil fit*. No doubt Harriot was influenced, directly or otherwise, by Bruno's lecturing in London (1583–5). At all events, the suspect Latin formula is to be found at several places in Harriot's manuscripts, not the least interesting being where it is said not to contradict *omnia fiunt ex nihilo*. What would Justice Popham have made of *that*? In his Preface to the *History of the World* (written and published during his many years in prison after the trial) Raleigh protested orthodoxy, and censured those who argued from "out of nothing nothing is made" to the world's eternity.[23] Their problem, in brief, is well expressed by Jean Jacquot: "Either you followed Lucretius and denied God and Creation, or you denied Creation in time and deprived God of his freedom".[24] This, at least, is how the problem was seen. Atomism went hand in glove with these two sorts of impiety, more by historical association than by logic; and when, in a correspondence in 1606–7, Harriot tried to persuade Kepler of the virtues of the concepts of atom and void, Kepler fought shy. Even Harriot's friend Nathaniel Torporley tried his hand at a refutation of atomism, fearing that the Christian creation dogma was at stake. No wonder that Harriot complained to Kepler that he could not philosophize freely.[25]

Descartes was closer to Bruno than were Locke and Newton, and if I may be excused for labouring what is obvious to some but not all historians of philosophy, the social pressures on the "philosopher of the infinite" were very different at the end of the century from what they had been in Harriot's day. Even the bold Thomas Hobbes was content to call the question of the duration of the world, and its twin, the question of its extent in space, finite or infinite, questions "unlawful to dispute of". He was, he said, content to leave them in the hands, not of philosophers, but of "those lawfully authorized to order the worship of God".[26] (Seth Ward's *In Thomae Hobbii exercitatio epistolica*[27] reveals a marked dislike of Hobbes' stance. This was all after mid-century.) Three or four generations of philosophers were still, however, destined to debate fervently the *ex nihilo* principle, and to probe the concept of causation in the process, without serious fear of civil redress – although there is a quavering in Leibniz's voice when in or around 1676 he shows concern for the prejudice to piety caused by abuse of "la philosophie nouvelle".[28] Bentley, in his third Boyle lecture, argued that something must have existed from all eternity because nothing can emerge from nothing; but then he went on to say that the "something" was of course God, and thus *ex nihilo nihil fit* was made to look innocuous. The same sort of thing must have been said many times over. It is, for instance, virtually the same as Locke's statement in the *Essay*, IV. 10.4, published a couple of years earlier. Others turned the debate in a more interesting direction. Could a creature be consistently said to be "coeternal with its creator"? You could sling mud at those who thought so, call them Arians and so forth, but they didn't write their books in the Tower of London. Locke, incidentally, prepared his answer to those who thought matter eternal because it could not be made from nothing, and it is in this connection that he hints at a mechanism for the creation of matter by God, a mechanism that seems to have been suggested by Newton, and one to which I shall refer later.

To return to Descartes: his account of the infinite might well have served to placate a few theologians, but it would be quite wrong to present that as its main philosophical purpose. It is in fact close to the very heart of Cartesian epistemology. To put the matter very simply, God and infinity come almost at the top of Descartes' list of certainties. Once again, how very different from Aristotle! We are tempted to suppose that we perceive the infinite as a negation of the finite. No, says Descartes, on the contrary, "I see that there is manifestly more reality in infinite substance than in finite, and therefore that in some way I have in me the notion of the infinite earlier than the finite – namely, the notion of God before that of myself".[29] It is almost as though, not content with putting God before things, in developing his philosophy from the *cogito*, he has begun to feel that somehow he should have put God higher on the list than the thinking self! We are still in a period of history when philosophers could play casuistical games with arguments as to which of two opposed concepts was the positive and which the negative. "Infinite" is grammatically negative, a fact which disturbed Descartes as it had disturbed other Christian writers before him. The remedy was to stress the limitation of the finite, by which it differs from the infinite, and is therefore "a non-entity or negation of being".[30] In the end, as well all know, Descartes made what was seen as an audacious claim, namely that we *can* have a clear and distinct idea of God, but that "our mind, being finite, cannot *comprehend* the infinite" – or all that is in it.[31]

Before going further consider what McGuire calls "Newton's comment on this kind of reasoning", namely Descartes' argument (against Hyperaspistes) that to negate presupposes positive knowledge of what is negated.[32] In fact when we consider the quotation from *De gravitatione,*[33] there is not only "much here that Descartes can agree with"; the drift of Newton's remarks is straightforwardly Cartesian. In playing with words, making *limit* a negative concept and therefore *infinite* a positive one, he is following in Descartes' footsteps. Locke, in his 1671 draft of the *Essay*[34] criticizes this line of argument, but rather oddly suggests that those "who contend for a positivie Idea of the Infinite" (and surely Descartes was one of those intended) were in this matter concerned with *quantity* (extension, duration, power etc.). In examining the question, Locke becomes involved in a discussion that had greatly occupied the thoughts of Aristotle and many scholastic philosophers, namely whether the "end" of a "duration" is the last moment of existence or the first of non-existence.

Even on the question of the relation of God to space, Descartes and Newton differ less than is apparent at first sight. We can place too much emphasis on the passage (from *De gravitatione*) in which Newton charges Descartes with having feared that infinite space "would perhaps constitute God because of the perfection of infinity".[35] Newton counters that infinity of extension is only as perfect as that which is extended – an estimable proposition, but one that misrepresents Descartes' views, not to say his fears. "I dare not call the world infinite", wrote Descartes to More, "for I perceive that God is greater than the world", but "not by virtue of his extension . . . which does not properly exist in God at all, but by virtue of his perfection".[36] Descartes does not set up extension as a rival to God. He avoided doing so by having two separate concepts of infinity. He is quite prepared to *admit* that the extension of the world, the number of parts into which matter may be divided, and so on, might in principle be infinite, and *then* "infinite" seems to mean only "without limit". To emphasize again the letter to More: Descartes confesses that he does not *know* whether these things are infinite or not. To Chanut he admitted that *God* might know limits to the world.[37] The indefiniteness is emphatically a reflection of the failure of our own faculty of knowing. The important thing is that we should not try to match up the infinity of God with that of extension – which "does not properly exist in God at all". Newton must have agreed. For all Leibniz's misunderstandings, due in large measure to Samuel Clarke's having mixed Newton's views on an omnipresent Deity with his own on space as an attribute of God, Newton nowhere (so far as I am aware) speaks of God as extended – infinitely or otherwise.[38]

EXTENSION, PERFECTION, AND DIVINITY

So much for Descartes' fears – which are too easily exaggerated. As for his way of speaking of perfections, here again he and Newton are closer than appears at first sight – despite the (fundamentally scholastic) way he has of occasionally using the word "perfection" as a synonym for "reality".[39] Like Newton, he thought himself able to pronounce on what are to be classed as perfections. They say similar things about the perfection of the intellect. Leaving aside a comparison of their pronouncements, however, consider what Descartes says about the limitation of perfection – which he often speaks of as having degrees. In a letter to Clerselier (23 April 1649) he quotes from the Latin edition of the *Meditationes*, treating of defects in substances, that is, limits to their perfection.[40] These limits are said to be accidents; but in the infinite substance (God's is clearly the only one envisaged) the infinitude of perfections is treated as something other than an accident of the infinite substance which "has" it. What else can it be? It must be the very essence of the substance![41] The convolutions of Cartesian metaphysics are seductive, and there is a temptation to play the neo-scholastic game. Is it not a tenet of the *Meditationes*, for instance, that all lesser perfections (such as I have, and including my free will) are a part of the supreme perfection of God? If a limit to perfection is an accident, is not absence of limit also an accident? In the case under discussion, however (the nature of God) infinitude must consequently be something other than absence of limit, since according to Descartes infinitude is not an accident. In fact his "positive" view of infinity-as-essence represents infinity as "a real thing . . . *incomparably* greater than all those that have some end".[42] Here he lays down a route which Newton will not be able to persuade himself to follow. Descartes may claim that he *never* uses the word "infinite" to signify "having no end-point", but note such remarks as are made in the letter to Chanut, where he spoke of God's knowing limits to the world. There are several comparable statements in his works. For much of the time he writes as though he is obliged to draw a sharp line between two categorically distinct concepts of the infinite – the positive, "essential", non-mathematical, and the negative, absence-of-limit concept. And yet there is a tension imposed not only by his mathematical vocabulary but by his mathematical way of looking at problems.

I want to suggest that there is no real conflict between the views of Newton and Descartes on the relationship between the concept of extension and that of divinity, for the simple reason that neither man was anxious to allow the two concepts to interrelate significantly. In McGuire's words, for Newton "the infinity of extension in no way competes with the reality of the Divine being".[43] As explained, Descartes would surely have agreed (although he would not have put it like this without first being vouchsafed the knowledge that extension is indeed infinite). As I have said, the two infinites were for Descartes supposedly categorically distinct. Faced with the statement that the world is indefinite, and that we do not know enough to say anything more positive, we naturally assume that Descartes would have allowed that it is *conceivable* (in some sense of that word) that the world be infinite – that is, infinite in extension. Is this so? Is there, in other words, what we might call a "positive concept of infinite extension", in *some* sense resembling the infinity of God? McGuire argues at length that "extension contradicts the nature of infinity, because extension is an imperfection, and infinity cannot be admitted where there is not perfection".[44] Why then should Descartes have introduced the concept of *indefinite*? Did he really wish to say that when he described the world as indefinite, meaning that he did not know whether or not it was infinite, in one sense of the word "know", he could adduce *a priori* arguments to show the impossibility of an infinite world, namely arguments proving the self-contradictory nature of the concept *infinite extension*? Would one ever say "I simply do not know whether or not the window is both broken and unbroken", or the like?

The argument as spelled out by MicGuire is essentially this: imperfection is incompatible with infinitude, and extension, a created and dependent being, is imperfect. The text quoted (reply to the Second Objections) in fact talks only of *body*, in which *all* perfections are found, rather than of infinite extension. In the paragraph preceding, Descartes has actually been talking of ideas we may have of "infinite attributes"; and one of the infinite attributes he will admit to be (eminently as opposed to formally) contained in the idea of God is length. Not, admittedly, extension; but if infinite length is possible, so is infinite extension, at least in thought. There is much mystery spun out of Descartes' conception of extension, but following its development in his writings it is clearly possible to say that the *basic* part of it is something like "the volume (a function of lengths, breadths and heights) [of a body] in combination with the distribution of that volume".[45] Thoughts and extensions, he explains, are modes found in substance – and he explains what he means in regard to thoughts by reference to extension, which he obviously considers clear enough. In *Le Monde*, extension is plainly meant to be understood as measurable: it is the property matter has of occupying space, not an accident of matter but its "true form or essence".[46] Extension is a mathematical concept. In the tenth of his Principles of Philosophy he talks of conceiving the same extension as space, emptied of body, and as constituting body. (His arguments elsewhere against a void need to be reconciled with this, but that is not our job for the moment.) I know of no text which would support the claim that the "characteristics of unity and simplicity cannot be found in the nature of extension"[47] or that "material extension in itself is an imperfection". In Principle *X* Descartes speaks of the "generic unity" of extension considered as space. Considered on the other hand as the nature of body, extension was admittedly considered "as particular" and conceived to "change just as body changes". Presumably he would have said that extension is divisible in this sense; but does that mean that the corresponding concept of infinite extension is, for him, self-contradictory? I think not; but had Descartes given more serious attention to non-theological concepts of infinity, as Aristotle had done, we might have been more confident.

ON UNDERSTANDING AND NOT UNDERSTANDING THE INFINITE

The difficulties in the Cartesian position are not made any smaller by Descartes' distinction between two sorts of knowing – between *intelligo* and *comprehendo*. This is paralleled in many scholastic writers, not to say in Augustine. By "comprehension" Descartes meant to stress the *completeness* of our grasp of something. This idea of totally enclosing something (knowledge) within ourselves, an idea which sprang from a metaphorical usage, was codified by medieval philosophers (thus Ockham gives five meanings) without any overt appeal to metaphor.[48] When Descartes writes that "it is a manifest contradiction that, when I comprehend anything, that thing should be infinite",[49] it is natural to think in terms of a container metaphor of mind. You can't put an infinite number of pints in a finite pot, so to speak. In fact Descartes' words at this particular point seem to me to suggest that his argument was a different one and that he was saying rather that it is by reason of the fact that the definition of "infinite" contains the concept "incomprehensibility" that the infinite cannot be comprehended. (He certainly uses the other argument elsewhere – in the third meditation, for instance –

but in the same place he *will* allow that we have the *idea* of infinite substance, not from our finite selves, but proceeding from some truly infinite substance with "manifestly more reality" than is in finite substance.) If this reading is correct, we can ignore the *non sequitur* only because it does not occupy a central place in his argument. (If, as he suggests, it is impossible to comprehend "infinite", that is, a concept "incomprehensible and F, G, H, . . ", where F, G, H, . . offer no difficulty, then the concept "incomprehensible" – and hence also "comprehensible" – is incomprehensible, and the discussion cannot even begin.) The upshot of it all is only that our ideas – even of God – are inherently imperfect. Descartes is far from clear on these questions. He calls our idea "complete", for instance, but the tenor of the paragraph in question (the 7th in his Reply to Objections V) is that our idea of the infinite *represents* the whole of infinity. Consider his final sentence. " . . . it is enough to understand a thing bounded by no limits in order to have a true and complete idea of the whole of infinity". He goes on to make a similar point in regard to our idea of God. We shall very soon get into deep water if we try to impose total consistency on Descartes' several utterances on these themes. The "true" in the last quotation, for instance, is confusing in view of his repeated emphasis on the imperfection of our idea; the suggestion that our idea is one of the *whole* of infinity does not immediately seem to square with his emphasis on the incomprehensible character of the infinite; and finally there is something rather puzzling about the statement as a whole, bearing in mind his remarks (quoted earlier) to More, to the effect that he never uses the word "infinite" to signify only "having no end-point". Descartes was replying to objections raised by Gassendi (V, 7), and I cannot help feeling that he would have benfitted by the infusion of some of Gassendi's thoughts in his own. Gassendi, briefly, can be said to have argued that our idea of infinite substance can come from the elaboration of limited principles concerning the visible world – in the language of the schools, by "composition and amplification".[50] This looks at once more medieval and more modern than Descartes' talk of the idea of infinite substance proceeding from the infinite. Whatever the merits of the two sides of this debate within a debate, it is enough for my present purposes to point out that Descartes did in a reasonably strong sense *agree* with Newton (if we may be allowed to turn the god of chronology on his head) that infinity can be understood. We can have a complete idea of it, but not comprehend it, in the peculiar sense of that word.

In saying that infinity can be understood, does this apply only to the concept in its solitary proven application, namely its application to God? We must come back to the supposition that infinite extension is a self-contradictory notion. In arguing for this view, McGuire reminds us of the utter dependence and contingency of created things (because of their re-creation by God at successive times), and passes straight on to extension, as though the same thing holds for that. "It is in this strong sense, then," he writes," that Descartes conceives of extension as a contingent and dependent being".[51] In sum, we are told that extension is for Descartes (1) divisible and hence imperfect; (2) contingent and dependent; and hence that on both scores it is impossible that it be positively infinite. My own view is that the style of argument is not entirely true to Descartes'; but since he clearly does *not* want to say that extension is positively infinite, I will be as brief as possible in saying why the style is wrong.

Certainly extension is divisible – and in fact Descartes makes its divisibility even basic to the divisibility of matter.[52] Certainly infinite extension cannot be found respectability along theological lines, for it is not a property of God, as we have already seen. Certainly it is divisible, but to say that it is – on this account – imperfect seems to me to misrepresent Descartes' way of speaking about imperfection. Imperfection may be revealed through limitation – as when we say a piece of matter is bounded, or our knowledge is limited – but the imperfection of the bounded thing derives from its reduced value, its loss of power to do what it is its function to do. It is *inappropriate* to talk of extension's limitation in this way. I realise that this is a rather impressionistic reconstruction of the case, and that it is always possible to make the rejoinder "whatever you say about the matter has to be said of extension, in the sense of occupied space". This will be the rejoinder of one who will not allow my argument to continue along the lines " in limiting extension I am no more rendering it imperfect than is a geometer who marks off a segment on a line" – an argument that will be denied because (although as quoted earlier, Descartes does talk of empty space) extension without matter does not exist in the Cartesian world. Descartes does talk freely enough of imagining limits to extension. Why then should he not speak of imagining the *absence* of such limits? Consider that very clearly-and-distinctly explained imagining of *Principles* II.21: if, he says, we imagine a limit to extension, we shall find that we are still "able to imagine beyond that limit spaces indefinitely extended", and these we perceive to be filled with corporeal substance. Indefinitely extended. We *are*, it seems, able to imagine a situation where there are no limits to extension. Indefinitely, not infinitely, extended: but – to repeat an earlier quotation – "it is enough to understand a thing bounded by no limits in order to have a true and complete idea of the whole of infinity",[53] so although we cannot be expected to imagine the whole of infinity, it does not seem that we are meant to question the consistency of the concept "infinite extension". We are driven back, not by an inconsistency, but by the sheer limitations of our knowedge.

THE CREATION OF EXTENSION, AND THE INFINITY OF IT AND THE WORLD.

On the question of the creation of extension, Descartes is not as explicit as he might have been, and indeed in all such matters we have to remember that we are on the verge of indulging in a species of philosophical necromancy in order to promote a dialogue between Descartes and Newton. Extension is not in God, and therefore must either be (a) outside God, and have been so from before the Creation – we ought really to say from before the *first* creation, on Descartes' view of continual creation; or (b) created by God *ex nihilo*.[54] The first would detract from God's omnipotence. The second must be accepted, thought Descartes; but why did he not consider a third possibility, namely that (c) extension is created necessarily *with* matter, as its essence?

Would this have been to presume to limit God's freedom? One might circumvent such an objection along the same lines as Descartes tried to circumvent his trouble with the eternal verities. (God did not necessarily will them, but he willed them to be necessary.)[55] Of course Descartes' religious style will not allow him to have extension competing with God, but if for a moment we allow infinity in different *respects* (e.g. in length as well as perfection) there is no reason why infinity in a limited number of respects should imply and challenge to God, who is infinite in all perfections.

We are now back to Descartes' distinction between indefinite and infinite. Inevitably he must admit that infinitude is only specifiable in certain respects, although, as we have seen, he finally settles on a unique infinite, God himself, the sum of all perfections. But it does not seem to me that his philosophy forces this position on him. I suggest that in the last analysis his infinite/indefinite distinction is rather trivial and that it was an obstacle to an easy dialogue involving later philosophers on this topic. Consider the somewhat abject tone of *Principles* I.26: we must not try to dispute about the infinite, for it would be absurd to comprehend it in our finite minds. Is the half of an infinite line infinite, he asks, or an infinite number odd or even? Two perfectly good mathematical questions, you might think; but he will not allow those without infinite minds to consider them. How disappointing, in view of the start already made by scholastic philosophers and Galileo to think constructively about a potential mathematics of the infinite. Not, of course, that they had made much progress, but they seem to have sensed the riches in store, riches which do not seem to have interested Descartes very seriously. "For our part, while we regard things in which, in a certain sense, we observe no limits, we shall not for all that state that they are infinite, but merely hold them to be indefinite". We observe no limits; but might we not speculate on the way "limits" might behave? Descartes mentions indefinite extensions, subdivisions of bodies, and stars, and he goes on to explain his distinction between an apparent absence of limits due to our lack of understanding, and one due to the nature of what is considered.

Now of course there is nothing wrong with such a cognitive distinction, as long as it is not used to prevent us from asking awkward questions. What would it have meant for Descartes to have said, quite categorically, that the world *is* limited (in regard to addition, in Aristotle's sense)? Since empty extension is not conceivable,[56] he would have been obliged to work out a geometry with a concept of limit very different from any previously developed. (His would not have been a finite and unbounded world, but a world finite and bounded by non-extension, whatever that may be made to mean.) We do not need to imagine the horror he would have had of a *bounded* world, because on more than one occasion he expressed it himself, in response to More's Lucretian-style challenge.[57] It involves a contradiction, thought Descartes, to suppose that the world is finite or that it is bounded. In the discussion with Burman he tentatively acknowledged that we may perhaps have to say that the world is infinite.[58] In the face of all this evidence I fail to see how Descartes can be said to have thought infinite extension a self-contradictory notion, and I think we must assume that he would have chosen from the following alternatives:

(1) The world, like the totality of extension, is infinite. There is no inconsistency involved in saying as much. We are, as it happens, unable to say whether the assertion about the world is true, and therefore, being cautious, we call it "indefinite", but it *is* infinite in the (authentic Cartesian) sense of "bounded by no limits".

(2) The world, and extension, are finite. As it happens, we do not know whether this is so, and so call the world "indefinite".

(3) The world and extension are finite. As it happens, we do not know whether this is so, and what is worse, we cannot even conceive of the possibility. We call the world "indefinite".

He would presumably have wished to say *either* "either (1) or (2)" *or* "either (1) or (3)". The problem with (3) is that it requires the word "indefinite" to have a different meaning from that of (1) and (2). (Briefly, "inconceivable" rather than "unknown".) As it happens, everything Descartes wrote, where relevant, points to his having preferred (1). Had this not been so, the potential ambiguity of the word "indefinite" might have attracted more attention. As things stand, the word is subject to another sort of ambiguity, the roots of which are hidden in the origins of the word "definition" (or rather its Latin ancestor). If *P* is a proposition which conveys information about the world and the information happens to concern the spatial limits to the world, and if I say "I know *P*", then I might say that the world is defined; and if I say "I do not know *P*", then I might say that the world is indefinite in the sense of "not defined" – here, as before, "defined" meaning that a definition is available in regard to the limiting of the world, I might, however, equally well translate "not-*P*" with the phrase "the world is not defined", meaning by this that *limits do not exist*. A case might be made out for saying that by his word "défini" (and its equivalents) Descartes was subconsciously trying to catch both meanings of my "defined" simultaneously; but of course he could not do so without making a category mistake.

NEWTON AFTER DESCARTES

There is a measure of irritation in Newton's charge that Descartes slides, in effect, from an epistemological thesis of the indeterminacy of (the limits of) extension to the thesis that it is an indefinite nature.[59] If this is a correct interpretation of Newton, then perhaps he was making the point I have just made about the category mistake. Perhaps Newton was not so much saying that the "indefiniteness thesis" amounts to an ontological claim as that he is impatient with Descartes because he will *not* make a genuine ontological claim, and insists on sitting on the fence.

Despite the fact that I have said so much about them, I find Descartes' writings on infinity (and related questions) disappointing. He might not have been afraid of the theologians, but something certainly petrified his thoughts. Newton's comments show that he did not always represent Descartes' views accurately; but there is a fresh breeze blowing through his comments – and these comments so often seem to be what Descartes himself should have said, on his own premises. For both, the notion of limit involves there being something to both sides of the limit. They would have agreed that they could not conceive of a point in space with space to one side and not the other.[60] Both assume that "infinity" has a basic meaning of "not limited". Newton can simply conclude that space is infinite, because he is not racked by Descartes' doubts:

> . . . God understands that there are no limits, not merely indefinitely but certainly and positively, and because although we negatively imagine [space] to transcend all limits, yet we positively and most certainly understand that it does so.[61]

There is something comforting about a plain application of the principle of the excluded middle: "either [real being] has limits or not, and so is either finite or infinite".[62] The imagination may be unequal to the task of coping with the infinite, but the understanding – and *a fortiori* God's understanding – can do so.

I would agree with McGuire that Newton's distinction between the imagination and the understanding is more or less Descartes' – although I think there are important differences. Descartes' position is, after all, much more elaborately worked out, philosophically speaking. The two ways of speaking of thoughts, as representations and as contents of the mind, were common property, and are incidental to the imagination/understanding dichotomy. The danger here is in making Descartes (and, for that matter, Newton) too Kantian. Descartes refers to the faculties of imagination and pure intellection (in the sixth meditation), the first a "power and inward vision of my mind", requiring a particular *effort* of mind for its use, the second (*intellectio* in Latin but *conception* in the French translations) something the mind must always be doing. The important thing is not to make too sharp a distinction betwen faculties, in the traditional sense of the term. Thinking has many modes: doubting, understanding, willing, imagining, feeling, conceiving (intellection), and so forth. He makes intellect and volition the two essential modes, and intellect the more fundamental. (A mind must always be thinking – not so always judging or imagining or willing.)[63] In judging, the understanding conceives ideas of things, which may be affirmed or denied, and if they were really clear false affirmation would not arise. Free will is also involved in judgement, a faculty of choice, and error arises when we fail to control our volitions.[64] Whatever its antecedents, this cannot be called a "traditional" doctrine, and so far as I can see, was not taken over by Newton in any but a superficial sense. (It has had followers, of course, between Descartes' day and our own, but that is another matter.) And in passing, allow me to note that Descartes talks of the *infinity* of the human will: "As a matter of fact I am conscious of a will so extended (*sic*) as to be subject to no limits . . . [a] faculty . . . even infinite". Seeing, he goes on to say, that he can form the idea of it, he recognises "from this very fact that it pertains to the nature of God".[65] Again, I am not aware of any Newtonian parallel, although, as McGuire has pointed out, for Newton the understanding can apprehend that *extension* is "an external, infinite, [and] uncreated" entity.[66]

How does it do so, and is Newton's epistemology really "closer to Descartes" than it is to Locke, especially on the question of 'how the mind grasps the notion of the infinite"?[67] To make Newton sound more like Descartes, McGuire[68] sets side by side Newton's remarks about "the understanding grasping the infinite" and Descartes' about "having the notion of the infinite in me before the finite". The two contexts are, however, utterly different. Newton is talking of *extension*, while Descartes (writing to More) is talking about *being*, in a theological setting. One has to remember that (in the third meditation, which is what he is discussing with More) Descartes is extending his philosophy from the *cogito*, yielding the thinking self, to prove God's existence (i.e. the existence of a complete and independent being).[69] There is a sense in which both Descartes and Newton are prepared to talk of "the finite as the negation of the infinite", as opposed to Locke, who can be said to put it the other way around; but that by itself does not seem to me to be a particularly deep point of agreement or disagreement, philosophically speaking.

SOCIAL LINKS IN THE INTELLECTUAL CHAIN

Before making a closer comparison between the three writers it is worth considering some of the social aspects of the network of scholars to which they belonged. Thomas Digby was in the Paris group surrounding Mersenne in 1637, the year of publication of the *Discours*, and some of his admiration was conveyed to Hobbes, with whom he corresponded. Most of the Digby correspondence has disappeared. Sancroft copied letters relating to Hobbes, Herbert of Cherbury, and Descartes, and these survive.[70] Hobbes was actually in Paris when the manuscript of the *Meditationes* arrived, and Mersenne gave it to him. A set of objections Hobbes prepared was sent by Mersenne to Descartes, without telling him who the author was, and Descartes thought the objections implausible.

The objections were received, and the book published, in the same year, 1641.[71] The two philosophers were so very far apart in their basic tenets that no dialogue of importance was likely to ensue, but with Henry More it was different. His *Infinitie of Worlds* (1647) shows the influence of Descartes' thinking; but from the position of disciple he grew progressively more critical. Like those other Cambridge Platonists Culverwel and Cudworth, he made much use of Cartesian philosophy, and especially its physico-mechanical aspects – which influenced Newton in ways too well known to require more than mention. On the other hand, they saw in Cartesianism a number of tendencies which they judged to be hostile to religion. By a piece of extraordinarily bad exegesis the Cambridge group managed to run together Cartesian philosophy and Democritean atomism – the latter, as I have mentioned in connection with Harriot, generally feared as atheistic. The Cambridge Platonists feared an alliance of Hobbesian principles with Cartesianism; and in what they saw as the answer to materialism, they coined the philosophy of an *anima mundi*, the soul of nature, the immanent cause of organic and inorganic nature. From the affable tone of the correspondence with Descartes from which I have been quoting (1648–9), More has so changed his tone by 1668 that in the Preface to his *Divine Dialogues* of that year he actually charges Descartes with atheism. Spirit, said Cudworth and More, is not "substance without extension". Extension does not have to be material; and if spirit were without it, God would be nowhere. Space is extension, not of matter, but (therefore) of spirit, and in infinite space we have the *extension of infinite spirit*.

The link with Newton has long been recognized, although McGuire has added several qualifications to the accepted account, and has argued that "Newton is largely a Gassendist on the problems of space, time, and existence".[72] Whatever the line of influence (McGuire suggests that Walter Charleton might have been the intermediary between Gassendi and Newton),[73] the broad outline of Newton's arguments resembles More's: (1) God's existence is eternal and uncreated; (2) everything – including God – must exist with respect to space and time if it is to count as an existent; (3) infinite space and time are consequences of God's external existence (Newton) as intrinsic attributes of God's nature (More).

The rivalry between Hobbes and the Cambridge philosophers could well be at the root of his expressed determination to leave the problem of infinity in the hands of "those that are lawfully authorized to order the worship of God".[74] The Platonists opposition to Hobbes extended to moral and political philosophy. They preached a personal religion, opposed to religious and civil centralization, and opposed therefore to the puritans and various arms of the official English church (which came and went with small predictability at about this time). Although in fact Hobbes does have a number of interesting things to say, despite his protestation, on the subject he said he would avoid, it is worth remembering how easily a philosopher's views may be distorted by sectarian antipathies.

AND LOCKE

Influence did not always run along social lines. John Locke's general agreement with the Cambridge Platonists on questions of political liberty and religious toleration (and note his long friendship with the family of Cudworth's daughter, Lady Masham) did not do much to bring their epistemological positions together. But to return to a comparison of Locke's "philosophy of the infinite" with Newton's and Descartes': on the face of things, their views, too, could hardly have been more different. Leaving God out of the picture, we have seen that Descartes was agnostic while Newton claimed that the understanding can get a clear idea of extension (eternal, infinite, and uncreated though it is), based as it is on a nature "real and intelligible". This was easy to say, but it all seems rather vague when we look for an explanation of the *workings* of the understanding – along the lines of Descartes' or Locke's accounts, for example. Newton's positive idea of actually infinite space looks at times almost like an innate idea, a datum so fundamental that there is no point in showing how it is built up. To Locke this is anathema, and he hopes to show how we synthesise the idea. Portions of extension are deemed to affect our senses (*sic*), carrying with them the idea of the finite. (The same goes for durations.) We have the idea of repeated addition (or doubling) of extensions, and the idea of doing this without stopping, and this, according to Locke, is how we are led to an idea of infinite space.[75]

In the same spirit Locke writes of utmost *divisibility*, that our thoughts cannot reach an end in the boundless progression. In all this there is no very clear sign of any influence from Newton's quarter. Locke was of course ten years older, and had more or less completed the *Essay* by 1686, three or four years before he first met Newton and several months before he first read Newton's *Principia.*[76] [See "The empiricism of Locke and Newton", by G. A. J. Rogers, in G. C. Brown (ed.), *Philosophers of the Enlightenment*, 1980, pp. 1–30] The *Essay* finally appeared in 1690. Before considering the Newton-Locke relationship further, I shall look further into the *Essay*, explain very briefly why Locke comes out against an actual idea of a completed infinite, and examine his theology.

I have already noted Locke's "idea of infinity" considered as "an endless growing idea".[77] This he contrasts with "the idea of any quantity", a "terminated" idea (at the time the mind has the idea). It is not, he thinks, "an insignificant subtilty" to say that we must carefully distinguish between "the idea of the infinity of space and the idea of a space infinite". The first is a supposed *endless progression* in the mind, the second an actual idea in the mind of a *completed endless repetition*; in short, a contradiction in terms. Berkeley commended the distinction in the closing words of his youthful essay on the infinite.[78] It is occasionally said to have been adopted from Locke by Kant, who does indeed seem to make use of it for the proof of the thesis of the First Antinomy; but there is ample evidence that Kant did not *accept* it, that he thought that those who rejected the "actual mathematical infinite" did so "in a very casual manner", and that he had a trick up his sleeve rather reminiscent of certain passages in the writings of Descartes, namely in introducing an intellect which may exist – "though not indeed a human intellect" – capable of perceiving a multiplicity in a single intuition without the successive application of a measure.[79]

Whether or not Locke's account should be called "Aristotelian" is a question of emphasis. For Aristotle the infinite was something that could not be gone through; for Locke in his published version of the *Essay* it was something which required you to keep adding. There is more to Locke's account, but at this fundamental level he is less precise, less careful, than Aristotle, to the extent that he does not consider the type of endlessness exemplified by the ring. Nevertheless, both rely heavily on metaphor when they discuss what we might call "going on without stopping", for both speak of it as though it was a process requiring time, as walking, or calculating, requires time. As for Aristotle, number was basic for Locke – it "furnishes us with the clearest and most distinct idea of infinity we are capable of" – , but even when we number intervals of space and time and make use "of the ideas and repetitions of numbers", the "clearest idea any mind can get of infinity, is the confused incomprehensible remainder of endless addible numbers, whch affords no prospect of stop or boundary".[80] The infinity of number "lies only in a power still of adding any combination of units to any former number, and that as long and as much as one will".[81] This phenomenological approach to the subject is appreciably different from Descartes'. When Descartes spoke of an absence of limits he treated the case as one that might be judged instantly: we simply "perceive no limitation in the thing".[82] His other kind of infinite, unlimited perfection – in a word, God – , was likewise something that could be the object of "clear and intuitive knowledge". "In some way I have the notion of the infinite prior to the notion of the finite".[83] Those who are looking for a mathematics of the infinite will consider neither approach very promising. A completely different approach can be detected in some scholastic writers, and in Leibniz it is clearly expressed: it requires a complete separation of the conceptual apparatus of the infinite from that to which it may be applied – or not, as the case may be. (The real may be governed by the ideal, according to Leibniz, and "infinites and infinitesimals are grounded in such a way that everything in geometry, and even in nature, takes place *as if* they were perfect realities")[84]

Accepting that Locke's programme was of a different sort, we shall still understand it properly only if we first understand his notion of *positive idea*. We saw how Descartes and Newton in their different ways made the infinite "positive", and the finite its negation. Locke takes the common-sense reversal of labels. In an example of the mariner who fails to find the sea-bed with his sounding line (cf. Ockham), he says the mariner *knows* that the depth is so much as the line is long (a positive idea) and more (another positive idea).[85] On the other hand, the idea of "so much greater as cannot be comprehended" is not positive but negative. The act of numbering is fundamental to an understanding of his "primary" meaning of finite and infinite – referred to, following tradition, as "modes of quantity". His argument so far leads him to say that we can have no positive idea of infinite number.[86] He later extends the doctrine to numerable things. Besides his primary meaning for "infinite" Locke gives a secondary meaning relating to God's power, wisdom, and so on.[87] As he explains, this is *figurative*, and is arrived at by analogy with number or extent of the acts or objects of God's power, and the like. In this matter the *Essay* is poles apart from Descartes, who is made to look all the more a scholastic by contrast.

Starting from the basic principle that there is no acceptable positive idea of actual infinite number, Locke extends his argument to numerable things. Some ideas are repeatable and some are not. Two idea of white, for instance, when combined "run into one". Space, duration, and number alone have parts. (So much for the intension and remission of forms!).[88] Locke therefore draws the same conclusion as to the unavailability of positive ideas of infinite *duration* (eternity being the "number of duration" looking forwards or backwards from the present),[89] and of infinite *space*.[90]

Like Aristotle, then, Locke rejected the idea of the completed infinite, and also turned his back on the idea of the infinite as a "separate, independent, thing" and an "indivisible thing". These were phrases beloved of theologians and metaphysicians – and yet was Locke's empiricism not influenced by Christian theology, by his Christian (latitudinarian) belief? We know that Descartes reserved positive infinity to God, and that he made much of God's *unity* and *independence*. We have seen the Cambridge Platonists' view of God as an infinitely extended entity, "One, Simple, and Immovable", and also "not only Eternal but Independent . . . of anything whatever".[91]

We might have expected Locke to have adopted some similar attitude to such widely diffused 17th century ideas. What sort of God was Locke's?

In the chapter of the published *Essay* dealing with the idea of infinity, using almost identical phrases to those of the 1671 draft, he says that he thinks it "unavoidable" that every "considering rational creature" should have a notion of an eternal wise Being who had no beginning: "and such an idea of infinite duration", he adds, "I am sure I have".[92] This is at first sight an astonishing admission, for is not the duration assigned to God of the same kind as is experienced on earth? There were other theological opinions than this; did Locke have them in mind in writing this common-sensical passage? Turning back to the 1671 draft we find him referring to "the unconceivable *punctum stans* of the schools" which is the idea some might claim to have of duration of eternal beings having "noe succession".[93] Note how almost all leading philosophers inveigh against the "*nunc stans*". In the published *Essay*, however, he faces up to the problem of God's power to see "all things past and to come". "They all lie under the same view ... To conclude, Expansion and Duration do mutually imbrace, and comprehend each other; every part of Space, being in every part of Duration; and every part of Duration, in every part of Expansion".[94] Such an omnitemporal and omnipresent God sounds rather like the God of Newton's *Principia*.

Putting aside for a moment Locke's difficulties with the causal argument for God's existence, what about the concept of God's "having no beginning"? Surely this means that it is *innappropriate* to speak of God's beginning in time. This is not suggested by the following *volte face*, however:

> But this negation of a beginning, being but the negation of a positive thing, scarce gives me a positive idea of infinity; which, whenever I endeavour to extend my thoughts to, I confess myself at a loss, and I find I cannot attain any clear comprehension of it.[95]

Elsewhere we find him slipping back into forbidden ways. "We easily conceive in God infinite duration";[96] and God is immovable "not because he is immaterial, but because he is an infinite spirit".[97] The second remark is reminiscent of a proof offered by Henry More that "no Infinite extended entity which is not co-augmented from parts, or in any way condensed or compressed, can be moved ... ".[98] Locke's favoured argument for the existence of God is the causal argument, though he occasionally has the argument from design. The cause of my existence must be other than a material being – it must be a "cogitative being"; and since it is inconceivable that there was a time when nothing existed, it being impossible that the real being of God came from a non-entity, therefore "from eternity there has been something".[99] He does not describe this Being as infinitely wise and powerful, but as "most powerful and most knowing", and indeed "most perfect", but on the score of extension, "it is hard to find a reason why any one should doubt that [God] ... fills immensity". There is again a touch of More in the next sentence: "His infinite being is certainly as boundless one way as another; and methinks it ascribes a little too much to matter to say, where there is no body, there is nothing".[100] Where there is no body there is God; but does he have Newton's omnipresent God in mind, or More's extended deity? Space was taken earlier to be a possibility for extended beings to exist, but that is not enough to allow us to place Locke alongside More on this question and to do so would very probably be mistaken, if we are to judge by Locke's 1671 draft of the *Essay*. Let us ask someone who thinks he has a positive idea of infinity whether it is of infinite local extension or infinite duration. "If it be local extension that cannot belong to God, who is a spirit &is therefor uncapable of local or corporeal extension, & is therefor proper only to body".[101]

In connection with the extension/omnipresence question there is a curious theory for the creation of matter at which Locke hints at one point in his *Essay*. The theory is that God by his power prevents anything from entering a particular region of space, and thereby (endowing it with impenetrability) creates matter there. To move the matter he simply creates, at a later time, the same impenetrability in another region. Again this tells us very little about God's relation to space; but since it seems that the theory was Newton's brain-child, perhaps we may assume that Locke and Newton both remained inclined to the view that God is omnipresent in space, without being extended.[102]

It would lead too far afield to consider Locke's application of the causal argument for God's existence, but its involvement in the problem of infinite regress is worth mentioning. The argument, briefly, is that the cause of my existence cannot be a mere nothing, since *ex nihilo nihil fit*, and by the same principle its own cause cannot be a mere nothing – wherefore the cause of my existence is eternal. On the principle (much used by the Cambridge Platonists) that there must be more reality in the cause than in its effects, and indeed more of all powers, the cause, which we call God, is the most real, the most powerful, the most intelligent, and so forth. Locke spends the later part of the relevant chapter[103] demonstrating the immateriality of God. His causal proof was not likely to attract much notice, since it was already part of an established Christian (neo-Platonic) tradition. The problem, for Locke's philosophy of the non-availability of a positive idea of the infinite, is that by the same token he should never claim to have an idea of the first term in his causal series; and of course the first term is none other than God. But he does so claim. In the words of the 1671 draft, "I acknowledge I have the notion of a first being or a being without a begining & I thinke it unavoidable for every considering rational creature that will trace his owne thoughts to have such an one". But, he goes on to say, this is "note a positive Idea of Infinitie" rather "the negation of a positive thing (as the beginning of any thing is)".[104] He is clearly not above playing Descartes' game of positives and negatives, but I cannot see how it can have been thought to help his case.

The three vertices of the McGuire triangle are Descartes, Locke, and Newton – all of them compared on the basis of their writings, but not in any "strongly historical" style of the sort that might reveal the paths by which the ideas of one developed from (or in dialectical opposition to) those of another. Of course he has often written in such a style, and one paper in particular – "Existence, actuality and necessity: Newton on space and time" – provides much of the background to his present essay, so to some extent my strictures may seem unfair. Consider, however, the last sentence of the essay: "If I am right about how Newton claims to know infinitely extended space, any kinship with Lockean empiricism must be denied". This may be read in different ways. A philosopher might have a very clear idea of that timeless entity "Lockean empiricism", and care little for chronological relationships. It might be argued that kinship does not imply interdependence, but merely a common inheritance. A historian might point out that the very early works of Newton could hardly be expected to show strong Lockean influence, for they were written at a time when Locke's epistemological interests were barely formed, let alone committed to print. Regarded as a comment on early (i.e., before the late 1680s) influence from Locke to Newton, the statement does not look very bold; but as a denial of influence in the opposite direction, it adds a little weight, certainly, to current criticisms of the old view that Locke was simply the philosophical interpreter of the Newtonian physics. No-one is likely to regard Newton and Locke as identical twins, but it does seem to me so very obviously right to think of them as kindred philosophical spirits that I think we should resist Ted McGuire's over-subtle separation of their natures. By the same token we shall have to say that the early Newton and the late Newton were not kindred spirits. Both may be called Cartesian dualists. The qualification is entirely justifiable in the 17th century context, for in both writers we find passages lifted virtually straight out of the work of Descartes. Mind and matter were for them separate substances, related in a causal theory of perception. Both of them accepted a broadly Cartesian language of ideas, and both drew the line at innate ideas. Their views of space and time were in their elementary aspects rather similar – although Locke changed his mind on that subject, for he had flirted with a relational doctrine during his travels on the continent. And of course there are the differences (dare I call them small?) on the question of "positive ideas of infinity". The number of passages from Newton's writings with a bearing on what might be called the "metaphysical infinite" *is* rather small. Both were avowedly anti-scholastic over the various doctrines of substance, and both combined a continuum view of space and time with a corpuscular view (in fact with a variety of corpuscular views) of matter. There are differences in plenty, but are these points of fundamental agreement evidence enough for kinship?

NEWTON AND LOCKE

Should we be looking for something deeper? I think Newton's geometrical fragments on the "infinite in fact and not only imagination" are extremely interesting, but they do not seem to me to put any more distance between him and Locke than we have already found. Before commenting on them, let me quote Locke again, from the *Essay*:

> Some mathematicians perhaps, of advanced speculations, may have other ways to introduce into their minds ideas of infinity. But this hinders not but that they themselves, as well as all other men, got the first ideas which they had of infinity from sensation and reflection, in the method we have here set down.[105]

Such generosity, on Locke's part, puts the status of his earlier argument in doubt. In developing his account of what he calls the "primary meaning" of infinity, was he providing nothing more than a naive "commonsense" account, consistent, perhaps, but weak, and in principle capable of "mathematical" refinement? There is a sense in which he will have fulfilled his purpose if he can get his opponents, believers in our innate idea of the infinite, to acknowledge that indeed the mind works as he says, and receives its (possibly naive) ideas of infinity from sensation and reflection; but is this only a contingent truth about the ordering of experience in the seventeenth century mind? Might we not catch our philosophers young, and instil more refined ideas in them before they have had a chance to think along naive lines? After all, the typical child does not hear of the infinite at its mother's knee. We may grant Locke the contradictory character of the notion of an actual idea of an incomplete completed repetition; but what guarantee have we – and he himself can see none – that there are not other, more powerful, and consistent theories of the infinite? None at all, of course. The most that Locke can hope for is that his opponents will grant him his opening definition of "infinite".

How then do the "geometrical" quotations from Newton square with Locke's meanings?[106] One side of a triangle is swung round a vertex until eventually it is parallel to the side opposite that vertex. In the first quotation, overlooking a phrase where he seems to be suggesting that there exists a last point of intersection on the line opposite the vertex, he is in fact arguing that *there is no last point* on the line. It is a curious example, since it introduces elements (other lines, angles, the notion of monotonic increase, intersection, existence, parallelism) any of which might be the first point of attack by a modern geometer, aware of geometries alternative to Euclid. Could he not have spoken simply of proceeding to greater and greater distances along a line? Why did he not do so? I think the answer is that he wanted to make the obscure concept "actually more than finite" understandable, by making it a function of more readily intuited concepts. (In particular, the angle increases to a finite and intuitable limit.) In his other triangle example (from the earlier *De gravitatione*), where he talks about the point towards which the two sides of a triangle are directed and where they meet (although this meeting point might have to be pretended to lie "beyond the limits of the corporeal universe"), Newton is again ultimately referring simply to the points of a line. Why not talk about the line and nothing else? It seems to me that he is offering yet again what a recent generation of philosophers would have called an "operational definition" of the line in terms of things supposedly easily understood. Rather as before, the "actual" line "extends beyond all distance". Unlike Locke's infinite process, a process without an end, Newton's is a process in which the infinite *is* specified, at least "*fingatur*". It would be anachronistic to expect him to offer an existence proof. By the lights of the time, the point existed. Can it be that when he wrote of "many things concerning numbers and magnitudes which to men not learned in mathematics will appear paradoxical" he was thinking of the words I quoted from the end of Locke's chapter on the infinite, and was consciously offering "ways of introducing into the mind ideas of infinity" of a sort that Locke suspected might be open to mathematicians of "advanced speculations"?

WHAT THE SEVENTEENTH CENTURY CONTRIBUTION WAS NOT

It was a commonplace in philosophy and mathematics before the nineteenth century that what was infinite was *quantity*. It is to Locke's great credit that he saw the confusion bred by a failure to distinguish between infinite quantity regarded (1) as a unique, completed, thing; and (2) as something of which we have an "endless growing idea".[107] Bolzano would later make the same point more effectively, and – like Locke – emphasize the logical priority of *number* over quantity. From number, Bolzano and Dedekind would shift the focus of attention to *classes*, equivalent classes sharing the property of having the same power or cardinality. The seventeenth century can offer at least a handful of hesitant attempts to come to terms with the problem of comparing infinite classes – setting up a one-to-one correspondence between their members in the now familiar way. There is nothing new about this problem, which has in any case been aside from the main themes I have been discussing. It was quite commonly discussed in fourteeth century Oxford, for example, and was subsequently never totally forgotten by European scholarship. Neither Descartes nor Locke, even so, seems to have made any attempt to grapple with the problem of the "inequality of infinites", and the apparent paradox which results, for instance, when each term in the series of natural numbers is made to correspond with its double. Galileo had given the problem publicity and Harriot had struggled with it at much the same time. Descartes shows from a letter to Mersenne of 28 January 1641 that he *knew* perfectly well of the problem, for in his letter he commented on the work *Quod Deus sit mundusque ab ipso creatus fuerit in tempore* (1635) by the mathematics professor at the Collège Royal in Paris, J. -B. Morin. Morin had recently attacked Descartes, despite which fact Descartes' reply was restrained:

> I have read M. Morin's book. Its main fault is that he always discusses the infinite as if he had completely mastered it and could comprehend all its properties. This is an almost universal fault, which I have tried carefully to avoid. I have never written about the infinite except to submit myself to it, and have not written to determine what it is not . . . He supposes also that there cannot be an infinite number – which he could never prove.[108]

Fragmentary enough; but if it is to be believed, this suggests that Christian humility overcame the mathematician's curiosity. Hobbes was probably referring to Morin – or at least to the sorts of argument Morin introduced – when he too agreed to leave such problems in the hands of the established

Church; but Hobbes' language is so loose as to make it seem unlikely that he would have made much progress had he tried.[109] Leibniz makes an amusing aside about Hobbes' attempts to apply the exact methods of geometry to morals and physics: "Il y a un mélange chez Hobbes d'un esprit merveilleusement pénétrant et estrangement foible incontinent à pres. c'est qu'il n'avoit pas assez profité des Mathematiques pour se garantir des paralogismes".[110] As it happens, one of Leibniz's earliest disputations (Leipzig, 7 March 1666), known only from a list of theses to be defended, was on the theme "One infinite is greater than another",[111] so there at least we have a mathematician who did not brush aside the problem.

One could mention others who touched on the problem of the equivalence of an infinite set with a proper sub-set, but the story would be always the same, namely that there was never any suggestion that the infinite be *defined* in terms of the paradoxical property. Where the infinite does enter seventeenth century mathematics significantly is in connection with infinite series (seen clearly by mid-century to define a limit), often as the basis for quadratures (some introducing their own "paradoxes of the infinite", as when an infinite line contains a finite area). There are mathematical uses of the word in connection with almost every tract on the differential and integral calculus, and to abstract the many nuances of its use would induce something approaching infinite tedium, but one rough generalization is in order here. The mathematician almost invariably thought of his infinite collections (lines, spaces, numbers, times, etc.) as continuous magnitudes rather than as sets; and insofar as he tried to set up a "mathematics of the infinite" it is in the form of a calculus of 0 and ∞. Number was seen as a ratio of quantities (an obscure idea, but thought meaningful enough by those who used it). Leibniz was one of many who regarded the "infinite quantity" as a quotient of 1 and 0; and taking 0 as analogous (or even equal to) an infinitesimal quantity, it became perfectly natural in the fulness of time for Euler to speak of orders of infinity corresponding to orders of differentials, namely their reciprocals. The notion has respectable modern descendents, as have so many historical characters who behaved in their day with wild abandon.

I mention these tendencies only to underline the fact that the Descartes-Locke-Newton stream of thought we have been considering is a separate stream from what might be crudely called the "inverse infinitesimal" stream (with its vast network of tributaries), and different again from the Bolzano-Dedekind-Cantor network. This is all put very crudely, I know, but is needed by anyone who is tempted to regard the concept of infinity as monolithic. (No-one is likely to treat it as *simple*, I imagine.) Whether or not "number is prior to quantity", is it possible to frame a meaningful doctrine of the infinite without reference to the concept of number, and if so, does it resemble anything from 17th century metaphysics? And if not, is it true to say, with E. T. Bell, that "The 'infinites' of religion and philosophy are irrelevant for mathematics"? The biography of Cantor alone is enough to controvert Bell as far as historical influence is concerned.[112] What about their relevance to physics? Space, matter, void, time, motion, . . . Does it matter to physics that we include an "infinite" God in the list, as all our characters from the 17th century wished to do? I think such counterfactual hypothetical questions are an important part of philosophical history; and in this case I think my answer would be that metaphysics did *not* help physics very much; but I am equally sure that I am not qualified to give an answer to this particular question.

Finally, what did the 17th century metaphysics of the infinite do for the rest of philosophy? We have seen how it prompted Descartes to think carefully about the difference between comprehending, having an idea, and knowing, in different senses. Descartes' words in this vein reverberated through the world of philosophy. Take Hobbes: we have in our mind no idea of the infinite and eternal except that of our insufficiency to comprehend them. Take Malebranche's "intelligible extension", what we are supposed to think about when we think of the infinite. And Newton – understanding rather than imagining infinite space. And Locke – for whom the infinite was a testing ground for his way of ideas. And Leibniz – whose thoughts on infinity led to some influential ideas on necessity. A few examples only, but not a bad score sheet.

University of Groningen

NOTES

1 *Phys.*, II.4; 202b.30–.
2 204a.6.
3 L. Baudry, *Lexique philosophique de Guillaume d'Occam*, Paris, 1938, p. 122.
4 *Met.*, 1066b.17–18.
5 *Phys.*, 212a.20.
6 *Ibid.*, 212b.20–21.
7 *Met.*, 1048b.9–17.
8 *Phys.*, 204b.5–6.
9 *Phys.* III.6, 7 is the source of the remainder of this paragraph and the next.

10 *Phys.*, 206b.20–24.
11 *Ibid.*, 207b.18–20. These propositions are accepted at several places in Aristotle's writings.
12 Some remarks were added here which are open to several interpretations, but they are beside my present point. See *Phys.*, 207b.27–34. One interpretation is that if a geometer is embarrassed by a figure which bids to stretch beyond the universe (in his thoughts) then he may use in his proofs a smaller, similar, figure.
13 Cf. *Met.*, 1051a.21–33.
14 "Aristotelian infinity", *Phil. Rev.* 75 (1966), 197–218.
15 *Ibid.*, pp. 217–8.
16 [C. Adam and P. Tannery (eds.), *Œuvres de Descartes*, Cerf, Paris, 13 v., 1897–1913 (reprinted Vrin, Paris, 1957–8) will be cited in the usual way as "AT".] AT. V, p. 274.
17 AT.V, pp. 50–1.
18 For references, and more detail, see A. Koyré, *From the Closed World to the Infinite Universe*, Baltimore, 1957, Chapter 1 and the notes to it.
19 Notes, however, the case of the endless ring, previously discussed. The ring is not limited, but it can be "gone through" in the sense that all its parts can be surveyed in a finite operation.
20 *Op. cit.*, 1970 edition, p. 11.
21 AT.V, p. 167, for the Burman letter. Henry More, to take an example of confusion, *approves* the indefiniteness concept of Descartes but speaks of God (and thus, for him, space) as infinite and the world as indefinite, meaning by this *finite*. See Koyré, *op. cit.*, p. 140. Joseph Addison, in *The Evidences of the Christian Religion*, 6th ed., 1776, p. 107 (reprinted from the *Spectator* (c. 1714)), writes of the extent of the throne of God as follows: "Though it is not infinite, it may be indefinite; and though not immeasurable in itself, it may be so with regard to any created eye or imagination". Comment would be superfluous.
22 Jean Jacquot, "Thomas Harriot's reputation for impiety", *Notes and Records of the Royal Society* 9 (1952), 164–87 at p. 167.
23 *Ibid.*, p. 179.
24 *Ibid.*, p. 180.
25 *Ibid.*, pp. 181–2.
26 *Elements of Philosophy*, IV.26.1; W. Molesworth (ed.), *The English Works of Thomas Hobbes*, London, Vol. 1, 1839, pp. 410–14. The date of the epistle dedicatory is 1655.
27 Oxford, 1656, pp. 116 ff.
28 [*Die philosophischen Schriften von G. W. Leibniz*, Gerhardt (ed.), 7 Vols, Berlin, 1875–90 = "G".] G. VIII, pp. 64–5; G. IV, pp. 343–9.
29 *Meditatio* III; AT.VII, pp. 45–6.
30 To "Hyperaspistes", August 1641; AT. III, pp. 426–7.
31 This is often reiterated. See, for instance, to Mersenne, AT.III, p. 233, and Gassendi's objection to the third meditation, AT.VII, pp. 296–7, for a contrary view.
32 Cf. pp. 92–3 above.
33 *Ed. cit.*, pp. 101–2.
34 P. H. Nidditch (ed.), *Draft A of Locke's Essay Concerning Human Understanding. The Earliest Extant Autograph Version*. University of Sheffield, 1980, pp. 163–5.
35 *De grav.*, pp. 102–3; quoted p. 94 above.
36 AT.V, p. 344.

37 AT. V, pp. 51–2.
38 A passage from Des Maiseaux's preface to the 1720 edition of the *Clarke-Leibniz Correspondence* (see H. G. Alexander's edition, Manchester U.P., 1956, pp. xxviii–xxix) shows that Clarke was not altogether happy with his own theory, which also appears in his Boyle lectures. For a typical occurrence of it see *ed. cit.*, p. 23.
39 For example at AT. VII, p. 40.
40 AT. V, pp. 355–6.
41 *Loc. cit.*
42 *Ibid.*, p. 366. My italics. Note the language of comparison, and compare the mathematical language of ratios.
43 This is one of several references which I can no longer give, since the printed version of the paper to which I am replying differs somewhat from that version to which I actually replied.
44 See n. 43 above.
45 *Princ.* I, pr. 64; AT. VIII, p. 31.
46 AT. XI, p. 36.
47 See n. 43 above.
48 Baudry, *op. cit.*, pp. 49–50, lists Ockham's meanings.
49 AT. VII, p. 368; see n. 43 above.
50 AT. VII, p. 295.
51 See n. 43 above.
52 Cf. AT. VII, p. 163.
53 For the context, see above. Note that Koyré, *op. cit.*, pp. 104–6, turns *Princ.* II.21, 22 to wrong use, saying that the infinity of the word "seems thus to be established beyond doubt and beyond dispute", but that "as a matter of fact Descartes never assers it". *Princ.* II.22 is in fact aimed only at showing the unity of the world – and that it admits of no gaps between one "world" and another.
54 See the interview with Burman, AT. V, p. 155.
55 AT. IV, p. 118 (to Mesland).
56 Cf., for instance, *Princ.* II.16.
57 AT. V, pp. 311–12 (More), p. 345 (Descartes).
58 J. Cottingham (tr.), *Descartes Conversations with Burman*, Oxford U.P., 1976, p. 33. For the Latin originals see AT. V, especially p. 167, and cf. the letter to Chanut, where he admits that although the putative limits to the world are *incomprehensible* to him, they may be known to God. AT. V, pp. 51–2.
59 McGuire, but see n. 43.
60 *De grav.*, pp. 100–01.
61 *Ibid.*, p. 102.
62 *Loc. cit.*
63 AT. VIII, p. 25.
64 AT. VIII, p. 17.
65 AT. VII, p. 56.
66 See n. 43 above.
67 See n. 43 above.
68 See n. 43 above.
69 AT. VIII, pp. 16, 53.
70 See Marjorie Nicolson, 'The early stage of Cartesianism in England', *Studies in Philology* 26 (1929), 356–74.

71 See L. J. Beck, *The Metaphysics of Descartes*, Oxford U.P., 1965, especially pp. 17, 41.
72 "Existence, actuality and necessity: Newton on space and time", *Annals of Science* 35 (1978), 463–508. The Gassendi connection was explained and defended in R. S. Westfall, "The foundation of Newton's philosophy of nature", *British Journal for the History of Science* 1 (1962), 171–82.
73 *Op. cit.*, p. 469, n. 25.
74 See pp. 120–21 above, where I quoted other words to the same effect, from the same place.
75 *Essay*, Bk. II, Chapter 17.
76 See G. A. J. Rogers, "The empiricism of Locke and Newton", in G. C. Brown (ed.), *Philosophers of the Enlightenment*, London, 1980, pp. 1–30.
77 *Essay*, II.17.7.
78 A. C. Fraser (ed.) *The Works of George Berkeley*, Vol. 1, 1901.
79 For the most apt text see N. Kemp Smith, *A Commentary to Kant's "Critique of Pure Reason"*, 2nd ed., Macmillan, London, 1923, p. 486.
80 *Essay*, II.17.9. In the 1671 draft Locke expressed himself only marginally differently. On number, there also taken as basic, he says the same thing: there is no other infinity than that of numbers, "which is never actual but is always capable of addition". As for infinite power and knowledge he expressed the hope that on a future occasion he may show that thereby we mean no more than "such a power or knowledge which cannot by any thing that doth or can possibly exist be limitted or resisted. Which is not any notion of a positive actual Infinity, but only potential as of numbers, to the bounds whereof even in thinking we cannot arrive; & soe that is Infinitum to us ad cuius finem pervenire non possumus". *Ed. cit.* p. 165, my punctuation. Apparatus omitted. Compare this with my next paragraph, p. 136.
81 *Essay*, II.17.13.
82 AT.VII, p. 113.
83 AT.VII, p. 45.
84 Letter to Varignon, 2.02.1702.G (Math.) IV, pp. 91–5. Loemker (tr.), pp. 342–6. My italics.
85 *Essay*, II.17.15.
86 *Essay*, II.17.8.
87 *Essay*, II.17.1.
88 *Essay*, II.17.7.
89 *Essay*, II.17.10–13.
90 *Essay*, II.17.18.
91 Quoted from Koyré, *op. cit.*, pp. 148–50. (More, *Enchiridium metaphysicum*, c. VIII, 9.) Note More's proud boast that Aristotle made immovability the highest attribute of the First Being. Such ideas continued to be influential. Joseph Raphson, for instance, although he criticized Spinoza generally, followed the essentially Spinozist position in taking God to be extended; and Berkeley (*Of Infinites*) took him to task for this. Raphson quotes More in his own support. (See his *De Spatio Reali seu Ente Infinito*, 1697.)
92 *Essay*, II.17.17. *Ed. cit.*, p. 168, for "Draft A".
93 *Ed. cit.*, p. 167.
94 *Essay*, II.15.12.
95 Compare his use of the word "comprehension" with Descartes".
96 *Essay*, II.15.4.

[97] *Essay*, II.23.21.
[98] Koyré, *op. cit.*, p. 149.
[99] *Essay*, II.10.4.
[100] *Essay*, II.15.3.
[101] *Ed. cit.*, p. 166, my punctuation.
[102] I am not aware of any recent comment on the passage in question, or on its elucidation by P. Coste, the French translator of the *Essay*, a man who had lived with Locke during the last seven years of his life. I owe the reference to W. Hamilton, *Discussions on Philosophy and Literature, Education and University Reform*, 3rd. ed., Edinburgh and London, 1866, pp. 199–200.
[103] *Essay*, IV.10.
[104] *Ed. cit.*, p. 168. Cf. my earlier quotations from the same.
[105] The closing words of II.17.
[106] Leibniz, Harriot, and a number of medieval writers use the same figure to a rather similar, if not identical, purpose.
[107] *Essay*, II.17.7; see above, p. 136.
[108] AT.III, p. 292.
[109] *Op. cit.* (*English Works*), Vol. 1, p. 413.
[110] L. Couturat (ed.), *Opuscules et Fragments inédits de Leibniz*, Paris, 1903, p. 178. The date was c.1686.
[111] He refers to Cardan's *Pract. Arith.*, c.66, nn. 165 & 260. Leibniz was confusing Ward with John Wallis when he said that Seth Ward dissented, in his *Arithmetic of Infinites*. G.IV, p. 42; Loemker, p. 75.
[112] *Development of Mathematics*, 2nd ed., New York, 1945, 545.

PART III

PAUL WEINGARTNER

THE IDEAL OF THE MATHEMATIZATION OF ALL SCIENCES AND OF 'MORE GEOMETRICO' IN DESCARTES AND LEIBNIZ

The purpose of this paper is to give the key-principles which are involved in the ideal of the mathematization and of the 'more geometrico' – idea in Descartes and Leibniz. This is done in the following way. In Chapter 1 some of the main-principles of the philosophical doctrine of rationalism are given. These principles it is thought underly the philosophy of both thinkers Descartes and Leibniz. And so they underly also their doctrines concerning the ideal of more geometico and of mathematization of all sciences. In Chapter 2 the method 'more geometrico' will be analyzed. In Chapter 3 the ideal of mathematization will be dealt with. In respect to 'more geometrico' and to the ideal of mathematization Descartes and Leibniz differ considerably and it is Leibniz who has to offer the main principles here.

For several reasons I have chosen the method of stating explicitely the respective principles rather than giving a long description of the many different claims in the philosophies of Descartes and Leibniz. I hope to be able to show that these principles have also a good support in the writings of both authors and can be supported by the references given. However the principles given cannot be understood as being just a direct interpretation of some passages of the text. They are interpretations in the light of the research in our time. Thus they result from interpretation, reconstruction and construction. An attempt was made to avoid the latter whenever interpretation and reconstruction was sufficient for getting a consistent and adequate picture.

1. RATIONALISM

The Rationalism of the 17th and 18th century is a rather complicated philosophical doctrine or perhaps better a number of related doctrines. However there are some common and necessary characteristics of it under which the following seem important:

(1) There are first evident principles which are the sources of human knowledge.

(2) There are logical and mathematical principles of deduction by which theorems (further knowledge) can be derived from these first evident principles.

(3) There is a truth-criterion by which principles of the sort (1) and (2) can be selected from other sentences.

William R. Shea (ed.), Nature Mathematized, 151–195.

(4) There are primitive innate ideas (concepts) which are the building blocks of the first evident principles.

(5) There are logical and mathematical principles of definition by which a reduction of complex ideas to primitive ideas is possible.

(6) There are criteria to select the primitive innate ideas from other ideas.

The Rationalism of the 17th and 18th century has a lot of additional features. For example there is a certain devaluation of empirical knowledge or a stronger or less strong rejection of tradition etc. But these and other chracteristics are at least not so important for the topic developed in this paper as those which are listed above.

If one would ask who are representatives of philosophical doctrines of rationalism which satisfy these six conditions a fair and sound answer would be: Descartes, Leibniz, Spinoza and Christian Wolff. There are of course other thinkers who held either all or some of these conditions to be true but probably not in that general sense as with the above mentioned four. Thus for instance Pascal would have held all conditions to be true but for an area of applicability which is much narrower than the one Descartes, Leibniz, Spinoza and Wolff had in mind. And outside this area he had quite different 'anti-rationalistic' views.

To state the six conditions of rationalism in more detail and with more accuracy the first three are given in Chapter 1.1 and the second three in Chapter 1.2.

1.1. *Scientific Knowledge*

1.1.1. The proposition p is scientifically known (SK) if and only if (iff) either p is a first principle (FP) or p is deduced from a first principle (by logical principles of deduction)

$$(p)\ [SK(p) \longleftrightarrow FP(p) \vee (\exists q)\,(FP(q) \wedge q \vdash p)]\,.$$

What is here called 'first principle' Descartes would call a statement (judgement)[1] known by intuition.

Thus for Descartes' terminology one could reformulate 1.1.1 also in the following way: The proposition p is scientificially known iff either p is known by intuition or by deduction. These two – intuition and deduction – are the two fundamental operators of the mind,

"two mental operators by which we are able, entirely without any fear of illusion, to arrive at the knowledge of things"[2] . . . "two operators which are the most certain routes to knowledge".[3]

It has to be observed that these two operators are not the scientific method since the scientific method consists in a set of rules for correct application of the operations of intuition and deduction.[4]

Leibniz would call 'primitive truths' what is here called 'first principles' which can be – like the derived truths – truths of reason or truth of fact:

"Ideas and truths can be divided into such as are primitive and such as are derivative".[5]

"Veritates absolute primae sunt inter veritates rationes identicae et inter veritates facti haec, ex que a priori demonstrari possent omnia experimenta . . . ". "Omne veritas quae non est absolute prima, demonstrari potest ex absolute prima. Omnis veritas aut demonstrari potest ex absolute primis (quas indemonstrabilis esse, demonstrabile est) aut ipsa est absolute prima."[6]

1.1.2. If p is a first principle then it is known by intuition:

$$(p)\ [FP(p) \rightarrow INT(p)]\,.$$

The principle 1.2 is clearly held by both, Descartes and Leibniz:

"But the first principles themselves are given by intuition alone . . . "[7]

"The primitive truths, which are known by intuition, are like the derivative, of two kinds."[8]

Examples for first principles according to Descartes and Leibniz would be the principle of non-contradiction (or identity) – as a truth of reason – and the "cogito ergo sum" as a truth of fact.

Therefore it holds according to both Descartes and Leibniz that there are first principles:

1.1.3. $(\exists p)\, FP(p)$.

"Thus each individual can perceive by intellectual intuition that he exists, that he thinks, that a triangle is bounded by three lines only, a sphere by a single surface, and so on."[9]
"The primitive truths of reason are those which I call by the general name of identicals . . . Those which are affirmative are such as the following: everything is what it is, and in as many examples as we may desire, A is A, B is B"[10]

"As for the primitive truths of fact, they are immediate internal experiences of an immediacy of feeling. And it is here that the first truth of the Cartesians or of St. Augustin occurs: 'I think, therefore I am, i.e. I am a thing which thinks' . . . "[11]

1.1.4. If p is known by intuition then p is known "clare et distincte" (i.e. one has a truth-criterion (TC) for knowing that p or shortly p is a clear and distinct knowledge)

$$(p)\ [INT(p) \rightarrow TC(p)]\,.$$

For Descartes the truth-criterion is 'clare et distincte'. He speaks of certain and evident intuition in Rule 3 of RD:

"intuitûs evidentia et certitudo".

"By intuition, therefore, is meant a purely intellectual activity, an intellectual seeing or vision which is so clear and distinct that it leaves no room for doubt." [12]

According to Leibniz "Knowledge is either obscure or clear. Clear knowledge is confused or distinct. Distinct knowledge is adequate or inadequate, and is also either symbolical or intuitive".[13] From this division it follows that intuitive knowledge is always distinct and clear.[14]

1.1.5. If p is known clare et distincte then p is true (T).

$$(p)\ [TC(p) \to T(p)].$$

There are a lot of passages in Descartes which support principle 1.1.5:

"I came to the conclusion that I might assume as a general rule that the things which we conceive very clearly and distinctly are all true".[15]

"to accept nothing as true which I did not clearly recognize to be so: that is to say, carefully to avoid precipitation and prejudices in judgements, and to accept in them nothing more than what was presented to my mind so clearly and distinctly that I could have no occasion to doubt it." [16]

Leibniz had a much more detailed view on the problem of a truth criterion which is connected with his more detailed division of the different kinds of knowledge.[17] As is clear from Leibniz's *Meditationes de Cognitione, Veritate et Ideis* and from one of his letters to Foucher[18] he didn't accept Descartes principle 1.1.5:

"La Regle Generale que plusieurs posent comme un principe des sciences, quicquid clare distincteque percipio est verum, est sans doute fort defectueuse." [19]

However Leibniz assumed a weaker principle:

1.1.5.1 If p is known clare et distincte and if p is known to be possible ($K \Diamond p$)) then p is true.

$$(p)\ [(TC(p) \wedge K \Diamond(p)) \to T(p)].$$

1.1.5.1 is weaker than 1.1.5 in the important sense that a proof which shows the possibility (the consistency) is required. An idea may be clear and distinct and even the marks may be stated in a *nominal definition*, but one is not assured to derive knowledge of the truth from the definition as long as it is not established in a *real definition* that the thing defined is possible:

"Je viens à vostre Examen du grand principe des Cartesiens et de Dom Robert, que j'ay déja touché: sçavoir que nos idées ou conceptions sont tousjours vraies. Et comme j'ay déja dit, je suis bien eloigné de l'admettre, parce que nous joignons souvent des notions incompatibles, en sorte que le composé enferme contradiction. J'ay examiné plus distinctement ce principe dans une remarque sur les idées vrayes ou fausses que j'ay mise dans le journal de Leipzig. Et je tiens que pour estre asseuré, que ce que je conclus de quelque definition est veritable, il faut sçavoire que cette notion est possible. Car si elle implique contradiction, on en peut conclure en même temps des choses opposées."[20]

From the principles stated so far it follows for Descartes that every first principle is true (from 1.1.2, 1.1.4, and 1.1.5). For Leibniz it follows that every first principle is true provided the ideas put together in the principle are possible (consistent).

1.1.6. Everything derived from a true premiss is true.

$$(p)\,(q)\,[(T(p)\quad p \vdash q) \rightarrow T(q)]\,.$$

Deduction is Descartes' second operation (besides intuition)

"by which we are able, entirely without any fear of illusion to arrive at the knowledge of things."[21]

It is also described by him as

"all necessary inference from other facts which are known with certainty."[22]

"But the first principles themselves are given by intuition alone while the remote conclusions, on the contrary, are furnished only by deduction."[23]

Though Descartes describes the operation of deduction he didn't state explicitely logical rules of deduction. And although he says that there is certainty when we arrive at knowledge by deduction and that deduction contains necessity[24] he does not say very explicitely to what kind of knowledge the principles of deduction belong.

This is different with Leibniz. First of all, principles of deduction belong certainly to the truths of reason. Secondly they are hypothetical truths, not existential truths. Examples of principles of deduction are according to Leibniz the modes of syllogism; but also other valid principles not included in the theory of syllogisms:

"Further it should be known that there are good asyllogistic conclusions . . . e.g.: Jesus Christ is God, therefore the mother of Jesus Christ is the mother of God . . . If David is the father of Solomon, without doubt Solomon is the son of David. And these consequences do not fail to be demonstrable by truths upon which common syllogisms themselves depend."[25]

As he says himself one of the main principles of deduction is the principle that what implies contradiction is false, i.e. a principle of reductio ad absurdum which was known in logic and mathematics since antiquity. The second principle he describes is the principle of sufficient reason which will be dealt with later (cf. Chapter 2.5).

"In demonstration I use two principles, of which one is that what implies contradiction is false, the other is that a reason can be given for every truth."[26]

1.2. *Scientific Analysis*

1.2.1. A (scientific) term (concept) t is analyzable (AN) iff t is a primitive term or else t is reducible to a primitive term with the help of (a chain of) definitions.

$$(t)\ [AN(t) \longleftrightarrow [PRIM(t) \vee \exists t'\ (PRIM(t') \wedge t = \mathrm{df}\ t')]] .$$

Descartes understands by method

"a set of certain and easy rules such that anyone who observes them exactly will never take anything false to be true and, without any waste of mental effort but by increasing his knowledge step by step, will arrive at a true understanding of all those things which do not surpass his capacity."[27]

This method has two parts, the first part is analysis or resolution and the second is synthesis. With the method of resolution Descartes reduces obscure propositions and obscure ideas step by step to those which are simpler: The purpose is

"to divide up each of the difficulties which I was to examine into as many parts as possible and as seemed requisite."[28]

In fact the goal of this analysis or resolution is to end up at genuine knowledge by intuition (first principles) and at the apprehension of simple natures (primitive terms or concepts) which are contained in the first principles.[29]

A detailed example of a resolution he gives in his commentary to rule 12 Chapter 5 and 6 with the idea of figure. Another one we find in Chapter 13 where the analysis of "a body has extension and figure" leads to the simple natures 'body', 'extension', 'figure' where

"these elements have never existed in isolation from each other. But relatively to our understanding we call it a compound constructed out of these three natures."[30]

As a last point it has to be mentioned that Descartes didn't use very much definitions. In fact he had a prejudice about them – similar to that concerning syllogistics – because he was afraid of the definitions used by scholastic philosophers who, as he thinks, use definitions for something which cannot become clearer, i.e. for primitive ideas.[31] Thus the part "$t = \mathrm{df}\, t'$" fits better into Leibniz's philosophy. At least it cannot be understood as a direct interpretation of Descartes writings, reconstruction and abstraction from Descartes prejudices is necessary here.

Leibniz on the other hand understands by analysis or analytical method a reduction with the help of definitional replacement in order to reach primitive ideas or first principles in the sense of identicals:

"Analysis is as follows: Let any given term be resolved into its formal parts, that is, let it be defined. Then let these parts be resolved into their own parts, or let definitions be given of the terms of the (first) definition, until (one reaches) simple parts or indefinable terms."[32]

"When definition pushes analysis until it reaches primitive notions, without presupposing anything whose possibility requires an *a priori* proof, the definition is perfect or essential."[33]

As Rescher puts it "Leibnizian analysis is a logical process of a very rudimentary sort, based on the inferential procedures of *definitional replacement* and *determination* of *predicational containment* through explicit use of logical processes of inference."[34]

1.2.2. All (scientific) terms (concepts) which are primitive and derived from the mind itself are innate.

$$(t)\ [(PRIM(t) \wedge INTEL(t)) \rightarrow INN(t)].$$

Descartes speaks of innate ideas in several places: when he discusses the different types of ideas (those which are innate, those which come from sense experience and those which are constructed by the own phantasy)[35] and when he states that certain ideas (and truths) are implanted in our intellect by nature.[36] The principle 1.2.2 however he states explicitely in Rule 12, Chapter 14 where he says:

"14. Dicimus secundò, res illas, quae respectu nostri intellectûs simplices dicuntur, esse vel purè intellectuales, vel purè materiales, vel communes. Purè intellectuales, illae sunt, quae per lumen quoddam ingenitum, & absque ullius imaginis corporeae adjumento ab intellectu cognoscuntur: tales enim nonnullas esse certum est, nec ulla fingi potest idea corporea quae nobis repraesentet, quid sit cognitio, quid dubium, quid ignorantia, item quid sit voluntatis actio, quam volitionem liceat appellare, & similia; quae tamen omnia revera cognoscimus, atque tam facilè, ut ad hoc sufficiat, nos rationis esse participes."[37]

From this it is clear that Descartes divides first the primitive ideas into those which are purely intellectual (rely upon our understanding alone) – like knowledge, ignorance, doubt etc. – and those which are purely material, i.e. can be recognized as being in a material body (res extensa) like figure, extension, motion; and thirdly into those which are both intellectual and material (like existence, unity and duration). And then he says unambigously that the purely intellectual ones are known by a kind of inborn light and without the assistence of a material picture.

Leibniz had a very similar view. He uses the expression innate truth (vérité innée) in the New Essays,[38] to denote a truth in which all the ideas are innate, i.e. not derived from sense; but he explains that there is a different use of the word.

In the sense in which he uses it, "the sweet is not the bitter" is not innate, because *sweet* and *bitter* come from the external senses. But "the square is not the circle" is innate, because *square* and *circle* are ideas furnished by the understanding itself.[39]

1.2.3. There are simple ideas which are coming from the mind itself.

$$(\exists t)\ [PRIM(t) \wedge INTEL(t)]\,.$$

The examples which are given by Descartes and Leibniz for primitive ideas and for such primitive ideas which are intellectual seem to be different. Whereas Descartes gives as examples knowledge, doubt, ignorance, i.e. epistemological examples[40] Leibniz uses either mathematical, like square, triangle, circle[41] or metaphysical examples like the ideas of being, of possible, of equality. Of the latter he says that they are essential elements of our mind:

"L'idée de l'être, du possible, du même, sont si bien innées, qu'elles entrent dans toutes nos pensées et raisonnemens, et je les regarde comme des choses essentielles à nostre esprit."[42]

On the other and Leibniz agrees with Descartes that those inborn and primitive ideas cannot come from sense experience:

"The soul comprises being, substance, unity, identity, cause, perception, reason, and many other notions which the senses cannot give."[43]

1.2.4. All clear and distinct ideas are innate

$$(t)\ [CD(t) \rightarrow INN(t)]\,.$$

It is very probable that Descartes held this principle although I wasn't able to find a literal reference.[44]

It is however uncertain whether Leibniz agreed to the principle 1.2.4. There is a clear passage in the *NE* saying that the opposite of 1.2.4, i.e. the principle

"All innate ideas are clear and distinct"

does not hold:

> "What is innate is not at first known clearly and distinctly as such; often much attention and method is necessary in order to perceive it. Students do not always do so, still less every human being."[45]

Moreover it has already been said that an analoguous principle of 1.2.4 concerning propositions (i.e. 1.1.5) is not accepted by Leibniz, he accepts only the weaker principle 1.1.5.1.

2. MORE GEOMETRICO

The principles stated so far under the title "Rationalism" (Chapter 1) could be defended – at least partially – for principles of the method more geometrico too. On the other hand the principles stated so far are principles not only of 17th century rationalism but of many rationalistic doctrines, even if they are embedded in other philosophical frameworks. Thus the principles in the chapter "Scientific knowledge" (1.1.1 – 1.1.5) are accepted also by phenomenologists, even if they would use some other terms for "clare et distincte". More accurately phenomenologists would agree with rationalists that we have some knowledge by intuition (first principles which are evident) and some knowledge by deducing theorems from these principles. But those philosophers who accepted the method more geometrico had the much stronger view that the statements which we know by intuition and by deduction have to be arranged in a deductive system of axiomatic character, the first principles being the axioms and the statements "known by deduction" being the theorems. Moreover they had a certain view on the properties of such a deductive system, for instance that it is (as a whole or as a part) complete etc. Thus the following principles go beyond the principles of rationalism.

But Descartes and Leibniz differ in their views and especially in the explicite description of the more geometrico-ideal. Whereas Descartes' "more geometrico" means roughly the axiomatic system, Leibniz goes further to explicitely state and require very general and important features of deductive systems (like axiomatizability, decidability, soundness, completeness etc.). Though one cannot say that Descartes did not assume such features for the axiomatic systems he had in mind he didn't formulate them explicitely.

2.1. All scientific knowledge (*SK*) (or: everything what is scientifically known) can be integrated in a deductive system (*DS*)

$$(p)\ [SK(p) \rightarrow (\exists\alpha)\ (DS(\alpha) \wedge p \in \alpha)].$$

According to Tarski one can define a deductive system with the help of its consequence class (*CN*):

$$DS(\alpha) \longleftrightarrow CN(\alpha) \subseteq \alpha.$$[46]

Principle 2.1 is described in several passages of the works of Descartes. Moreover it is stated as a rule or general norm of scientific methodology:

"Method consists wholly in the ordering and disposing (literally, in the order and disposition) of those objects to which the attention of the mind must be directed if we are to discover any truth. We shall observe this method exactly if we reduce involved and obscure propositions step by step to those which are simpler, and if we then start with the intuitive apprehension of the simplest propositions and try by retracing our path through the same steps to ascend to the knowledge of all the others."[47]

"Apres avoir bien fait entendre ces choses, j'aurois voulu mettre icy les raisons qui servent a prouver que les vrays Principes par lesquels on peut parvenir à ce plus haut degré de Sagesse, auquel consiste le souverain bien de la vie humaine, sont ceux que j'ay mis en ce Livre: & deux seules sont suffisantes à cela, dont la premiere est qu'ils sont tres-clairs, & la seconde, qu'on en peut deduire toutes les autres choses: car il n'y a que ces deux conditions qui soient requises en eux."[48]

In many of his works Leibniz describes the structure of a deductive system and holds that every truth can be incorporated in it:

"Omnis veritas quae non est absolute prima, demonstrari potest ex absolute prima. Omnis veritas aut demonstrari potest ex absolute primis (quas indemonstrabiles esse, demonstrabile est) aut ipsa est absolute prima."[49]

"Connection is of two sorts: the one is absolutely necesssary, so that its contrary implies contradiction, and this deduction occurs in eternal truths like those of geometry; the other is only necessary *ex hypothesi* and, so to speak, by accident and it is contingent in itself, when the contrary does not imply contradiction."[50]

Leibniz of course doesn't mean here that a consequence wouldn't logically or necessarily follow from its premiss (no matter what kind of propositions are involved); what he wants to express is that as far as the axioms (premisses) are not necessary in itself the conclusion cannot get the necessity by being a correct consequence (but will also be contingent).

2.2. All propositions have exactly one of the definite truth-values true or false (where 'false' means 'not-true').

$$(p)\ [V[p] = T \vee V[p] = F].$$

where F = not-T or: $V[p] = F \longleftrightarrow V[\sim p] = T$ and p is a proposition. Thus we can write the principle also in this form:

$$(p)\ [V[p] = T \vee V[\sim p] = T].$$

This principle is very restrictive: It is a bivalence principle which rules out any many-valued logic. Although Leibniz accepted the principle 2.2 he had a differenciated view on it. He says that one consequence of the principle that every proposition is either true or false is that a proposition cannot be both true and false:

"Le principe de contradiction est en general: *une proposition est ou vraye ou fausse;* ce qui renferme deux enonciations vrayes, l'une que le vray et le faux ne sont point compatibles dans une même proposition, ou *qu'une proposition ne sauroit estre vraye et fausse à la fois*."[51]

This is correct and important: If this consequence would be taken to be the principle many-valued logics wouldn't be ruled out. This can be seen from the following consideration: All what the above consequence forbids is that there would be at least one line in a matrix where p takes both values T and F. But it is of course not forbidden by this consequence that there may be lines in the matrix of p where p takes a different value (for example: indefinite) i.e. a value which is neither T nor F. Thus the usual 3-valued logics, for instance those of Łukasiewicz, Bochvar, and Kleene are not excluded. Even a logic with infinitely many values between true and false would not be ruled out.[52]

The above consequence "a proposition cannot be both true and false" implies that a proposition and its negation cannot both be true and cannot both be false. This is probably in accordance with Leibniz who does not explicitely mention that.[53]

The second consequence of the principle of contradiction as stated by Leibniz is that there is no middle between true and false or that it is impossible that a proposition is neither true nor false:

"l'autre que l'opposé ou la negation du vray et du faux ne sont pas compatibles, on qu'il n'y a point de milieu entre le vray et le faux, ou bien: *il ne se peut pas qu'une proposition soit ny vraye ny fausse*. Or tout cela est encor vray dans toutes les propositions imaginables . . . "[54]

This second consequence tells us unambiguously that Leibniz understood a very restricted principle by the law of contradiction. Namely a principle which includes a strong kind of a tertium non datur.

This strong version of tertium non datur seems to exclude two things: First it clearly excludes many-valued systems in the sense that a proposition p cannot take any other value except true and false.[55]

Secondly it also excludes truth-value gaps, i.e. the principle implies that a proposition – which according to Leibniz has either subject-predicate form or the form "*A* exists"[56] – take always a definite truth-value.

Thus principle 2.2 is not acceptable for systems or for theories of truths which allow truth-value-gaps like those proposed by van Fraassen and Kripke.[57] On the other hand principle 2.2 is in accordance with Tarski's theory of truth and with his truth-definition. Since this definition is an interpretation of the classical Aristotelean theory of truth and since Leibniz is very much in the tradition of Aristotle and the scholastic philosophers in this respect it seems very natural that also he accepted the principle.

Although Descartes accepted the principle 2.2 it is hard to find passages where he formulates it explicitely. On the other hand there are many passages where he seems to presuppose it:

"By method I understand certain and easy rules such that anyone who observes them exactly will never take anything false to be true."[58]

"Nunc autem ordo videtur exigere ut prius omnes meas cogitationes in certa genera distribuam, & in quibusnam ex illis veritas aut falsitas proprie consistat, inquiram. . . ."[58a]

"Jam quod ad ideas attinet, si solae in se spectentur, nec ad aliud quid illas referam falsae proprie esse non possunt . . . Nulla etiam in ipsâ voluntate, vel affectibus, falsitas est timenda . . . Ac proinde sola supersunt judicia, in quibus mihi cavendum est ne fallar."[58b]

2.3. Every deductive system is finitly axiomatizable

$$(\alpha)\ [DS(\alpha) \rightarrow FAX(\alpha)]$$

where

$$FAX(\alpha) \longleftrightarrow (\exists\beta)\ [\beta = CN(\alpha)].$$ [59]

It is known that the principle 2.3 is not generally valid; i.e. there are deductive systems which are not finitly axiomatizable. But it seems that both Descartes and Leibniz thought of a deductive system as one that has the potential structure of an axiom system with a finite number of axioms. But an important difference has to be mentioned here:

Descartes seems to have thought that the axiomatization can actually be carried through or found by men in respect to both, truths of mathematics and empirical truths; though in both cases one has to have different axioms which are known by intuition, mathematical ones (like "things which are equal to the same thing are equal to one another") and factual ones (like the "cogito ergo sum").

"Hinc jam dubium esse potest, quare, praeter intuitum, hîc alium adjunximus cognoscendi modum, qui fit per *deductionem*: per quam intelligimus, illud omne quod ex quibusdam alijs certò cognitis necessariò concluditur. Sed hoc ita faciendum fuit, quia plurimae res certò sciuntur, quamvis non ipsae sint evidentes, modò tantùm à veris cognitisque principijs deducantur per continuum & nullibi interruptum cogitationis motum singula perspicuè intuentis: non aliter quàm longae alicujus catenae extremum annulum cum primo connecti cognoscimus."[60]

We shall observe this method exactly if we reduce involved and obscure propositions step by step to those which are simpler, and if we then start with the intuitive apprehension of the simplest propositions and try by retracing our path through the same steps to ascend to the knowledge of all the others."[61]

Leibniz on the other hand made an important distinction here: In respect to the *vérités de raison* such axiom systems can be found; in this case the *vérités innées* are the axioms from which the other truths of reason can be derived. Also in respect to those contingent truths which are about ones own inner states (i.e. that I think, that I will) including ones own existence a finite axiomatization by man is possible. However this does not hold of the other contingent truths. Although they are incorporated in a deductive system which is finitly axiomatizable it is not the case that man can find such an axiom system. The reason is that man would need infinitly many steps to trace back these contingent truths to simple axioms. Only God knows the axiom-system with one act of thinking (without thinking in finitly or infinitly many steps). – One has to observe here that the concept of finite axiomatizability is not an epistemic concept (for example it does not mean: "*knowing* a finite set of axioms for a set of propositions") but a semantic concept concerning deductive systems which is defined with the concept of *consequence class* (which itself is not epistemic).

The passages which support principle 2.3 can be found in many of Leibniz's writings. First we find it in his proposals for a scientia generalis:

"Elementa Veritatis aeternae, seu de forma argumentandi qua per modum calculi omnes controversiae demonstrative tollantur, et vel absolute determinetur veritas, vel quando sufficientia non sunt data . . . "[62]

"Absolutely first truths are, among truths of reason, those which are identical and among truths of fact those, from which all experiments can be proved à priori."[63]

Although Leibniz does not explicitly say that these absolutely first truths, i.e. the axioms are finite in number he implicitely assumes this. This is supported by the lists of definitions he adds[64] and by notes such as the one added to the expression "veritates absolute primae" in the above passage:

"Veritas definitio realis est. Verum est quod ex identico demonstrabile est per definitiones. Quod demonstratur ex definitionibus nominalibus verum est hypothetica, quod ex realibus absolute . . ."[65]

Sometimes he reduces the number of axioms – for instance to derive the theorems of arithmetic and geometry – to a very narrow set, even to one axiom plus definitions:

"The great foundation of mathematics is the pricniple of contradiction or identity, that is, that a proposition cannot be true and false at the same time, and that, therefore, *A* is *A* and cannot be non-*A*. This single principle is sufficient to demonstrate every part of arithmetic and geometry, that is, all mathematical principles."[66]

How this can be done he tries to show us for instance with the following example:

"TH. Je dis, que je vous attendois là bien preparé. Ce n'est pas une verité tout à fait immediate que deux et deux sont quatre, supposé que *quatre* signifie trois et un. On peut donc la demonstrer et voicy comment:

Definitions
(1) *Deux*, est un et un.
(2) *Trois*, est Deux et un.
(3) *Quarte*, est Trois et un.

Axiome. Mettant des choses egales à la place, l'egalité demeure.

Demonstration

2 et 2 est 2 et 1 et 1 (par la def. 1)	$2+2$
2 et 1 et 1 est 3 et 1 (par la def. 2)	$2+1+1$
3 et 1 est 4 (par la def. 3)	$3+1$
Donc (par l'Axiome)	4
2 et 2 est 4. Ce qu'il falloit demonstrer.	

Je pouvois, au lien de dire que 2 et 2 est 2 et 1 et 1, mettre que 2 et 2 est egal à 2 et 1 et 1, et ainsi des autres. Mais on le peut sousentendre par tout, pour avoir plustost fait; et cela, en vertu d'un autre axiome qui porte qu'une chose est egale à elle même, ou que ce qui est le même, est egal."[67]

2.4. All finitly analytic (*FA*) propositions are (necessarily) true.

$$(p)\ [FA(p) \rightarrow T(p)].$$

This principle is more a principle of Leibniz than of Descartes but when the terms are explained one can see that it is also applicable to Descartes.

According to Leibniz a finitly analytic proposition is one that is either an 'identical' (i.e. one of the form *A* is *A* or *AB* is *A* like "a rational animal is an animal") or one that can be reduced to an identical proposition with the help of a finite number of steps using definitions:

"Quoniam vera propositio est quae identica est, vel ex identicis potest demonstrari adhibitis definitionibus . . . "[68]

Copleston and Rescher say that principle 2.4. is Leibniz' principle of contradiction (or identity).[69] But they do not give a passage where Leibniz calls this principle the principle of contradiction. On the other hand there are several passages where Leibniz speaks explicitely of the principle of contradiction, but they read differently:

"The principle of contradiction is in general: A proposition is either true or false."[70]

". . . the principle of contradiction or identity, that is, that a proposition cannot be true and false at the same time."[71]

Following the first citation Leibniz speaks of two consequences of the principle of contradiction which have been discussed already in Chapter 2.2. In addition to that Leibniz uses the term 'principle of contradiction' also for the inference-principle *reductio ad absurdum*, i.e. for the principle that every proposition is false if a contradiction is derivable from it.[72]

From these passages it seems clear that the principle 2.4. can hardly be called "Leibniz's principle of contradiction". I turn now to a different question, namely whether principle 2.4. is a true and even important principle in the philosophy of Leibniz. This is indeed so. And there is even a good reason for calling 2.4. a principle of (non-)contradiction although there is little evidence that Leibniz called it so.

My claim is this: The principle 2.4. is the principle of the consistency of the axiomatic system of the truths of reason. Thus it is a principle of (non-)contradiction (i.e. of consistency) but it does not apply to arbitrary propositions but to those which are integrated in the axiomatic system of truths of reason. In this sense the principle 2.4. amounts to a principle of soundness, i.e. like *every proposition which is derivable in the axiomatic system is (necessarily) true*.

If this interpretation is correct Leibniz at least used – in addition to different forms of the traditional principle of contradiction including the reductio ad absurdum principle – a principle for the consistency of axiomatic systems or a soundness-principle in the form of 2.4. That Leibniz indeed used such a principle is supported by many passages in which he describes the method of analysis.

By this method first of all the axioms can be shown to be true: Since they are identical propositions of a simple form (first principles) they are evident insights of the human mind. In most cases if not in all they are vérités innées according to him. Then the theorems have to be shown to be true. This is possible since every step (of the finitly many) can be watched and the transformations with the help of definitions and inference rules[73] can be checked. But since the principle 1.1.6. – that the (correct) logical consequences of true propositions are true – is accepted one can prove that the theorems (which have been derived in the axiomatic system) must be true too:

"Itaque cujuscunque veritatis reddi potest ratio, connexio enim praedicati cum subjecto aut per se patet, ut in identicis, aut explicanda est, quod fit resulutione terminorum. Atque hoc unicum summumque est veritatis criterium, in abstractis scilicet neque ab experimento pendentibus, ut sit vel identica vel ad identicas revocabilis."[74]

According to the last passage the principle 2.4. can also be expressed in this way: If a proposition (which can be incorporated into a deductive system) has a (sufficient) reason, i.e. if it has a proof by which it can be traced back to identical propositions by using definitions then it is true (and consequently also consistent).

The principle 2.4. seems to be included in Descartes' concept of method. By method he understands a set of rules which guide the two fundamental operations of the human mind – intuition and deduction in such a way that one will never take anything false to be true:

"By method I understand certain and easy rules such that anyone who observes them exactly will never take anything false to be true . . . "[75]

The above text together with the fifth rule can be taken to say that whatever proposition we arrive at as long as our intuition and deduction are guided by our method (set of certain and easy rules of deductive ordering) we can be sure that this proposition is not false (i.e. true).

2.5. Every true proposition is finitly or infinitly analytic. Or: Every true proposition has an a prior proof.

$$(p)\ [T(p) \rightarrow [FA(p) \vee IA(p)]\,].$$

This is one of the two main-forms of Leibniz's principle of sufficient reason. As Russell says correctly[76] there is a group of earlier formulations which date around the year 1686. Of these the principle 2.5. is a more or less direct intepretation. This can be seen from the following passages:

"that nothing is without a reason, or that every truth has its à priori proof, drawn from the notion of the terms, although it is not always in our power to make this analysis."[77]

"Generally, every true proposition (which is not identical or true *per se*) can be proved *à priori* by the help of axioms, or propositions true *per se*, and by the help of definitions or ideas. . . . It is certain, therefore, that all truths, even the most contingent, have an *à priori* proof, or some reason why they are rather than are not. And this is itself what people commonly say, that nothing happens without a cause, or that nothing is without a reason."[78]

"In demonstration I use two principles, of which one is that what implies contradiction is false, (while) the other is that a reason can be given for every truth (which is not identical or immediate), that is, that the notion of the predicate is always contained, explicitly or implicitly, in the notion of its subject, and that this holds good no less in extrinsic than in intrinsic denominations, no less in contingent than in necessary truths."[79]

But there are also formulations of a later period which differ somewhat from those given above:

"Our reasonings are founded upon two great principles, that of contradiction . . . and that of sufficient reason, in virtue of which we judge that no fact can be found true or existent, no statement veritable, unless there is a sufficient reason why it should be so and not otherwise, although these reasons usually cannot be known to us. There are also two kinds of truths, those of reasoning, and those of fact. Truths of reasoning are necessary, and their opposite is impossible; truths of fact are contingent, and their opposite is possible. When a truth is necessary, the reason of it can be found by analysis. . . . But there must also be a sufficient reason for contingent truths or truths of fact, *i.e.* for the sequence of things which are dispersed throughout the universe of created beings, in which the resolution into particular reasons might go on into endless detail."[80]

2.5.1. It seems to me that though these later formulations are different they can be treated as a more special or more detailed fomulations of the earlier ones.[81] This can be seen from two considerations:

(1) One has to keep in mind what Leibniz means by analysis. As Rescher puts it analysis is "a logical process of a very rudimentary sort, based on the inferential procedures of definitional replacement and determination of predicational containment through explicit use of logical process of inference."[82]

An important point is further that in the realm of truths of reason the proof-process is terminating after a finite number of steps, i.e. the proposition proved is finitly analytic. On the other hand in the realm of contingent truhs the proof-process is not terminating after a finite number of steps, i.e. the proposition is infinitly analytic. In both cases however we have a genuine proof-process intended at least as an inferential procedure of the most rigorous kind, i.e. a proof of a proposition within a deductive system. In both cases therefore Leibniz doesn't hesitate to speak of an a priori proof even if in the case of contingent truths the necessity is hypothetical (ex hypothesi).

(2) One has to realize that Leibniz uses both a more ontological (or metaphysical) way of speaking and a more logical one when he describes the principle of sufficient reason. He himself announces sometimes this or that way of speaking as he does in the following passage for the metaphysical way:

"Thus far we have spoken only as mere physicists: now we must rise to metaphysics, by making use of the great principle, little employed in general, which affirms that nothing happens without a sufficient reason."

Thus a short cut of the metaphysical version of the principle is just that nothing happens without a sufficient reason. But it is important to notice that the continuation of the above citation turns again to an epistemological or logical version of the principle:

"i.e. that nothing happens without its being possible for one who should know things sufficiently to give a reason sufficient to determine why things are so and not otherwise."[83]

One may translate the latter citation into the following more logical way of describing the principle. Let's assume a state of affairs is happening (is obtaining) iff the proposition representing it is true. Then instead of saying: if a state of affairs obtains then there is a sufficient reason for the state to obtain (and to obtain in this particular way) one can say also: If a proposition (representing a state of affairs) is true then one (who should know things sufficiently) can give an explanation for it. Or: if a proposition (representing a state of affairs) is true then one can give an a priori proof for this proposition from suitable premisses (explanatory propositions) in a deductive system.

This leads to my claim for the interpretation of Leibniz's principle of sufficient reason:

2.5.2. The principle of sufficient reason is a statement about the completeness of a deductive system.

First of all it has the general form of a completeness result: Every true (valid) proposition is provable in the system. Secondly the deductive system here can be understood as a comprehensive theory with an axiomatic structure of any of the following areas: logic, mathematics, physics, jurisprudence (including ethics), metaphysics and even natural theology (including theodicy):

"In philosophy, I have found a means of accomplishing in all sciences what Descartes and others have done in Arithmetic and Geometry by Algebra and Analysis . . . "[84]

Applied to these different areas the principle of sufficient reason can be reinterpreted in the following way:

(1) Provided logic and mathematics are built *more geometrico*, i.e. as a deductive system *LM* with the properties given in the principles 2.2. – 2.4. (the principles of rationalism presupposed) then every true (necessarily true) proposition of logic or mathematics has an a priori proof in *LM* with a finite number of steps. That is: *LM* is (deductively) complete.

(2) Provided metaphysics and natural theology (Leibniz thinks here especially of theodicy and proofs of the existence of God) are built *more geometrico*, i.e. as a deductive system *MT* with the properties given in the principles 2.2. – 2.4. (the principles of rationalism presupposed) then every true (necessarily true) proposition of metaphysics or natural theology is provable a priori in *MT* with a finite number of steps. That is: *MT* is (deductively) complete.

(3) Provided physics is built *more geometrico*, i.e. as a deductive system *PH* with the properties given in the principles 2.2. – 2.4. (the principles of rationalism presupposed) then every true (contingently true) proposition of physics has an a priori proof in *PH* with an infinite number of steps. That is: *PH* is (deductively) complete although it goes beyond the capacity of humans to carry through effectively this proof.

(4) Provided the moral sciences including jurisprudence are built *more geometrico*, i.e. as a deductive system *MJ* with the properties given in the principles 2.2. – 2.4. (the principles of rationalism presupposed) then every true (necessarily true for some and contingently true for other) proposition of the moral sciences or of jurisprudence has an a priori proof in *MJ* with an infinite number of steps. That is: *MJ* is (deductively) complete although it goes beyond the capacity of the human mind to carry through effectively this proof.

The first two versions (1) and (2) of the principle concern sciences which deal with truths of reason.[85] Since metaphysics deals with substances the second version is very much connected with certain important completeness-properties which are possessed by every substance according to Leibniz. This point, though interesting, would be beyond the scope of this paper.[86]

The third version (3) concerns physics or in a broader sense the natural sciences. These truths are (at least partially as far as they presuppose existence assumptions and do not deal just with abstract concepts) not finitly but infinitly analytic:

"The laws of motion which actually occur in nature and which are verified by experiments are not in truth absolutely demonstrable, as a geometrical proposition would be."[87]

What has been said to the third version (3) applies similarly to the fourth version (4): These truths are at least partially (for instance if the free will is taken into account) not finitly but infinitly analytic.

Putting together these four versions of the principle of sufficient reason into a general principle one may say:

Provided that everything what can be scientifically known is integrated *more geometrico*, i.e. built into a comprehensive deductive system *CU* (with the properties given in the principles 2.2.–2.4., the principles of rationalism presupposed) then every true proposition which can be scientifically known has an a priori proof in *CU* either with a finite or with an infinite number of steps. That is: *CU* is (deductively) complete although it goes beyond the capacity of the human mind to carry through effectively the proof in case of contingent truths.

2.5.3. Did Descartes hold the principle 2.5.? He certainly did not have this principle in the following two senses: (1) Descartes doesn't have a principle of sufficient reason in the sense that he had a version which would fit very well to Leibniz's earlier or later formulations of the principle. (2) Descartes has not stated such (or a similar) principle as one of the most important and fundamental of his philosophy (as Leibniz did with the principle of sufficient reason). However, if my interpretation of the principle of sufficient reason as a completeness-principle is correct one can find a kind of completeness-idea implicitly in some of the rules for the direction of the mind. In the commentary to the fourth rule, after describing how he understands *method,*[88] Descartes says that one has to observe here two different points: first that he who observes these rules will never take anything false to be true and secondly that he (when observing the rules) by increasing his knowledge step by step will *arrive at a true understanding of all those things* which do not surpass his capacity. The second point is the one which is important here. And he goes on to explain this in more detail in saying that if the method clearly and distinctly tells us how we have to use intuition in order not to commit an error and how to use deduction in order to achieve the knowledge of all things then this method doesn't lack anything in respect to its completeness:

"Notanda autem hîc sunt duo haec: nihil nimirum falsum pro vero supponere, & ad omnium cognitionem pervenire. ... At si methodus rectè explicet quomodo mentis intuitu sit utendum, ne in errorem vero contrarium delabamur, & quomodo deductiones inveniendae sint, ut ad omnium cognitionem perveniamus: nihil aliud requiri mihi videtur, ut sit completa, cùm nullam scientiam haberi posse, nisi per mentis intuitum vel deductionem, jam antè dictum sit."[89]

A similar view is expressed in rule 7:

"Ad scientae complementum oportet omnia & singula, quae ad institutum nostrum pertinent, continuo & nullibi interrupto cogitationis motu perlustrare, atque illa sufficenti & ordinatâ enumeratione complecti."[90]

Though these passages do not give us the clear completeness idea as it is expressed in different versions of the principle of sufficient reason, they still give the impression that Descartes had the idea that if the scientific approach is directed by his rules one can end up with a deductive system which allows us to derive step by step anything true about all things.

2.6. Concerning the *more geometrico idea* as a whole I may add that the principles given in 2.1.–2.5. can be viewed firstly as metatheoretic principles of a very general methodology of sciences. In this sense and under this aspect they are not axioms of a particular deductive system. However, there is a second understanding of some of these principles, namely as *first principles* of a comprehensive deductive system. Not all the principles can be understood in this second sense however (for example 2.1., 2.3., 2.4. and 2.5. cannot).[91] On the other hand principle 2.2. (the principle of contradiction or identity) has at least an analogue in the object language according to Leibniz, insofar it belongs to the socalled *identicals*, i.e. to the identical propositions and therefore has instances which can be used as axioms. For instance "what is A cannot be non-A" or "$A = \sim\sim A$" as an example of a negative identical is according to Leibniz an application of the principle of contradiction. But since he identifies the principle of contradiction with the principle of identity he also says that the positive examples are instances of identicals going back to the principle of contradiction:

"vulgaria autem axiomata resolutione vel subjecti vel praedicati vel triusque ad identica revocantur sive demonstrantur, ut contrarium supponendo apparent idem simul esse et non esse. Unde patet ostensivam et apagogicam demonstrationem in ultima analysi coincidere, et recte Scholasticis quoque notatum, omnia axiomata intellectis terminis revocari ad principium contradictionis."[92]

Finally there is another point to be added to the principles 2.1.–2.5. which are essential for the *more geometrico idea*:

They may be interpreted not only as propositions but can also be read as first normative principles according to the idea *more geometrico* of a general methodology of all sciences[93] or even of a universal science. In this case they would read thus:

Everything what is (or can be) scientifically known should be integrated in a deductive system (cf. 2.1.).

All propositions (which are used in science) should have exactly one of the definite truth-values true or false (where 'false' means 'not-true') (cf. 2.1.).

Every deductive system (used for scientific inquiry) should be finitly axiomatizable (cf. 2.3.).

In an axiomatic system used for scientific inquiry all finitly analytic propositions should be (necessarily) true (since otherwise the system would be inconsistent or incompatible with facts) (cf. 2.4.).

In an axiomatic system used for scientific inquiry every true proposition should have an a priori proof (should be finitly of infinitly analytic) (cf. 2.5.).

3. THE IDEAL OF MATHEMATIZATION

3.1. *Mathematical Language of Sciences*

There are passages in Descartes and Leibniz in which both present arithmetic and geometry as an ideal case of scientific knowledge and also as an ideal which should be imitated by the other scientific disciplines.

In these passages different aspects are stressed. Descartes for instance underlines the aspect of certainty and security which is one of the most important ideals in his whole philosophy:

"Jam verò ex his omnibus est concludendum, non quidem solas Arithmeticam & Geometriam esse addiscendas, sed tantummodo rectum veritatis iter quaerentes circa nullum objectum debere occupari, de quo non possint habere certitudinem Arithmeticis & Geometricis demonstrationibus aequalem."[94]

"Ces longues chaînes de raisons, toutes simples et faciles, dont les géomètres ont coutume de se servir, pour parvenir à leurs plus difficiles démonstrations, m'avaient donné occasion de m'imaginer que toutes les choses, qui peuvent tomber sous la connaissance des hommes, s'entresuivent en même façon, . . . "[95]

As to the question how this certainty is reached Descartes has to offer his Rules for the Direction of the Mind and his Discours on Method.

Though both Discours and Rules contain directions of how to proceed, they are too general to be an effective help to build up the other sciences like mathematics.

Also Leibniz underlines certainty and the possibility for deciding controversial problems when he speaks of the ideal of mathematical reasoning:

"Projet et Essais pour arriver à quelque certitude pour finir une bonne partie des disputes et pour avancer l'art d'inventer. Les hommes ont sçu quelque chose du chemin pour arriver à la certitude: la logique d'Aristote et des Stoiciens en est une preuve. Mais sur tout l'exemple des Mathematiciens . . .

Cependant on n'a pas suivi ce chemin, parce qu'il est un peu incommode, et parce qu'il y faut aller lentement et à pas comptés. Mais je croy que c'est, parce qu'on n'en a pas sçu les effets. On n'a pas consideré de quelle importance il seroit de pouvoir establir les principes de Metaphysique, de Physique et de Morale avec la meme certitude, que les Elemens de Mathematique."[96]

But in contradistinction to Descartes Leibniz had a lot of detailed proposals to reach the precision and certainty of mathematics also in other scientific disciplines. First he realized – like Frege – that the natural language is not sufficient for that purpose. Like in mathematics one has to develop a special kind of language:

"Linguae vulgares etsi plurimum prosint ad ratiocinandum, attamen innumeris aequivocationibus sunt obnoxiae, nec officium calculi facere possunt, nempe ut errores ratiocinationis ex ipsa vocabulorum formatione et constructione detegi possint, tanquam soloecismi et barbarismi. Quod sane admirabile beneficium hactenus solae praestant notae Arithmeticorum et Algebristarum, ubi ratiocinatio omnis in usu characterum consistit et idem est error animi qui calculi."[98]

Secondly he gave a set of related proposals which contain concrete rules to build a formal and universal language according to the ideal of mathematics. This formal and universal language (usually called *characteristica universalis*)[99] had different parts: An alphabet of basic terms, rules how to build compound terms out of basic terms, rules for the combination of terms in general (i.e. for subject and predicat) to build true propositions, rules for the derivation of propositions from others, rules for finding new true propositions (*ars inveniendi*) and rules for defending, proving and deciding propositions (*ars judicandi*).

The following principles state important characteristics of Leibniz's proposals for an universal mathematical language:

3.1.1. Every primitive term can be represented (*REP*) by a characteristic basic number (*BN*).

$$(t)\ [PRIM(t) \rightarrow (\ \ x)\ [BNx \wedge REP(xt)]\,].$$

3.1.2. Every compound (*COM*) term can be represented by a characteristic number (*N*) which is equal to the result of applying a certain mathematical function (*F*) to numbers which represent primitive terms.

$$(t)\,[COM(t) \rightarrow (\exists x)(\exists y_1 \ldots y_n)(\exists t_1 \ldots t_n)\,[Nx \wedge REP(xt) \wedge x = F(y_1 \ldots y_n) \wedge PRIM(t_1) \wedge \ldots \wedge PRIM(t_n) \wedge REP(y_1 t_1) \wedge \ldots \wedge REP(y_n t_n)]\,].$$

"Sed nescio an quisquam mortalium veram rationem hactenus perspexerit, qua cuique rei numerus suus characteristicus assignari postest."[100]

Although – according to this and the following passage – Leibniz believes that it is not quite arbitrary to choose these numbers, i.e. that the characteristic numbers have to be found (discovered) one may use other numbers as substitutes until one can find the right ones. The only requirement is that these substitutes have to satisfy certain relations in order that the relations between the things or concepts (which are represented by numbers) hold too:

"Cum vero ob admirabilem rerum connexionem paucarum rerum ab aliis diversarum Numeros characteristicos dare difficillimum sit, ideo elegans ni fallor artificium excogitavi, quo ostendi possit, quod ratiocinationes per numeros comprobare liceat. Fingo itaque Numeros characteristicos illos, tantopere mirabiles, jam dari, observataque illorum generali proprietate quadam, tales numeros qualescunque ei proprietati congruentes interim assumo, iisque adhibitis statim mirabili ratione omnes regulas Logicas per numeros demonstro et ostendo, quomodo cognosci possit an argumentationes quaedam sint in forma bonae. An vero argumenta vi materiae bona sint aut concludant, tum demum sine ullo labore animi aut errandi periculo judicari polerit cum ipsi veri Numeri Characteristici rerum habebuntur."[101]

From 3.1.1. and 3.1.2. it follows that also every Subject-term and every Predicat-term of a categorical proposition is representable by a characteristic number.[102]

In his last and successful application to syllogistics Leibniz represented each term of a categorical proposition not by just one number but by an ordered pair of numbers of which one is positive the other negative and both have no common divisor (cf. 3.1.3.2).[103]

3.1.3. The characteristic number of the compound term has to be chosen in such a way that in addition to condition 3.1.2. the following condition is satisfied: If the predicate (P) is contained in the subject (S)[104] the number representing the predicate term (t_p) has to be contained mathematically (i.e. where 'containing' is interpeted by a certain mathematical function F^*) in the number representing the subject -term (t_s).

$$[REP(x, t_P) \wedge REP(y, t_S)] \rightarrow [P \text{ inest } S \longleftrightarrow x = F^*(y)].$$

Regulae usus characterum in propositionibus categoricis sunt sequentes: *Si propositio Universalis Affirmative est vera*, necesse est ut numerus subjecti dividi possit exactè seu sine residuo per numerum praedicati."[105]

From this it is clear that one of Leibniz's basic ideas was that the mathematical function representing the intensional inclusion, i.e. the inclusion of the predicate in the subject is the function of division. In his successful proposal the division has to be applied to the first and second members of the ordered pairs of numbers representing the subject and predicate-term (cf. 3.1.3.2).

3.1.3.1. Leibniz has given quite a number of different proposals for a mapping from the forms of the different categorical propositions to characteristic numbers.[106] The difficulties are not only related to the categorical propositions alone but also to their role in syllogistic modes so as to be able to distinguish valid arguments from invalid ones by a method of mathematical calculation (cf. 3.1.4). It seems that only Leibniz's last proposal was successful.[107] According to this proposal the terms of categorical propositions are represented by an ordered pair of two numbers, one positive, the other negative ($+x_S^1$, $-x_S^2$ for the subject term and $+x_P^1$, $-x_P^2$ for the predicate term) and with the important condition that they are relatively prime to each other, i.e. they have no common divisor. Then the following holds for the four categorical propositions:

(1) "All S is P" is true iff P inest S iff $+x_S^1 : +x_P^1$ and $-x_S^2 : -x_P^2$ (divisible without remainder)
"All S is P" is false iff this is not the case.

(2) "No S is P" is true iff there is a (common) divisor z of at least two non-corresponding numbers, i.e. iff

$$\begin{array}{ll} \text{either} & +x_S^1 : z \text{ and } -x_P^2 : z \\ \text{or} & -x_S^2 : z \text{ and } +x_P^1 : z \end{array}$$

"No S is P" is false iff this is not the case.

(3) "Some S is not P" is true iff "All S is P" is false.

(4) "Some S is P" is true iff "No S is P" is false.

3.1.3.2. This is substantiated by the following passages:

"I come now to the numbers by which terms are to be expressed and to that end I shall give the following rules and definitions:

(I) Any term of any proposition, whether a subject or a predicate is to be written as two numbers, the one being modified by a plus sign and the other by a minus sign . . .

(II) A true universal affirmative proposition, for example
Every wise man is pious

+20, −21 +10, −3

is one in which any symbolic number of the subject (e.g. +20 and −21) can be divided exactly (i.e. in such a way that no remainder is left) by the symbolic number of the same sign belonging to the predicate (+20 by +10, −21 by −3) . . . Conversely when this does not hold the proposition is false . . .[108]

(IV) A true universal negative proposition, for example
No pious man is unhappy

+10, −3 +5, −14

is one in which two numbers of different signs and different terms have a common divisor. (For example −10 and −14, since the former has the sign + and the latter the sign −; the former is taken from the subject, the latter from the predicate . . ."[108a]

One is reminded on Boole's Algebra and on Gödel's idea of representing the wffs by Gödel-numbers.

3.1.4. The proposition q follows from the premisses $p_1 \ldots p_n$ iff there is a certain mathematical function such that the ordered sequence of numbers representing q is equal to the result of applying this function to the ordered sequence of numbers which represent the premisses $p_1 \ldots p_n$.

In the more specific application to syllogistics the general principle 3.1.4 reads:

3.1.4.1. The conclusion *SP* (i.e. one of the four categorical propositions) follows from the two syllogistic premisses involving the terms *S, P, M* iff the ordered pairs of numbers for the subject- and the predicate-term satisfy the conditions given in 3.1.3.1 (1)–(4) for one of the four categorical forms. For example:

Every wise man is pious	+20, −21; +10, −3
No pious man is unhappy	+10, −3; +5, −14
Therefore: No unhappy man is wise	+5, −14; +20, −21

By this method all laws of conversion, of the square of opposition, and all the valid moods of assertoric syllogism can be proved to be valid. Moreover the method exhibits as valid Lukasiewicz' four axioms of assertion and also his axiom of rejection.[108b]

3.1.4.2. There is however some difficulty which seems to have bothered also Leibniz. There are invalid syllogistic forms for which one can find numbers in such a way as to exhibit them as valid. But, on a closer look, this fact does not make Leibniz's mathematization of syllogistic incorrect. As Marshall points out correctly, the argument for rejection is as follows:

"If invalidating instantiations can be produced, the mood is invalid; it is valid if they cannot."[108c]

3.1.5. Provided the principles 3.1.1. to 3.1.4. hold (on the basis of the principles of rationalism and of *more geometrico*) there is an *ars inveniendi* i.e. there is a mechanical and combinatorial method with the help of which one can find every (or almost every) truth.

There are many passages where Leibniz describes his ars inveniendi; most of them are embedded in texts in which he describes the whole approach of the mathematization of the scientific language (which is sometimes called "characteristica universalis" and if especially the calculating process is meant "calculus ratiocinator"):

"In Philosophy, I have found a means of accomplishing in all sciences what Des Cartes and others have done in Arithmetic and Geometry by Algebra and Analysis, by the *Ars Combinatoria*. . . . By this all composite notions in the whole world are reduced to a few simple ones as their Alphabet; and by the combination of such an alphabet a way is made of finding, in time, by an ordered method, all things with their theorems and whatever is possible to investigate concerning them."[109]

"Methodus inveniendi perfecta, si praevidere possimus, imo demonstrare antequam rem aggrediamur, nos ea via ad exitum perventuros; perfecta magis illa quae nullis utitur theorematis apud alios demonstratis, vel problematis ab aliis solutis."[110]

Although Descartes has not detailed principles like 3.1.1. – 3.1.4. he had also strong convictions that one could have a method (consisting mainly of the rules he gave in his *RD*) to find all the truths.[111]

I agree with Hermes that an interpretation of Leibniz's idea of *ars inveniendi* in the light of modern mathematics can be given with the concept of *recursive enumerability*.[112] The claim of Leibniz that he has found an ars inveniendi to find out all (scientifically investigated) truths amounts then to the claim that he has found a certain mathematical function, i.e. a recursive function the values of which are all true propositions. Or to the claim that the set of true propositions is a recursively enumerable set. And this again means that the elements of this set – i.e. the true propositions – can be found by a mathematical-mechanical procedure beginning with an initial (countable) function and continuing with some principle which leads only from countable functions to new countable functions.

3.1.5.1. There is an interesting further question concerning the interpretation of those passages given for the ars inveniendi of Leibniz. Did he have the stronger claim in mind that every proposition – of those dealt with in a scientific system (with the supposition of principles of rationalism more geometrico and 3.1.1. – 3.1.4.) – is decidable?

Hermes thinks that decidability would be a good candidate to interpret Leibniz's *ars iudicandi.*[113] In this sense Leibniz's claim that there is an ars iudicandi for all (scientifically investigated) truths would amount to the claim that there is a procedure by which one could decide within finitely many steps whether a certain proposition p belongs to that set of truths or not. In other words this would amount to the claim that there are both a recursively enumerable set of all truths and a recursively enumerable set of the complement set, i.e. of all falsehoods.

However, it goes beyond the scope of this paper to investigate further this interesting question which would require a further study of passages in Leibniz which are hardly clear enough to answer this question without difficulties (cf. Chapter 3.2.7.).

3.2. *Mathematical Order of the Universe*

My claim is that the mathematical order of the universe according to Leibniz is guided by one comprehensive principle. I mean the principle which he himself called[114] the *conspiration universelle*:

3.2.1. In all states and processes of the universe some property is at a maximum or a minimum.

The principle 3.2.1. is not explicitely stated by Leibniz. He states in fact more specified consequences of it:

"And indeed there is always in things a principle of determination which is based on consideration of maximum and minimum, such that the greatest effect is obtained with the least, so to speak, expenditure. And here the time, place, or in a word, the receptivity or capacity of the world may be considered as the expenditure or the ground upon which the world can be more easily built, whereas the varieties of forms correspond to the commodiousness of the edifice and the multiplicity and elegance of its chambers."[115]

"D'autant plus que par ce moyen il se satisfait à leur egard à une autre principe qui succede au precedent et qui porte qu'au defaut du moindre, il faut se tenir au plus determiné, qui pourra estre le plus simple, lors même qu'il est le plus grand."[116]

Rescher describes the *conspiration universelle* as follows:

"Inspired by its success in optics, Leibniz sought ardently to extend the applicability of minimax' principles by means of a general principle that in all natural processes some physical quantity is at a maximum or a minimum. He felt he had found a principle which cuts across the particular laws of physics, and clearly demonstrates the general interconnection of things. Here was a powerful unifying rule for the multitude of particular natural laws, giving coherence to natural science and showing clearly the economy in nature. He could draw upon pure mathematics, mechanics, optics, and dynamics for illustration of his principle. It makes possible the deduction of a multitude of particular laws from a simple general rule, and gives some insight into that remarkable unity of the phenomena on which Leibniz dwelt so fondly."[117]

One important reason for the higher generality of the *conspiration universelle* in contradistinction to other principles like that of perfection or of continuity or of harmony seems to me to be the following: The *conspiration universelle* says *that* (in every state or process) some property is at a maximum or a minimum. The other principles say *what* property is at a maximum or a minimum. Thus the principle of perfection tells us that perfection (the quantity of essence) is maximized, the principle of plenitude assumes that existence is maximized, the principle of continuity says that *liaison* gets a maximal value whereas gaps get a minimal value and the principle of harmony claims a maximization of the "concomitance or of the accord of substances between themselves."[118]

Having in mind that Leibniz is one of the inventors of the infinitesimal calculus it is evident that the mathematical invention or discovery influenced the metaphysical invention or discovery:

"Aussi est il remarquable que dans l'Analyse de maximis et minimis, c'est une même operation pour le plus grand ou pour le plus petit sans qu'on les distingue que dans l'appliquation aux cas divers, parce qu'on cherche tousjours le plus determiné en grandeur, qui est tantost le plus grand, tantost le plus petit dans son ordre, l'analyse n'estant fondée que sur l'evanouissement de la difference ou sur l'unicité des jumeaux reunis, et nullement sur la comparaison avec toutes les autres grandeurs."[119]

That the application of the infinitesimal calculus in the light of the global principle 3.2.1. is really most general can be seen from the following consideration:

(1) The universe is (according to Leibniz) a universe of moving and acting things. Rest is interpreted as a special extreme case of movement.[120]

(2) The laws of moving things can be understood (if one forgets 20 century physics, i.e. microphysics or certain astronomical processes for the moment) as differential equations (Hamiltonian diff. equations) which are based essentially on the infinitesimal calculus.

(3) Thus the laws of moving things – i.e. of the universe according to (1) – are based essentially on minimax principles.

3.2.2. In all states and processes of the universe the perfection is maximized.

3.2.2.1. perfection = df the quantity of essence

"Everything possible ... tends with equal right towards existence according to its quantity of essence or reality, or to the degree of perfection which it involves, perfection being nothing but the quantity of essence."[121]

The principle 3.2.2. is one way of stating Leibniz's principle of perfection. It seems to me that this version fits very well to many passages in Leibniz:

"Absolutely speaking, it must be said that another state (of things) could exist; yet (it must also be said) that the present state exists because it follows from the nature of God that He should prefer the most perfect."[122]

"Among the infinite combinations of possibles and of possible series *that one* exists by which the most of essence or of possibility is brought into existence."[123]

Though Rescher admits that the principle of perfection is also a maximization principle[124] he states the principle in the different version 3.2.2.2.:

3.2.2.2. Every infinitely analytic proposition is (contingently) true.

This version reminds us on Leibniz's soundness principle 2.4.: Every finitely analytic proposition is (necessarily) true. In Chapter 2.4. I tried to show that this principle can be understood as the principle of the consistency of the axiomatic system of the truths of reason. My claim now for the principle 3.2.2.2. is that it can analoguously be understood as the principle of the consistency of the axiomatic system of the truths of fact. In order to show this I want first to give some reason for the connection of 3.2.2.2. with the principle of perfection stated in 3.2.2. and secondly to give support to the above claim.

As to the first the connection is seen better if one adopts a definition of contingent truths like that of Rescher (or some similar): A contingent proposition is true iff the state of affairs characterized by it allows a greater perfection for the world than any other state of affairs.[125] Thus if by the principle of perfection in the version of 3.2.2. in all those states and processes of the universe which are represented by contingent propositions the perfection is maximized then it follows (by the above definition) that these contingent propositions are all true.

As to the second one has first to remember the principles 2.1. and 2.3. which say that all scientific knowledge can be integrated in a deductive system and that all deductive systems are finitely axiomatizable. In fact, if it were just for Leibniz (but probably not for Descartes) principle 2.1. could be stated in a stronger form:

All true propositions about the universe (and one might add: about God) can be incorporated into a deductive (and finitely axiomatizable) system. Thus not only the truths of reason form an axiomatic system but also the truths of fact. But these systems differ in several respects: (1) First the axiomatic system of truths of fact has to have additional axioms. The axioms of the system of truths of reason are only identicals according to Leibniz. That all truths of reason have an a priori proof (deduction and resolution with a finite number of steps) from and back to these identicals satisfies the principle of sufficient reason. The axioms of the system of truths of fact are in addition (to the identicals) principles of maximization. That all truths of fact have an a priori proof (deduction and resolution not with a finite but with an infinite number of steps) from and back to these axioms satisfies again the principle of sufficient reason. (2) Whereas the axioms of the system of truths of reason can be principially known by humans the axioms of the truths of fact cannot generally be known by humans but only by God, i.e. the metaphysical or ontological point that all the truths of fact are incorporated in an axiomatic system is independent of the epistemological point that the human mind cannot by a finite number of steps trace back these contingent truth to these axioms.

"All contingent propositions have reasons for being as they are rather than otherwise . . . ; but they do not have necessary demonstrations, since these reasons are founded only on the principle of contingency, or of the existence of things, that is, on what is or appears the best among several equally possible things."[126]

"There is a difference between analysis of the necessary and analysis of the contingent. Analysis of the necessary, which is that of essences, goes from the posterior by nature to the prior by nature and ends in primitive notions, and it is thus that numbers are resolved into units. But in contingents or existents this analysis from the subsequent by nature to the prior by nature proceeds to infinity, without a reduction to primitive elements being ever possible."[127]

"The Principle of Perfection is not limited to the general but descends also to the particulars of things and of phenomena and that in this respect it closely resembles the method of optimal forms, that is to say, of forms which provide a maximum or minimum, as the case may be – a method which I have introduced into geometry in addition to the ancient method of maximal and minimal quantities. For in these forms or figures the optimum is found not only in the whole but also in each part, and it would not even suffice in the whole without this . . . It is in this way that the smallest parts of the universe are ruled in accordance with the order of greatest perfection . . . "[128]

That the principle 3.2.2.2. is a soundness principle concerning the axiomatic system of truths of fact can now be seen as follows: First it has the general form of a soundness principle: Every proposition derivable (where "derivable" is not an epistemic concept here!) from the axioms is true. Secondly an infinitely analytic proposition is a contingent truth which cannot be traced back to the axioms by finite steps, but would need a proof-process of infinitely many steps. And thirdly the axioms in question are not only identicals but in addition maximization principles. Thus the statement "All infinitely analytic propositions are (contingently) true" says that all propositions belonging to the consequence class of the axiom system of truths of fact are (contingently) true. And this again means that the axiom system of truths of fact is consistent.

3.2.3. In all states and processes of the universe there is a maximization of existence.

According to Leibniz existence is a mode of perfection and so principle 3.2.3. follows from principle 3.2.2. insofar as a maximization of a specific mode of perfection follows from maximization of perfection.

"The true cause why certain things exist rather than others is to be derived from the free decrees of the divine will the first of which is to will to do all things in the best possible way."[129]

"Perfection being nothing but the magnitude of positive reality taken in its strictest sense."[130]

"An existent is . . . which is compatible with more things than are incompatible with it."[131]

The last citations in Chapters 3.2.2. and 3.2.3. show on a closer look that Leibniz didn't make only general and global claims about maximization of perfection and existence. He was aware that such maximizations involve maximization not only of quantity but also of variety, of ordering, of simplicity. This can be expressed in the following principle:

3.2.4. The universe is a combination of a maximization of compossible monads together with a maximization of variety, together with a maximization of order, together with a minimization of laws.

"The true reason why this thing rather than that exists is to be subsumed under the free decrees of the divine will, of which the primary one is the decision to do everything in the best possible way as seems wisest. Therefore it is occasionally permitted that a more perfect be excluded in favor of a less perfect; nevertheless in the sum that way of creating the world is chosen which involves more reality or perfection, and God works on the model of the master Geometer who in problems brings forth the best construction. Therefore all beings, insofar as they are involved in the first Being, have, above and beyond bare possibility, a propensity toward existing in proportion to their goodness, and they exist by the will of God unless they are incompatible with more perfect existence candidates."[132]

Also here it holds that principle 3.2.4. is a consequence of the principle of perfection since the maximization of compossible monads, of variety, of order and the minimization and simplification of laws are specified modes of perfection.

A further important point has to be mentioned here. In the passage cited Leibniz says:

"Therefore it is occasionally permitted that a more perfect be excluded in favour of a less perfect; nevertheless in the sum that way of creating the world is chosen which involves more reality or perfection . . . "

This means that the expression "in all states and processes of the universe" which is used in the principles 3.2.1., 3.2.2., and 3.2.3. must not be interpreted in such a way that a maximization would take place in an isolated group of states or processes. On the contrary it is to be interpreted in such a way that there is a hierarchical order of maximizations and the maximizations in parts are only allowed to the highest degree which is compatible with maximizing the respective property – in general: perfection – for the whole universe.

Friedmann formulated a maximizing principle similar to 3.2.4. and compared it with v. Neumann's maximizing principle [133] which states that the universal class V is maximized:

3.2.4.1. V is maximized iff (X) $(X \not\approx V \rightarrow X \in V)$ (where $X \not\approx V$ means that X is not in one to one correspondance with V). This says that V is maximized iff V contains just as many sets as possible because any class which V does not contain is already in one-one correspondence with V and thus 'too big'. Friedmann's comparison of that maximization principle of set theory with the principle 3.2.4. throws an interesting light on Leibniz' maximization principles. Friedmann lists three important analogies:

(1) Just as 3.2.4. asserts that the compossible monads are maximized in the universe so 3.2.4.1. states that the set-theoretical universe has the maximum number of sets.

(2) Just as 3.2.4. assumes that the variety is maximized, 3.2.4.1. implies the axioms of replacement and of choice which yield indeed a great variety of sets.

(3) Just as 3.2.4. asserts that the order is maximized, 3.2.4.1. implies that the universe of set can be well-ordered (since it implies the axiom of choice).[134]

3.2.5. In all states and processes of the universe continuity is maximized and discontinuity is minimized.

"Principium quoddam generale non in mathematicis tantum sed et physicis utile, cujus ope ex consideratione sapientiae divinae examinantur naturae leges. . . . Principium hoc Ordinis Generalis ab infinito habet originem, magnique in ratiocinando usus est, quanquam non satis usurpatum nec pro amplitudine sua cognitum. Absolutae est necessitatis in Geometria, sed tamen succedit et in Physica, quoniam suprema Sapientia, quae fons est rerum, perfectissimum Geometram agit et Harmoniam observat, cujus pulchritudini accedere nihil potest. . . . Enuntiari potest hoc modo: Cum differentia duorum casuum infra omnem quantitatem datam diminui potest in datis sive positis, necesse est, ut simul diminuatur infra omnem quantitatem in quaesitis sive consequentibus quae ex positis resultant."[135]

"Lorsque les cas (ou ce qui est donné) s'approchent continuellement et se perdent enfin l'un dans l'autre, il faut que les suites ou evenemens (ou ce qui est demandé) le fassent aussi. Ce qui depend encor d'un principe plus general, sçavoir: Datis ordinatis etiam quaesita sunt ordinata."[136]

The principle 3.2.5. – usually called "lex continuitatis" or traditionally "natura non facit saltus" – is like 3.2.1. very much connected with Leibniz's invention of the infinitesimal calculus. Continuous series of intermediate substances, ordered physical states (which are concerned with aggregates of substances) ordered series of psychic states (perceptions) etc. can only be understood in the light of this new mathematical understanding: "If two substances differ by a finite difference there must be a continuous series of intermediate substances, each of which differs infinitesimally from the next."[137]

In addition to that it is connected with the principle of the definiteness of truth, i.e. that a proposition takes for every subject and predicate a definite truth value.[138]

The principle of continuity is applied in very many different areas within Leibniz's philosophy. It applies in metaphysics concerning monads. But there is a maximization of continuity in many different aspects, i.e. in respect to a continuous order of perfection, a continuous order of quantity of existence, of variety of hierarchy, of mirroring the world etc. This holds analoguously for other areas as for the physical realm and for the psychic realm.[139]

A consequence of the *lex continuitatis* is the principle that everything is linked with everything:

3.2.5.1. In all states and processes of the universe the universal connection is maximized and the occurrence of gaps and jumps is minimized.

"In every possible world everything is linked together. The universe – however it might be constituted – is a unified whole, like an ocean; even the smallest motion extends its influence to any distance however large."[140]

3.2.6. In all states and processes of the universe there is a maximization of mutual analogies, i.e. a maximization of isomorphisms and homomorphisms and a minimization of complexities.

This is a possible way, it seems to me, to formulate the principle of the Pre-Established Harmony. Leibniz's examples show that there is always an isomorphic (one-one) or homomorphic structure involved in the perceptions of different monads.

"The explanation of all phenomena proceeds solely through the perceptions of monads functioning in harmony with each other." [141]

"Enfin la troisieme maniere sera de faire d'abord ces deux pendules avec tant d'art et de justesse, qu'on se puisse asseurer de leur accord dans la suite; et c'est la voye du consentement préétabli. . . . Ainsi il ne reste que mon hypothèse, c'est à dire que la voye de l'harmonie préétablie par un artifice divin prevenant, lequel dès le commencement a formé chacune de ces substances d'une maniere si parfaite et réglée avec tant d'exactitude, qu'en ne suivant que ses propres loix, qu'elle a recues avec son estre, elle s'accorde pourtant avec l'autre . . . " [142]

Leibniz's description of the accord of two clocks, as his explanation of the mind-body relation, is a good instance of an isomorphic structure. When Leibniz says that the principle of pre-established harmony is "the hypothesis of the concomitance or of the accord of substances between themselves" [143] one should add in the light of 3.2.4. two things:

(1) In the principle of pre-established harmony 'accord' and 'concomittance' have to be interpreted as specific mathematical structures, i.e. as isomorphic or homomorphic structures,[144] as the examples and the description of Leibniz suggest.

(2) The principle of pre-established harmony asserts that in the relations between substances and in general in all states and processes of the universe these isomorphic and homomorphic structures are maximized. This means at the same time that the degree of complexity in relations between substances and in general in states and processes is minimized.

From what has been said it is clear that principle 3.2.6. is very much connected with principle 3.2.4.:

Isomorphic and homomorphic structures are specified modes of order and such structures yield a simplification and minimization of laws.

3.2.7. Maximization and Decidability.

In his paper about the connection between Leibniz's monadology and Transfinite Set Theory Friedmann made the following interesting conjecture: "maximization implies decidability in transfinite set theory, just as in

Leibniz's Monadology. Thus our conjectured thesis is to be understood in a strong sense, namely, that maximization implies decidability *in all cases*."[145]

This conjecture raises the interesting question whether Leibniz hold that all truths are decidable (which has already been touched in Chapter 3.1.5.2.). And more specifically whether decidability can be attained by humans only in respect to truths of reason or even in respect to truths of fact. Leibniz' famous "calculemus" seems to support such a view:

"Quo facto, quando orientur controversiae,non magis disputatione opus erit inter duos philosophus, quam inter duos Computistas. Sufficiet enim calamos in manus sumere sedereque ad abacos, et sibi mutuo (accito si placet amico) dicere: *calculemus*."[146]

As is clear from the principles 2.1. (all scientific knowledge can be integrated in a deductive system) from 2.3. (all deductive systems are finitely axiomatizable), from 2.5. (every true proposition has an a priori proof), 2.5.2. and 3.1.5. I make only a weaker claim: It is the claim that Leibniz' principle of sufficient reason can be interpreted as the assertion that true axiomatic systems about the universe (as for instance Leibniz' metaphysics) are complete (i.e. every true proposition about the universe has an a priori proof in the axiomatic system). And that the same is the case according to Leibniz for a comprehensive axiomatic theory about the areas of logic, mathematics, physics and jurisprudence (including ethics). And furthermore one has to distinguish very sharply between an ontological or logical and an epistemical interpretation of Leibniz' principles: Leibniz does not claim that man is able to finitely axiomatize the system of truths of fact and similarly his view is that man is not able to trace back every truth of fact to its axioms by a finite number of steps. This is evident from many passages in Leibniz' works already cited. Thus man cannot give effectively an a priori proof for every true proposition – although in the objective sense both the axiom system of truth of reason and also that of truths of fact (which include all truths about the universe) is complete by the principle of sufficient reason. Though there is some plausibility that Leibniz may have held the thesis that every mathematical and every logical problem is solvable (a thesis defended by a number of mathematicians before 1931) it seems to me almost certain that he has not held such a thesis in respect to truths of fact (though he holds it for God). And this means that it is at least uncertain whether he has assumed that his strong maximization principles distributed over all in the universe yield a complete decidability of every state of affairs.

University of Salzburg and
Institut fuer Wissenschaftstheorie
Internationales Forschungszentrum Salzburg

NOTES

1 The problem of the different meanings of 'proposition', 'statement' and 'judgement' cannot be discussed here. It would lead too far away from the topics of this paper. It may however be mentioned that the English term 'judgement' would be probably more appropriate to Descartes – cf. MD 3, 8–12 – where he gives his division of psychic phenomena. On the other hand the term 'proposition' will be more appropriate to Leibniz.

Here the term 'proposition' is used because it is more neutral in respect to an interpretation which would be too close to some psychologistic view.

For symbolization I use the usual first order predicate logic and quantification for propositions.

2 RD 3; AT 10, 368.

3 RD 3; AT 10, 370. Cf. Röd (DEP) p. 49ff.

4 Cf. RD 4; AT 10, 371–2.

5 GP 5, 21.

6 GP 7, 194–195. Cf. GP 7, 296.

7 RD 3; AT 10, 370.

8 GP 5, 343; NE 4, 2, 1.

9 RD 3; AT 10, 368.

10 GP 5, 343; NE 4, 2, 1. On some passages Leibniz says that the principle of non-contradiction asserts that every proposition is true or false (GP 5, 343). This is a rather restricted formulation which rules out any many-valued logic. Cf. Rescher (MVL) p. 143ff. But Leibniz knows also forms of the principle which are different and apply to an axiomatic system. Cf. Chapter 2.2. and 2.4.

11 GP 5, 347; NE 4, 2, 1.

12 Copleston (HPh) Vol. IV, p. 73.

13 Russell (CEP), p. 167.

14 Cf. GP 4, 423–424.

15 DM 4, 3; AT 6, 33. Cf. RD 3 and Descartes' commentary to that rule.

16 DM 2; AT 6, 18.

17 GP 4, 422–424. Cf. the citation of Russell, note 13.

18 GP 4, 422–426 and GP 1, 381–385.

19 GP 1, 384.

20 GP 1, 384. Cf. GP 7, 194 (footnote) where Leibniz gives a definition of truth (of reason) with the help of identical propositions. Cf. Chapter 2.3. note 63.

21 RD 3; AT 10, 368.

22 RD 3; AT 10, 369.

23 RD 3; AT 10, 370. Cf. RD 12, 23.

24 Cf. especially his commentary to Rule 3 in RD.

25 GP 5, 461.

26 GP 7, 199–200.

27 RD 4; AT 10, 371–372.

28 DM 2; AT 6, 18.

29 Cf. Röd (DEP) p. 13f. There is a discussion in historical works about Descartes whether 'simple natures' can only have the form of propositions or whether they may be understood as certain simple ideas as well. It would go beyond the scope of this paper

to deal with this problem. It seems that Descartes himself was not entirely clear about that matter and that he used 'simple natures' for both, propositions and ideas in such a way that simple natures as ideas like "existence", "thinking", "three lines" are disengaged from first principles known by intuition like "I think, therefore I exist" or "a triangle is bounded by three lines". Cf. Copleston (HPh), Vol. IV, p. 77f.

30 RD 12, 13; AT 10, 418.

31 Cf. PP 1, 10.

32 GP 4, 64–65 (= De arte combinatoria, 64). Cf. Kneale (DLg), p. 333.

33 GP 4, 450.

34 Rescher (LIP) p. 22. Cf. Röd (RAM). Concerning Leibniz' views on the subject-predicate form of propositions cf. § 8 of his Discourse on Metaphysics (= GP 4, 427–463) and Rescher (LIP) p. 21. It should be mentioned that the characterization of a proposition of subject-predicate-form as true if the predicate is contained in the subject – which can be found also in Thomas Aquinas (cf. (STh) I, 2, 1) – requires an intensional interpretation: In the true proposition "men (subject) is animal (predicate)" the subject (class of men) is extensionally included in the predicate (class of animals). But animal (the properties of being an animal) is intensionally included in man (the properties of being a man). Elsewhere I have given an interpretation of intensional inclusion which is able to explain the traditional doctrines concerning intensional inclusion in Aristotle, the Scholastics, Leibniz and De Morgan and which can be introduced in the modern logic of classes (or into first order predicate logic). Cf. Weingartner (IIS), (BIG), and (SPI).

35 MD 3, 13.

36 DM 6; AT 6, 64. PP 2, 3; AT 8, 42.

37 RD 12, 14.

38 NE 1, 1, 18.

39 Russell (CEP), p. 161. Cf. GP 5, 66 and 79. Cf. Rescher (PhL), p. 128; (LIP), p. 122.

40 Cf. RD 12, 14; cf. note 37.

41 NE 1, 1, 18.

42 NE 1, 3, 3.

43 NE 2, 1, 2.

44 Cf. Copleston (HPh), Vol. IV, p. 83: "All clear and distinct ideas are innate. And all scientific knowledge is knowledge of or by means of innate ideas". Cf. the citations in Chapters 1.2.2. and 1.2.3.

45 NE 1, 2, 12.

46 Cf. Tarski (LSM), p. 185, Def. 18. α is a set of sentences (propositions). It is further assumed that $\alpha \subseteq S$ where S is the set of meaningful sentences (propositions). This set of meaningful sentences has to be understood relative to a language which has been prepared and made precise in some sense to discuss scientific or philosophical problems. It would be unnecessary to require a symbolic language with a recursive definition of well formed formula. For Descartes and Leibniz a language in which books of geometry (for instance Euclids elements) are written would be ideal. But observe that not only the symbols of the deductive system are meant here but the whole language in which the axioms and theorems are discussed. Descartes would certainly view his Meditations as a case in point, and Leibniz his Nouveaux Essays.

47 RD 5. AT 10, 379.

48 PP, Letter to C. Picot, AT 9–2 (19). Cf. Röd (DEP), p. 47 and Scholz (PZM), p. 29.

49 GP 7, 195. Cf. GP 7, 295.

50 GP 4, 437.
51 NE 4, 2, 1. Cf. GP 7, 299: "Ante omnia assumo Enuntiationem omnem (hoc est affirmationem aut negationem) aut veram aut falsam esse, et quidem si vera sit affirmatio, falsam esse negationem; si vera sit negatio, falsam esse affirmationem." Cf. Kauppi (LLg), p. 82f.
52 There are some extravagant many valued calculi, in which a proposition p can take both the value T and F. Rescher calls them quasi-truth functional. Cf. Rescher (MVL), p. 166ff. Calculi of that sort would certainly be ruled out by Leibniz' weaker consequence of his principle of non-contradiction.
53 At least not in this and related passages. – For different versions of the principles of contradiction and tertium non datur see Rescher (MVL), p. 143–154.
54 NE 4, 2, 1.
55 Observe that the expression 'many valued' has two different meanings: a system can be many-valued in the sense that its propositions can take values which are different (for instance: indefinite) from true and false. On the other hand it can be many-valued in the sense that there is no value between or outside true or false but a proposition can take more than one value of true (or: designated value) and more than one value of false (or: antidesignated value) such that the antidesignated values coincide with the non-designated ones. The passage in Leibniz's Nouveaux Essays is formulated in such a way that I think one can assume that also the second form of many-valued systems are ruled out; i.e. it is assumed that there is only a single truth-value true and a single truth-value false.
56 That Leibniz defends the subject-predicate form of propositions is clear from many passages – cf. GP 4, 432, GP 7, 295/96. But as Brentano already noticed he was not of the opinion that this was the only possible form. Thus he says for instance:

"On peut tousjours dire que cette proposition: *j'existe* est de la dernière evidence, estant une proposition, qui ne sauroit estre prouvée par aucune autre, ou bien une verité immediate." (NE 4, 7, 7). Cf. NE 4, 5, 3.

57 v. Fraassen (STT) and Kripke (OTT).
58 RD 4, 2; AT 10, 371–2.
58a MD 3, AT 7, 36f.
58b MD 3, AT 7, 37.
59 The definition used here is due to Tarski: A deductive system a is finitely axiomatizable (FAX) iff there is at least one axiom system (i.e. iff there is a finite set β of sentences) such that β is identical with the consequence class of α. Cf. Tarski (LSM), p. 355.
60 RD 3, 8; AT 10, 369.
61 RD 5; AT 10, 379.
62 GP 7, 57. Cf. GP 7, 25, where he describes his new invention as "certaines vues toutes nouvelles, pour reduire tous les raisonnemens humaines à un espece de calcul, qui serviroit à decouvrir la verité." Cf. further Courturat (OF), p. 221.
63 GP 7, 194. Cf. the passage cited in note 49 (GP 7, 195) and GP 7, 295.
64 Cf. GP 7, 195.
65 GP 7, 194. Cf. Chapter 1.1.1., note 6.
66 GP 7, 355. But it is very probable that he includes here also the principle of contradiction as an inference rule, i.e. as the reductio ad absurdum principle. Cf. Chapter 2.4.
67 NE 4, 7, 10.
68 GP 7, 195. Cf. the passage cited in Chapter 1.2.1., note 32.

69 Copleston (HPh), Vol. IV, p. 277, Rescher (PhL), p. 26 and (LIP) p. 24.
70 GP 5, 343. NE 4, 2, 1.
71 GP 7, 355. Cf. Chapter 2.3., note 66. This passage is as a matter of fact also cited by Copleston, p. 276.
72 Cf. Monadology 31 (= GP 6, 612)
73 For such definitions see the example cited in note 67 above. In special cases the definitions have to be real, not only nominal according to Leibniz. This is so when the conclusion should be stated absolutely not only hypothetically. Cf. note 65. Concerning inference rules one should observe that Leibniz held that firstly the syllogisms of the first figure can be proved by the law of non-contradiction (in the sense of indirect proof or reductio ad absurdum) and secondly the syllogisms of the second and third figure can be derived from the first, also by using only the principle of non-contradiction. Cf. NE 4, 2, 1.
74 GP 7, 295–296. Cf. GP 7, 194–195.
75 RD 4, AT 10, 371. Cf. the fifth rule which was cited in note 61.
76 Cf. Russell, (CEP), p. 30ff.
77 GP 2, 62.
78 GP 7; 300, 301.
79 GP 7, 199–200.
80 GP 6, 612 (Monadology 31–36).
81 This seems to be also the opinion of Couturat – cf. his (LgL), p. 214f. – and of Kauppi in her detailed discussion of the principle of sufficient reason, cf. (LLg), p. 87–94.
82 Rescher (PhL), p. 23, cf. (LIP), p. 23, cf. Leibniz's description which was given in Chapter 1.2.1., note 32.
83 GP 6, 602.
84 GP 1, 57. Cf. GP 7, 301, where he lists metaphysics, physics, the moral sciences and proofs of the existence of God for the application of the principle of sufficient reason.
85 That this is so for logic and mathematics is evident, but it is also true for metaphysics according to Leibniz though metaphysics deals with different kinds of truths of reason (hypothetical truths about possible substances, categorical truths about abstract concepts, analytic truths and last but not least verités innées). Cf. Rescher, (PhL), p. 127; (LIP), p. 121. Cf. Kauppi, (LLg), p. 31, note 3.
86 Cf. Rescher, (PhL), p. 17.
87 GP 3, 400.
88 See the citation about *method*, Chapter 1.2.1., note 25.
89 RD 4, AT 10, 372.
90 RD 7, AT 10, 387.
91 Kauppi is therefore right to say that the principle of contradiction and that of sufficient reason are in the first place not axioms but metatheoretical principles of proof in general. Cf. Kauppi, (LLg), p. 80 and (OF), p. 184.
92 GP 7, 295. Cf. NE 4, 2, 1.
93 For a discussion of normative principles in the methodology of all or specific scientific disciplines cf. my (NCS).
94 RD 2, AT 10, 366.
95 DM 2, 11; AT 6, 19.
96 OF 175.
97 Cf. Frege (BGS), Preface and (WBB).

98 GP 7, 205.

99 This expression is not used unambiguously in the literature on Leibniz. Very often it is used in a broad sense for some or all of the different parts listed subsequently.

100 GP 7, 184.

101 GP 7, 189.

102 Cf. OF p. 49 and 50. This is also the opinion of Kauppi, cf. her (LLg), p. 146.

103 Cf. OF p. 78 and 79.

104 Observe that according to Leibniz a categorical proposition is true iff the predicate is contained in the subject:

"Propositio vera est cujus praedicatum continetur in subjecto seu ei inest. OF, p. 68. Cf. GP 7, 199–200; GP 7, 208.

I have shown elsewhere that the traditional expression "the predicate is contained in the subject" (predicatum inest subjecto) can be interpreted as a kind of intensional inclusion which can be defined as $P \sqsubseteq S \longleftrightarrow (M)\ [P \subseteq M \rightarrow S \subseteq M]$, where $\subseteq$ is the usual class-inclusion. When this definition plus two additional ones for intensional intersection and intensional union are introduced into the usual theory of classes (without the need of any set-theoretical axiom) one gets a dual theory which is an intensional class-theory. The theorems of this theory – as for instance the theorem $P \sqsubseteq S \longleftrightarrow S \subseteq P$ which expresses the antagonistic relation between extensional and intensional inclusion – give suitable interpretations of doctrines of traditional logic especially of those of Aristotle, the Scholastics, Leibniz and De Morgan. Cf. my (IIS) and (SPI).

105 OF, p. 42.

106 Cf. OF, p. 40–84.

107 Cf. OF, p. 77–84. The discovery that this calculation method is successful for syllogistics is due to Lukasiewicz (ASy). Cf. D. Marshall (LLA). I am indebted to P. Simons for drawing my attention to this discovery and to the paper of Marshall.

108 OF, p. 78.

108a OF, p. 79. Only a selection of Leibniz's description has been cited. Under (III) Leibniz describes the particular negative proposition and under (V) the particular affirmative proposition. He also states 8 theorems concerning the validity of different syllogistic laws like that of conversion and of square of opposition. (Translation by G. H. R. Parkinson, in: Parkinson (LLP).)

108b Cf. Lukasiewicz (ASy), p. 126–129. Kauppi and Rescher seem not to be aware of the success of Leibniz's last proposal. They discuss only earlier proposals where Leibniz represents terms by simple numbers (not pairs). In this respect they were correct with their doubts whether it works with all the syllogistic modes (though the simple method works for some, for example BARBARA, as they observe.)

108c Marshall (LLA), p. 241. For the historically difficult question whether Leibniz had doubts about the correctness of his last proposal see Marshall, ibid., p. 241f.

109 GP 1, 57.

110 OF, p. 161.

111 Cf. the citations I have given for the principle of completeness in Chapter 2.5.3., note 89 and 90.

112 Hermes (ILG), p. 98.

113 Hermes (ILG), p. 99.

114 That Leibniz called the principle so is a claim by Rescher – cf. his (PhL), p. 41, note 15: "He termed it 'la conspiration universelle'". But Rescher doesn't give a reference.

The Tentamen Anagogicum (GP 7, 270–279) which contains some passages concerning that principle does not however contain the name 'conspiration universelle'. It may be that Rescher has the term from Couturat.
115 GP 7, 303.
116 GP 7, 274.
117 GP 7, 303.
118 GP 2, 58. The conspiration universelle is stated by Couturat and Rescher almost only in connection with some of the other principles listed above, especially with the principle of perfection. Cf. Couturat, (LgL), p. 230ff. and Rescher, (LIP), p. 38–39. But its importance as a basic principle underlying all others seems not to be seen clearly. Rescher views the principle of perfection as the source-principle of those of plenitude, continuity, and harmony, cf. (PhL), p. 57. For Leibniz' principle of plenitude cf. Hintikka (LPR).
119 GP 7, 275.
120 Cf. GM 6, 130 (About the principle of continuity). The conception of the universe as a universe of moving things goes back to Aristotle. Cf. his Physics, 2, 1. Similarly the view that physics is concerned with moving being. This was accepted also by the great philosophers of scholastics. Cf. Thomas Aquinas (APC), 1, 1, 3: "... It follows that mobile being is the subject of natural philosophy. For natural philosophy is about natural things, and natural things are those whose principle is nature. But nature is a principle of motion and rest in that in which it is. Therefore natural science deals with those things which have in them a principle of motion."
121 GP 7, 303.
122 Grua (LTI), 1, 393.
123 GP 7, 303.
124 Cf. Rescher (LIP), p. 39 and (PhL), p. 30.
125 Cf. Rescher (LIP), p. 39, (PhL), p. 38.
126 GP 4, 438.
127 GP 3, 582.
128 GP 7, 272–273.
129 GP 7, 309–310.
130 Monadology §41.
131 OF, p. 360.
132 GP 7, 309–310.
133 Cf. Friedmann (RLM), p. 345f.
134 Cf. Friedmann (RLM), p. 346f. Friedmann discusses then a second analogy between Leibniz's principle that every monad represents the entire universe from its point of view with the greatest possible series of perceptions it may have on one hand and the generalized maximization principle that every local universe is maximized (which is a generalization of 3.2.4.1.) on the other. He again states three similar analogies which are illuminating in respect to the conception of a monad as a series of perceptions, in respect to the hierarchical order of the monads and to the logical relation of the two maximization principles of Leibniz. Cf. (RLM), p. 348f.
135 GM 6, 129.
136 GP 3, 52. These formulations of the principle of continuity support at the same time principle 3.2.6.
137 Russell (CEP), p. 64–65.

138 Cf. the discussion of principle 2.2.
139 Cf. Rescher (LIP), p. 63f. (PhL), p. 51f.
140 GP 6, 107. Leibniz presupposed of course that a process originating from a place can have an unlimited extension. This is unlike the theory of relativity, where a process originating from a point can have at any time only limited extensions.
141 GP 2, 450.
142 GP 4, 501.
143 GP 2, 58.
144 I have shown in (AnS) that one can give different definitions of analogy with the help of the mathematical concepts of homomorphism and isomorphism which can be applied both to traditional doctrines of analogy and to analogies in contemporary science (for instance physics).
145 Friedmann (RLM), p. 336–337. Friedmann's conjecture is not without reason. He showed in (GCH) that the generalized continuum hypothesis is equivalent to the generalized maximization principle.
146 GP 7, 200.

REFERENCES

Aristotle
(Phys) *Physics, Intr. Comm.* W. D. Ross, Oxford, 1955.

Copleston, F.
(HPh) 'A History of Philosophy', Vol. IV, *Descartes to Leibniz*, Burns & Oates, London, 1960.

Couturat, L.
(LgL) *La Logique de Leibniz d'apres des documents inédits*, Paris, 1901.

Descartes
AT *Oeuvres de Descartes*, C. Adam and P. Tannery (eds.), 13 Vols., Paris, 1897–1913. Librairie Philosophique J. Vrin, Paris.
DM *Discourse on Method*
MD *Meditations*
PP *Principles of Philosophy*
RD *Rules for the Direction of the Mind*

van Fraassen, Bas
(STT) 'Singular Terms, Truth-value Gaps and Free Logic', in *The Journal of Philosophy* 63 (1966), 481–495.

Frege, G.
(BGS) Begriffsschrift, Halle 1879, in I. Angelelli (ed.), *Begriffsschrift und andere Aufsätze*, Wiss. Buchgesellschaft, Darmstadt, 1964.
(WBB) 'Über die wissenschaftliche Berechtigung einer Begriffsschrift', in *Zeitschrift für Philosophie und philosophische Kritik* 81 (1882), 48–56. Reprinted in, I. Angelelli (ed.), *Begriffsschrift und andere Aufsätze*, p. 106–114.

Friedmann, J. I.
(GCH) 'The Generalized Continuum Hypothesis is Equivalent to the Generalized Maximization Principle', in *Journal of Symbolic Logic* 36 (1971), 39–54.
(RLM) 'On some Relations between Leibniz' Monadology and Transfinite Set

Theory (A Complement to the Russell Thesis)', in *Studia Leibnitiana Supplementa*, Vol. 15–3, K. Müller, H. Schepers, and W. Totok (eds.), Steiner, Wiesbaden, 1975.

Grua, G.
(LTI) *Leibniz, Textes inédits*, G. Grua (ed.), 2 Vols., Paris, 1948.

Hermes, H.
(ILG) 'Ideen von Leibniz zur Grundlagenforschung: Die ars inveniendi und die ars iudicandi', in *Studia Leibnitiana Supplementa*, Vol. 15–4, K. Müller, H. Schepers, and W. Totok (eds.), Steiner, Wiesbaden, 1975.

Hintikka, K. J. J.
(LPR) 'Leibniz on Plenitude, Relations, and the 'Reign of Law', in H. Frankfurt (ed.), *Leibniz. A Collection of Critical Essays*, University of Notre Dame Press 1976.

Kauppi, R.
(LLg) 'Über die Leibnizsche Logik. Mit besonderer Berücksichtigung des Problems der Intension und der Extension', *Acta Philosophica Fennica XII*, Helsinki, 1960.

Kneale, W. and Kneale, M.
(DLg) *Development of Logic*, Oxford, 1962.

Kripke, S.
(OTT) 'Outline of a Theory of Truth', in *The Journal of Philosophy* (1975), 690–716.

Leibniz
GM *Die mathematischen Schriften von G. W. Leibniz*, C. I. Gerhardt (ed.), 7 Vols., Berlin, 1849–63.

GP *Die philosophischen Schriften von G. W. Leibniz*, C. I. Gerhardt (ed.), 7 Vols., Berlin, 1875–90. Georg Olms Verlag, Hildesheim.

NE *Nouveaux essais sur l'entendement humain*. Wissenschaftliche Buchgesellschaft, Darmstadt.

OF *Opuscules et fragments inédits de Leibniz*, L. Couturat (ed.), Paris, 1903. Georg Olms Verlag, Hildesheim.

Lukasiewicz, J.
(ASy) *Aristotle's Syllogistic from the Standpoint of Modern Formal Logic*. London (Oxford UP) [2]1957.

Marshall, D. Jr.
(LLA) 'Lukasiewicz, Leibniz and the Arithmetization of the Syllogism', in *Notre Dame Journal of Formal Logic*, **18** (1977), 235–242.

Parkinson, G. H. R.
(LLP) *Leibniz, Logical Papers*, A Selection Translated and Edited with an Introduction by G. H. R. Parkinson. Oxford (Clarendon) 1966.

Rescher, N.
(LIP) *Leibniz. An Introduction to his Philosophy*, Oxford, 1979.

(MVL) *Many-valued Logic*, McGraw Hill, New York, 1969.

(PhL) *The Philosophy of Leibniz*, Englewood Cliffs, 1967.

Röd, W.
(DEP) *Descartes' Erste Philosophie*. Versuch einer Analyse mit besonderer Berücksichtigung der Cartesianischen Methodologie, Bouvier, Bonn, 1971.

(RAM) 'Die Rolle der analytischen Methode in der Grundlegung der Leibnizschen Metaphysik', in *Studia Leibnitiana Supplementa*, Vol. 15–4, K. Müller, H. Schepers, and W. Totok (eds.), Steiner, Wiesbaden, 1975.

Russell, B.
(CEP) *A Critical Exposition of the Philosophy of Leibniz*, London, 1900, new edition 1937.

Scholz, H.
(PZM) *Die Philosophie im Zeitalter der Mathesis Universalis*, Descartes, Pascal, Leibniz (Lecture at the University of Münster, 1933–34), ed. by Mathematische Fachschaft of the University of Münster (mimeographed).

Tarski, A.
(LSM) *Logic, Semantics, Metamathematics*, Oxford, 1956.

Thomas Aquinas
(APC) *In Octo Libros Physicorum Aristotelis Expositio*, P. M. Maggiolo (ed.), Marietti, Roma, 1954.
(STh) *Summa Theologica, Latin Text and English Translation*, Blackfriars, Cambridge, 1964.

Weingartner, P.
(BIG) 'Bemerkungen zum Intensionsbegriff in der Geschichte der Logik', in *Zeitschrift für Philosophische Forschung* **30** (1976), 51–68.
(IIS) 'On the Introduction of Intensions into Set Theory', in M. Przełęcki, K. Szaniawski, and R. Wójcicki (eds.), *Proceedings of the Conference for Formal Methods in the Methodology of Empirical Sciences*, Warszaw, 1974, (1976), D. Reidel Publ. Co., Dordrecht, Holland, pp. 287–296.
(NCS) 'Normative Characteristics of Scientific Activity', in R. Hilpinen (ed.), *Proceedings of the Conference on Ethics and Science, Dubrovnik, 1978*, 1980, D. Reidel Publ. Co., Dordrecht, Holland, pp. 209–230.
(SPI) 'A Simple Proposal for Intruducing Intensions into Science', in: *Archives de l'Institut International des Sciences Théoriques, Actes du Colloque "La Sémantique dans les Sciences" de l'Academie Internationale de Philosophie des Sciences Rixensaart, 1974*, Bruxelles, 1978, pp. 193–209.
(AnS) 'Analogy among Systems', in *Dialectica* **33** (1979), p. 1–23.

FRANÇOIS DUCHESNEAU

THE "MORE GEOMETRICO" PATTERN IN HYPOTHESES FROM DESCARTES TO LEIBNIZ.

In this paper I will follow two related trends. First, I will take issue with Weingartner's general and unique scheme for interpreting such different doctrines of scientific rationality as Descartes's and Leibniz's In a second move, of a more special kind, I will analyze Leibniz's reflections on the logic of scientific hypotheses during the timespan between *Hypothesis physica nova* (1671) and the formulation of the original principles of his dynamics (1678).[1]

I

Prof. Weingartner intends to interpret the ideal of mathematization in Descartes' and Leibniz's theories of scientific knowledge in the light of research of our time. Such an interpretation is bound to reconstruct 17th century philosophical and scientific issues along the lines of some more recent patterns for the semantics and syntax of scientific theories. However, Prof. Weingartner wishes to remain faithful to the philosophical doctrines he examines and he aims at drawing a 'consistent and adequate' picture. Such an attempt might throw unexpected light on the subject. But, I will defend the view that his 'systematization' tends to overcast specific points in Descartes's and Leibniz's doctrines that would prove of interest to historians of science and philosophy. The actual story, though it may be more complex and less consistent than Weingartner's reconstruction, has to do with presuppositions which formed the essential background for important scientific moves, mainly in physics and mechanics.

One of Prof. Weingartner's assertions in the first part of his paper is that for Descartes and Leibniz first principles are known by intuition and that the criterion for such knowledge consists in a clear and distinct apprehension by the mind of what is to be known. He adds that Leibniz weakens the Cartesian truth-criterion by adding the condition that p be known to be possible: "If p is known *clare et distincte* and if p is known to be possible, then p is true." *Intuitus* was the term used in the *Regulae ad directionem ingenii* to describe the fundamental modality of knowledge when the rational connection of concepts is actually present to the atttentive mind. *Deductio* is described as

William R. Shea (ed.), Nature Mathematized, 197–214.

a chain of intuitions (indirect connection of concepts). Part 2 of the *Discours de la méthode* stresses evidence as the proper criterion for the direct connecting of concepts. The phrase '*clare et distincte*' is used to mean the epistemic features of the act of knowing when it reaches a clear-cut rational content. As opposed to obscure, clear means that the object is directly present to one's awareness; and distinct, as opposed to confused, means that the object is unambiguously conceived upon the very act of awareness. So, the entire operation of true knowledge depends upon the quality of awareness enacted by the knower. It involves a critical survey of the objective contents of judgments: a methodical *dubitatio*, whose scope depends upon the voluntary endeavour of the knower to analyze *problemata* down to evident principles. The necessary connections that afford true conceptual knowledge get proper recognition through a constant effort of actual awareness. Obscurity means absence of true knowledge. And only confusion can be deceptive. Evidence ties up more with analytic awareness than with any kind of axiomatic construct. Prof. Weingartner states that though Descartes describes the operation of deduction, he does not enunciate rules of deduction explicitly. I guess the proper operation of knowledge consists in becoming aware of the *simple natures* as the distinct forms and elements of rational awareness, and to unfold the connections involved in the forms and elements in an orderly pattern, taking into account the dispositional order needed in order to extend the scope of awareness. 'Enumération', 'Dénombrement' are terms designating the overall process of deduction. I doubt whether this would correspond unequivocally with Prof. Weingartner's scheme for deductive inference.

I would like now to turn to Leibniz's so-called weakened truth-criterion. I have in mind the analytic attempt at describing the different species of ideas in the *Meditationes de cognitione, veritate et ideis* (1684) and *Discours de métaphysique* (1686). The distinctions are summarized by Prof. Weingartner by means of a brief quotation from Russell. As a matter of fact, Leibniz's analysis is anti-Cartesian: he wants to disprove the Cartesian truth-criterion. Evidence is dangerously subjective. Methodical doubt is but a psychological process if separated from counter-checking devices such as proofs in the arithmetical operations. It affords no substitute for formal logical processes. Clearness and distinctness in conceptual analysis are but the more superficial (i.e. psychological) features. A clear notion is one which reminds us of the relevant object; a clear and distinct notion is one which, in addition, helps distinguish the relevant object from analogous ones. Cartesian evidence ties up with a kind of concept analysis at the level of *prima facie* meaning experience. The Leibnizian conceptual analysis aims at clear distinct adequate ideas, i.e. ideas whose very ingredients are distinctly known. For the majority of our ideas at least, this means specifying the content of an idea by using other less complex ideas as meaning-signs (symbolic knowledge) and proving that the meaning thus spelled-out is both complete and consistent. This process involves a logical analysis of a combinatory type. The truth-criterion involved is that of determinateness, i.e. of distinctness analyzed down to conceptual consistency. Instead of providing awareness of 'phenomenal' rational connections, the ideal of knowledge consists in fostering real (*vs* nominal) definitions. And so, I doubt whether Leibniz would have considered axiomatization otherwise than as a logical means of approximating real knowledge. Indeed, the optimal category of knowledge is that of adequate intuitive knowledge, the adequacy of the known object being at once actuated for the knower's understanding. This would apply to primitive ideas (and there must be some if the analytic operation is to be effective). Leibniz does not draw any list of our primitive ideas. He mentions the fact that we get a notion of such ideas by considering the integral numbers in Arithmetic, which come close to meeting the requirement. Otherwise, he lists *termini simpliciores*, which form the ultimate categories required to ground a deductive process whether it be in metaphysics or in mathematics. The intuition of such notions remains in a way phenomenal. But they afford absolute certainty for the axioms which they constitute; such axioms, within the limits of human powers of cognition, may be taken as primitive indemonstrable truths. In a manuscript quoted by Couturat, *Introductio ad Encyclopaediam arcanam*, Leibniz states: "Non videtur satis in potestate humana pervenire ad notiones primitivas, seu ad ea quae per se concipiuntur; sed magis in potestate humana est analysis veritatum, multas enim veritates possumus absolute demonstrare et reducere ad veritates primitivas indemonstrabiles."[2] Indeed, it is enough to ensure the status of primitive truths that the propositions be analyzed down to "identicals" and that they be shown to conform to the one absolute axiom: the principle of contradiction. The appropriate process consists in substituting definitional equivalents in order to assess that the predicate is contained in the subject. Before Rescher, Couturat had clearly established that this is the canonical form of Leibnizian analysis.[3]

I believe some "nuances" should be brought to statements in section 1.2 on Scientific analysis. Weingartner suggests that for Descartes and Leibniz: "A (scientific) term (concept) t is analyzable (AN) iff t is a primitive term or else t is reducible to a primitive term with the help of (a chain of) definitions." Analysis of concepts is in fact achieved through resolution down to primitive truths. 'Primitive term' is an ambiguous phrase. Descartes's rule of analysis in *Discours de la méthode* states that one should divide up each of the difficulties to be examined into as many parts as possible and as seem requisite. This is quite a relativistic process insofar as reaching ultimate concepts is concerned. The only requirement is to reach true elements of solution, ascertained by the application of the "clearness and distinctness" criterion. The Cartesian "simple natures" have a rather complex epistemological status. If we may consider that they are "atoms of evidence", they need not however be isolated and separate entities: "they can be considered from the methodological points of view as the 'ultimate limits of formal analysis'", insofar as they can be intuited as the genuine characters of reality apart from any abstractive analysis. They resemble what Spinoza will call common notions by contrast with general notions. A close look at the lists of simple natures in *Regulae* 6 and 12 would show that, considered as the most immediate objects of intellectual intuition, they can involve further internal implications to be explicited in propositional form; but the connection of these implications is apprehended at once as consistent and real. As stated by L. J. Beck:

> Despite some ambiguous expressions, there is no reason to consider that Descartes's doctrine implies that simple natures are mere atoms of thought, to which correspond atoms of reality. They can be considered 'atomic' in the modern sense of atomic. The simple natures enclose and engender a number of relations which may be of an infinite complexity, the total constituting the whole of reality and the whole of knowledge . . . The 'atom of evidence' must be understood as 'simple' only in so far as it is the *minimum cognoscibile* for the knowing mind, but as *minimum cognoscibile* it is a whole of the nexus and what is connected; it is, to continue borrowing phrases, an implication with its *implicans* and *implicatum*, a simple proposition.[4]

Let me just add in connection with Descartes's classification of simple natures, that it seems doubtful whether we need statement 1.22 "All (scientific) terms (concepts) which are primitive and derived from the mind itself are innate". I do not want to pick up the issue of innate knowledge for discussion. I would like to suggest that the real metaphysical issue about first principles is whether they rule over the intellect of God (cf. Leibniz's so-called Platonism). the alternative being the Cartesian doctrine of the creation of eternal truths. Leibniz's criterion of truth, involving the requirement that the possibility of concepts be demonstrated ties up with the transcendent value of the axiom of contradiction, which is not only regulative of our capacity for analysis, but constitutive of the order of reality. As a matter of fact, the Couturat–Rescher definition of Leibnizian analysis as "a logical process . . . based on the inferential procedures of *definitional replacement* and *determination of predicational containment* through explicit use of logical processes of inference" confirms the view that the only true axiom of general science is the law of contradiction. The reducibility of scientific concepts to absolutely primitive terms is a limit which the human mind can only approximate confusedly. But the definitional terms in geometry and arithmetic, though still nominal and somewhat imaginary in comparison with the real essences as conceived in God's understanding, exemplify perfectly the logical requirement of analyticity. They may be considered as provisionally primitive terms, pending further resolution of the so-called axioms (= definitions under the control of the law of contradiction). The analyzability of concepts is a fonction of their reducibility in terms of definitions and identical propositions: "Manifestum est, omnes propositiones necessarias sive aeternae veritatis esse virtualiter identicas, quippe quae ex solis ideis sive definitionibus (hoc est terminorum resolutione) demonstrari seu ad primas veritates revocari possunt, ita ut appareat, oppositum implicare contradictionem, et cum identica aliqua sive prima veritate pugnare."[5]

In the second part of this section, I want to dissent somewhat from Prof. Weingartner's view that all scientific knowledge according to Descartes and Leibniz can be integrated in a deductive system, meaning that scientific systems would consist of finitely analytic propositions comprised within the bounds of a finite axiomatic. I do not question the fact that such may be the purport of a straightforward expression of the 'more geometrico' ideal. A selection of texts might countenance that version. But 'more geometrico' is a phrase suggesting some analogy of process with the mathematical methods. Scholars in the French tradition such as Gueroult, Belaval and Serres, have insisted on the diversity of understanding of what mathematics and mathematical physics can be for Descartes and Leibniz. I doubt whether the quite univocal pattern of Weingartner in his section "more geometrico" would fit the philosophies of mathematics and the processes, methods, and principles involved. But to return to my point, I would question 2.4 "All finitely analytic (FA) propositions are (necessarily) true", in view of the fact that truth is not tied with a finite analysis of concepts, nor even with such an analysis for propositions. My point is that for Leibniz, truth depends upon displaying the possibility of propositions and concepts, and that this can be done through a thorough analysis of a finite character (at least down to the level of primitive truths). But the same result can be achieved through a survey of present and past experience, whatever be the difference in criteria of experiential truth for Descartes and Leibniz (this is a different matter). The general consequence of the fact is that the criterion for scientific truth seems to be twofold in a way: binding together the requirement of analyticity and the 'synthetic' requirement of a sufficient reason for truths of fact. It is only in the realm of truths of reason that sufficient reason can be adequately rendered in the analytic fashion. I think this is the true meaning of one of Weingartner's quotations from Leibniz: "Itaque cujuscunque veritatis reddi potest ratio, connexio enim praedicati cum subjecto aut per se patet, ut in identicis, aut explicanda est, quod fit resolutione terminorum. Atque hoc unicum summumque est veritatis criterium, *in abstractis scilicet neque ab experimento pendentibus*, ut sit vel identica vel ad identicas revocabilis."[6] Let us infer therefrom that scientific explanation is a "resolutio terminorum" under the general ruling of sufficient reason, but that a resolution into identicals is not to be expected in the case of truths of fact. Now, let us turn to Weingartner's 2.5 "Every true proposition is finitely or infinitely analytic". If we glance through the list of Leibnizian formulations for the principle of sufficient reason, the problem seems to be connected with the notion of a priori proof. 'A priori proof' does not necessarily mean that the reason for a fact or a statement can or could be found by complete analysis (from the viewpoint of our finite understanding, and excepting the infinite capacity for actual computation in God's understanding). Sufficient reason for contingent truths or truths of fact is to be given *a priori* by what I would call systematic presuppositions, holding good in "extrinsic denominations" to use the Leibnizian terminology. Weingartner's opinion is that with infinitely analytic propositions, "we have a genuine proof-process intended at least as an inferential procedure of the most rigorous kind, i.e. a proof of a proposition within a deductive system." This might be the ideal formula for assigning a sufficient reason. In actual practice, the hypothetical necessity in the case of contingent truths is posited by some kind of adventitious move of the intellect, which at first sight has little to do with the analytic version of the 'more geometrico' ideal. One should try to reconsider the concrete patterns for building physics in the Cartesian and Leibnizian moods beyond the abstract structure proposed by Weingartner: "Provided physics is built *more geometrico*, i.e. as a deductive system *PH* with the properties given in the principles 2.2–2.4 (the principles of rationalism presupposed), then every true (contingently true) proposition of physics has an a priori proof in *PH* with an infinite number of steps."

First of all, the abstract pattern would hardly fit the scheme of Cartesian physics, where the sufficient reason for factual statements consists in finitely intuited simple natures used as elements for a rational reconstruction of principles and laws (+ the subsidiary use of hypotheses and experiential data in the course of unfolding the explanation of phenomena). Secondly, even in the case of Leibniz, I doubt whether physics is conceived to be achievable as "an axiom system of truths of fact". Let me refer to Weingartner's quotation from Leibniz: " ... toutes les propositions contingentes ont des raisons pour être plutôt ainsi qu'autrement ... mais elles n'ont pas des démonstrations de nécessité, puisque ces raisons ne sont fondées que sur le principe de la contingence ou de l'existence des choses, c'est-à-dire sur ce qui est ou qui paraît le meilleur parmi plusieurs choses également possibles."[7] The so-called "additional axioms" required for the "system" of truths of fact (but is there any system of truths for Leibniz, except in a purely metaphysical sense?) are in effect architectonic and regulative principles whose derivation is highly complex. Section 3.2 Mathematical Order of the Universe lists some such principles: "conspiration universelle" (cf. σύμπνοια πάντα, *Monadologie*, § 61), perfection, continuity, harmony or finality. But the question of their legitimacy is left open. I doubt whether the calculus *de maximis et minimis* would alone give legitimacy to a system of truths of fact. Rescher's statement that "inspired by its success in optics, leibniz sought ardently to extend the applicability of minimax principles by means of a general principle that in all natural processes some physical quantity is at a maximum or a minimum",[8] is a case of simplification. The process is more complex, as shown by Belaval when he analyzes Leibniz's use of the minimax principle in optics. The calculus is subservient for expressing rigorously a sufficient reason modelled after architectonic and regulative ideas. These architectonic and regulative ideas are taken to express the teleological order of the contingents initially discovered by reflecting upon the analogies of experience:

> Qu'est-ce que la voie 'la plus déterminée' (en optique)? Leibniz le répète sans cesse. (1) C'est dans une variation continue, l'*unique* valeur sans symétrique, ou, comme dit Leibniz, sans jumelle, c'est-à-dire sans identique; (2) C'est, par conséquent, la plus *simple*, car elle exclut la pluralité numérique et ne distingue pas entre maximum et minimum, faisant dans les deux cas "la même opération"; (3) C'est la plus *féconde* parce que cette même opération produit un maximum ou un minimum, mais jamais un indéterminé, et que sous un énoncé *unique* elle rend compte du maximum de phénomènes. Unicité; simplicité, fécondité donnent la définition d'une fin: car l'exclusion des identiques permet, dans le plus déterminé, de reconnaître le plus déterminant pour une volonté parfaite comme celle de Dieu.[9]

The link is analogical between the analysis of a specific physical feature *de maximis et minimis* and the multi-levelled computation of 'compossibles' needed in order to determine a sufficient reason for empiricial truths. The interesting question is: how does the norm of mathematization relate to the framing scientific explanations of given phenomena? Truths of reason are necessary by logical necessity; truths of fact are contingent. The truths which constitute physical necessity fall somehow in between: they are conditional and non-existential as the logically necessary truths, but contingent as the plain truths of fact. If the foundation for truths of fact is purely empirical, whence does the necessity originate for those physical truths, superimposed on truths of fact? My tentative answer would be that there is some possibility of comprehending facts under sufficient reasons with the special requirement that they conform to mathematical models of determination. As a scientist, Leibniz is skeptical concerning unqualified appeals to sufficient reason. One has to abide by the criteria of distinct, and as far as possible adequate, symbolic knowledge, in framing scientific explanations.

II

In support of my reservations on Professor Weingartner's interpretation of the more geometrico ideal for the Leibnizian theory of science, I shall now analyze Leibniz's doctrine of scientific hypotheses as it emerges in the crucial period spanning his revision of Cartesian mechanics and natural philosophy.

A good starting point is the exchange of letters with Conring since it offers essential data on the epistemological criteria for scientific hypotheses. Conring defends the view that nothing original in matter of doctrine has been obtained through researches in 'experimental philosophy'.[10] Because they neglect the deductive mode of explanation, the experimentalists and observers of the Royal Society try to work out a type of knowledge that will fail to reach the level of science. This criticism is uttered by an Aristotelian who cannot conceive how explanations inferred from factual data can be subject to revision. Conring would insist that the first premisses on which to build philosophy of nature be found a priori. We do not have a clear view on how he would envisage to implement such a project of natural philosophy. He would probably grant a system demonstrative status provided it conform with the metaphysical principles in Aristotelian scholasticism. Leibniz replies by developing interesting statements on the nature of scientific demonstrations and the hypothetical status empirical explanations can be given. The difference between standard demonstration and hypothesis forms the essential epistemological issue for Leibniz's critique.

Leibniz conceives of demonstration as structured according to a chain of definitions (*catena definitionum*), and he reduces the logic of demonstration to the operation of substituting equivalents. This scheme of demonstration is accurately described by Couturat and Rescher, and it agrees with the classification of ideas which the *Meditationes* (1684) and *Discours de métaphysique* will develop.[11] In his reply, Leibniz identifies hypotheses with anticipations of causal determination when the relevant phenomena have only been analyzed in part. A similar case is that of a clockwork whose internal agency one does not know but can give a sufficient reason for in mentally planning a system of wheels and structures which would produce the same result. However, the same artificial phenomena might be obtained by a variety of means; the analytic decomposition of the wheelwork alone could reveal without possible equivocation which process had been used by the clockmaker. This would be equivalent to giving a definition of the process generating the phenomenal results: "Ipsum praecise modum quo Artifex usus est, nisi dissoluto opere definire non possum".[12] Of course, when one deals with "machines de la nature" instead of "machines de l'art", there is infinite complexity in the analytic decomposition, and the hypothesis can only "express" the causal order. But Leibniz makes a significant move: he does not limit the process to expressing only sufficient reason for the presently available experiential data. He requests that the hypothesis sketch an architectonic plan for other possible effects that experience will have to reveal by investigating phenomena in a more systematic and extensive way.[13]

Because of this view on hypotheses, Leibniz accredits empirical investigations (= collecting and relating data following the method of 'histories'), but his remarks on the *methodus medendi* suggest that one has to get to sets of empirical concepts bound together by a systematic necessity, which arises from the empirically controlled hypothesis. In empiricist medicine (of the Sydenhamian type), this is the way to proceed in aiming at a physiological theory that would rule over practice ("*constituta alia plane corporis oeconomia*").[14] The necessity to appeal to hypotheses leads to underline the speculative structure in Hobbes's and Descartes's physics. Clearly, Leibniz has in mind a more open form of hypotheses, since the a priori principles which are presumed, cannot pretend to full apodicticity. The mode of explanatory unfolding in the *Hypothesis physica nova* (1671) had framed the background for such an epistemological stand.

The letters to Conring during the years 1677–8 develop with more accuracy: (1) fine lines of demarcation from the methodology of Descartes's *Principia*; (2) a possible conciliation between the definitional theory of science and the "pragmatic" doctrine of hypotheses.

Cartesian physics had been built on an insufficient basis in experience. In most cases, observations depending upon common experience are the only vouchers for the explanatory relevance of principles. Caught in vain controversies and acting as head of a party, Descartes had to content himself with gross correlations between *experimenta in vulgus nota*. The potential involved in the Cartesian type of hypothetico-deductive analysis ranks high. Some specimens extracted from his scientific work would afford positive instances: e.g., the *Météores*, the physiological writings, especially *L'homme* and *Les Passions de l'âme*. The problem with the Cartesian scheme is that an ingenious system of hypotheses is built on experiences which lack analytic qualification and so can be constantly objected to on the basis of a plurality of other relevant systems. Choosing between alternative hypotheses would depend upon experiments that could distinguish more intimately the 'compossibles' from the mere "possibles" within concepts which purport to 'reflect' the manifold of experience. And so, one would reach some kind of adventitious apodicticity that might combine expressing a causal order of the phenomena and hypothetically deducing the phenomenal connections. *De facto*, experience evidences more and more each day that the Cartesian construction in physics is fragile. But the plan itself of such an edifice seems to accomodate a true methodological requirement: the ingenious ordering of elements in the explanatory fiction can be used as a model by whoever wants to set a natural philosophy offering better guaranties of objective relevance: instances of the model could be found in the explanation of tides, rainbow, generation of salts, meteors, and magnetic activity.[15] When Leibniz describes the Cartesian and post-Cartesian fields of enquiry, he notably states that those explanations can be revised: the laws of impact are inadequate, specially for two hard bodies of unequal size 'concurring'; correlated effects in chemistry remain unanalyzed; methods of mathematical expression are made superficial use of. Literally taken, the Cartesian hypotheses are fruitless. But, as a matter of fact, one has to separe the rules for mechanical expression and the geometrical analysis on the one hand, the presumed solutions of empirical problems, on the other.

Initially, Leibniz would not have intended to overthrow the mechanistic methodology to which Descartes, Hobbes, or Boyle, were referring. The first criticisms focus on the elements of the physical hypothesis, and Leibniz would have defended the dictum: "Omnia fieri mechanice in natura" without much qualification. But about 1677–8, applying the dictum seems to involve some difference between the Cartesian and Leibnizian ways. Leibniz's notion of hypothesis, which was bound with the explanatory models in Cartesian physics, tends to diverge, or, to be more precise, to involve a new transcription of the hypothetico-deductive process. It seems clear that this recasting happens in conjunction with the phrasing of the new dynamics. Original elements are then available for building physical hypotheses and, as well, for setting the regulatory principles of the explanation procedure. Following this trend, a later letter to Conring underlines the shift in epistemological style which concurs with the advent of the dynamics:

In Physicis quoque demonstrationes aliquot habeo magni usus circa motum. Archimedes aequiponderantium scientiam constituit. Galilaeus longius progressus accelerationes sub regulam vocavit et pendulorum proprietates mirabiles detexit. Sed de vi elastica quam vocant corporum post flexionem se restituentium ac de corporum inter se concurrentium ac repercussorum legibus nunc primum certa Elementa me demonstrata habere arbitror Archimedeo plane rigore. Unde constabit non Cartesium tantum, sed et alios summos viros nondum ad intimam hujus argumenti pervenire nec proinde hactenus Scientiae mechanicae Elementa absoluta haberi.[16]

In the same period, the draft of a letter to Malebranche which was never sent, shows the decisive shift in the Leibnizian criticisms of Descartes's physics. This time, Leibniz raises doubt on the axioms of the Cartesian system: that matter and extension are the same thing, that all truth depends on the arbitrary will of God, that the same quantity of motion is constantly kept among bodies. In the same vein, he attacks the demonstrative arguments for the Cartesian law of sines. Those arguments were based on the analogy of optical refraction with the mechanical determination of impact along the Cartesian pattern: Leibniz finds that hypothesis to be a mere mental artefact, and he extends his criticisms to cover the whole of Cartesian physics, which he judges unexplanatory: "Je ne veux pas toucher à son hypothèse physique, car on ne la saurait prouver qu'en expliquant les phénomènes de la nature."[17] For instance, Newton's experiments suggest that Descartes had not really explained the rainbow. And the explanation of the magnet is deficient since it had not accounted for declination. Descartes's theories are "ingénieuses au possible, mais souvent incertaines et stériles".[18] However, full rejection of the Cartesian *model* is delayed for methodological, or at least epistemological, reasons:

Néanmoins je conseillerais toujours à un amateur de la vérité [d'] approfondir [le système cartésien], car on y voit une adresse d'esprit admirable et sa physique tout incertaine qu'elle est, peut servir de modèle à la véritable, qui doit pour le moins être aussi claire et aussi concertée que la sienne; car un Roman peut être assez beau pour être imité par un historiographe. Pour l'abréger: Galilei excelle de réduire les mécaniques en science, des Cartes est admirable pour expliquer par de belles conjectures les raisons des effets de la nature.[19]

If the metaphors in this passage were removed, one could probably infer that the Cartesian model, recast so as to deal with problems concerning the empirical understanding of nature, might represent a form of demonstration that allows for drawing experience into a rational system. Another way of interpreting this passage would be to suppose that the Cartesian explanations are mere conjectures, but can be referred to an analysis that may reveal the presuppositions involved and offer better criteria for analogical translations of the experiential data. To be accurate, this translation might depend upon a conceptual analysis of the categories involved in our understanding of phenomena. For Leibniz, any conjecture can be referred to analysis and verification, and so revision is to be expected. It is worth mentioning that he wishes a mind capable of rational conjectures would come and raise empirical medicine up to the level of a rational discipline. In similar areas, Cartesian-type hypotheses prime further demonstrative arguments, the initial motive consisting in the hypothetical anticipation.

It might prove interesting to resume the study of how Leibniz conceives the recasting of Cartesian physics. The available *corpus* is considerable, specially if one wishes to account for the rectifications involved in the dynamics proper. The analysis could focus on the critical reading by Leibniz of Descartes's *Principia* (cf. *Animadversiones in partem generalem Principiorum Cartesianorum*, 1692). This was a major concern for Leibniz, if we follow Belaval's analysis of an early version of the *Animadversiones*, dating back to 1675.[20] Since the advent of the dynamics can be located in the first half of 1678, such a set of critical reflections are developed in conjunction with the original elements of Leibniz's system of nature and are integrated in the conception of physical hypotheses.

Nevertheless, the origin of Leibniz's epistemological doctrines cannot be drawn from Descartes. Even if these doctrines are significantly inflected by the analysis of the *Principia*, concerning argumentative patterns and presuppositions, Leibniz himself states that his divergence from the Cartesian research program is quite antecedent to his effective study of Cartesianism. He explains to Malebranche:

> Comme j'ai commencé à méditer lorsque je n'étais pas encore imbu des opinions cartésiennes, cela m'a fait entrer dans l'intérieur des choses par une autre porte et découvrir de nouveaux pays, comme les étrangers qui font le tour de France suivant la trace de ceux qui les ont précédés, n'apprennent presque rien d'extraordinaire, à moins qu'ils soient fort exacts ou fort heureux; mais celui qui prend un chemin de traverse, même au hasard de s'égarer, pourra plus aisément rencontrer des choses inconnues aux autres voyageurs.[21]

From the epistemological point of view, this crossroad would probably represent the foreseen conciliation between the definitional theory of science and the pragmatic justification of hypotheses. In the same crucial period for the framing of dynamics, a convergent theme emerges when Leibniz addresses Foucher's rendering of the skeptical arguments and discusses hypothetical truths: "Dans le fond, toutes nos expériences ne nous assurent que de deux choses, savoir qu'il y a une liaison dans nos apparences qui nous donne le moyen de prédire avec succès des apparences futures, l'autre que cette liaison doit avoir une cause constante."[22] The truth of hypothetical arguments is therefore something external, which does not depend on us. Our capacity of conceiving the verisimilar order of nature drawn from the analogies of experience regulates itself by what can be considered analytic in the phenomenal realm, and this consideration itself depends upon the reconstructed continuity of particular sufficient reasons. Along this line, Leibniz rejects any metaphysical form of doubt and allows for progressive criticisms bearing on the untrustworthy schematisms of experience (confused notions). On the one hand, hypotheses are anticipations of sufficient reasons which evade any attempt at direct analytic unfolding. On the other hand, by setting an analytic clarification of hypotheses under cover of admissible presuppositions, such as the principle of continuity, one can show the power of given hypotheses to express and systematize a manifold of empirical relations.This 'geometrization' of empirical knowledge confers rationality upon hypothetical truths.[23]

From the status of hypothetical truths, one can easily infer and set on some features of the physical hypotheses. Thus, Leibniz concedes (to Conring) that natural science will be based on both *experientia* and *ratio apodictica* (or, if such reasons are wanting, probable reasons [δοξαστικῄ]) but his notion of *ratio apodictica* will enable him to set the epistemological model for a theory of well-formed hypotheses.

Ego semper putavi, Demonstrationem nihil aliud esse quam catenam definitionum vel prof definitionibus, propositionum jam ante ex definitionibus demonstratarum aut certe assumtarum. Analysis autem nihil aliud est quam resolutio definiti in definitionem, aut propositionis in suam demonstrationem, aut problematis in suam effectionem. Sed quando plures ejusdem rei effectiones fingi possunt, tunc nova quaerenda sunt data sive experimenta, quibus excludantur eae effectiones sive causae quae non sunt hujus loci. Si vero ejusmodi data nova (qualia Verulamius instantias crucis vocat) nont sint in promptu, tunc non possumus praecise designare causam effectus veram, sed contenti esse cogimur hypothesi sive causa possibili, quae quo simplicior et concinnior, hoc probabilior, ut in Astronomicis Copernicana, in quibusdam Physicis Cartesiana.[24]

It is at once noteworthy that the "catena definitionum" model lends itself to substitution schemes on behalf of propositional chains which include primary elements of presumptive certainty. For an argument thus patterned the validity criterion consists in our being able to resolve the subordinate links analytically. This extends to solving problems by an analysis that would rise to sufficient reasons. If there are alternative solutions, the difficulty can be cleared up by extending the analysis to include crucial instances that would direct the analytic unfolding to one of the alternatives. These crucial instances would specify empirical conditions further; to the initial terms of the problem, new terms would be added as determinations of the *experimentum crucis*. As it seems, Leibniz distinguishes two types of complex problem solving: (1) solutions analytically completed by experiences that discriminate more and more accurately between reasons: i.e. solutions which approximate chains of demonstrative equivalents; (2) solutions restricted to presuming determinate conditions for the available data (this applies to Copernicus's and Descartes's hypotheses): the elements of the problem are retranscribed in a presumed order. This second type of solution cannot pretend to apodictic truthfulness, but it may be granted a degree of probability in accord with its simplicity or formal style (coherence in the ordering expressed by the *explicatum*). Occasionally, Leibniz will suggest that though these second rank hypotheses bring on alternative *explicantia*, some determinations of probabilities through computation might connect the enquiry with the demonstrative definitional model. However, he allows that such a strategy is only at its start.

The main problem with the analytic solution by means of hypotheses consists in this that an *explicans* satisfying the conditions of the *explicandum* can be inferred from false principles. Conring stresses the obstacle vigorously: "Nova data quin possint effingi atque ex iis utut falsis aliquid demonstrari, nemo negaverit. Sed ejusmodi demonstrationes non tamen sun ἐπιστηκονικαὶ, nec per analysin colligere licet datorum veritatem ex veritate consequentium."[25] Conring's criticism bears on both the Cartesian hypotheses and the inferential processes that could be connected with *experimental philosophy*. However, Leibniz had specified formal validity conditions for the analytic process: the chain of propositions must imply reciprocal arguments, i.e. whose converse be equally true, on the model of definitions (doubtless, definitions that show the possibility of the object) and of equations. The truth of what can be demonstrated depends not on some inductive grounds, but on the substitution of conceptual equivalents when relating terms; which means that notions are to be used which show themselves to be adequate. Propositions which are undemonstrable at the end of the analytic reductions will be either identical or empirical. In both cases, one has to do with connections of notions (either identity or a connection in accord with the empirical record) which disclose their possibility directly.

Against Conring, Leibniz will defend his view of analysis and demonstration and assert that it expresses consistently the actual procedure among scientists while it shows the method "ad inveniendum atque judicandum."[26] The validity criterion for the whole process is expressed in the following way:

> Patet ... omnes veritates resolvi in definitiones, propositiones identicas et experimenta (quanquam veritates pure intelligibiles experimentis non indigeant) et perfecta resolutione facta apparere, quod catena demonstrandi ab identicis propositionibus vel experimentis incipiat, in conclusionem desinat, definitionum autem interventu principia conclusioni connectantur, atque hoc sensu dixeram Demonstrationem esse catenam definitionum.[27]

Clearly, Leibniz conceives a mode of hypothetical demonstration in which the elements for solving the physical *explicandum* are provided by empirical propositions. Even if the terms are irreducibly problematic in those propositions, this does not mean any aporia in the demonstrative process itself. This process consists in the full analyticity of the propositional links, which turns out to be the full analyticity of conceptual connections. Leibniz draws a straight parallel between analysis of ideas, demonstration, i.e. the analytic reducing of propositions to truths already known, and solving problems into easier problems within the limits of our capacity. With the classification of ideas in the *Meditationes* and *Discours de métaphysique* one finds that the ranking in types of distinct ideas expresses stages in the analytic process. So, a clear distinct inadequate idea can occasion demonstrative combinations, provided the notions involved be signified by symbols allowing the substitution of equivalents. If the analysis is provisional, such an idea and its equivalents might be taken as adequate for the sake of demonstration. It is then understood that the conceptual elements involved will have to be checked against experience. The use of symbols and the operations they allow set the stage for provisional adequacy and problematic solutions. The Leibnizian distinctions aim at showing that intuition is not to be trusted for complex notions. As a matter of fact, Leibniz questions Cartesian evidence on the basis that it implies a psychological fallacy concerning conceptual connections. Likewise, the epistemological analysis of hypotheses shows that an hypothesis if accepted without demonstration is of no value notwithstanding the fact that observed phenomena could be derived from it. Between *explicandum* and *explicatum*, the connection must be set in by means of convertible propositions or sequences of equations, since the logical principle in demonstrations rules that conceptual equivalents be substituted in place of each other. Of course, there remains the case of a principle or hypothesis (of the Cartesian type) which cannot adequately meet the analyticity requirement. The mere derivation from observed phenomena is not enough to warrant the truth of the matter. So, Leibniz reinforces the Cartesian critera which support the transformation of moral into physical certainty, and he modifies the meaning and scope of the operation.

Illud ... fatendum est, hypothesin tanto fieri probabiliorem quanto intellectu simplicior, virtute autem ac potestate amplior est, id est quo plura phaenomena et quo paucioribus assumtis solvi possunt. Et contingere potest ut hypothesis aliqua haberi possit pro physice certa, quando scilicet omnibus omnino phaenomenis occurrentibus satisfacit, quemadmodum clavis in cryptographicis. Maxima autem (post veritatem) laus est hypotheseos, si ejus ope institui possint praedictiones, etiam de phaenomenis seu experimentis nondum testatis, tunc enim in praxi hypothesis ejusmodi pro veritate adhiberi potest.[28]

Evidently, this refers to *Principia*, IV, § § 203–206, where Descartes justifies his doctrine of hypothesis by showing that it conforms to the intellect's primary notions concerning corporeal nature *qua* known *a priori*. The fiction of imperceptible material parts which are given appropriate sizes, figures, and motions suffices to yield an explanation of perceptible phenomena. This hypothesis has only practical value (*sufficit ad usum vitae*); one does not need more, if it matches all natural phenomena exactly. The moral certainty which it possesses is of the same order as that which arises from inventing a cipher to formulate the tentative meaning of a cryptogram. The truthfulness of the interpretation is in direct ratio with the confirmed belief issuing from the fact that the phenomena have been unified under a coherent representation. This moral certainty of corroborated hypotheses can be further ensured by reference to God's veracity, since this metaphysical principle warrants applying the evidence criterion in demonstrations, and hypothetico-deductive inferences as well. Knowledge of the existence of material entities and also, all arguments establishing the system of nature are amenable to evidence by means of deductive consistency and/or direction from the "lumière naturelle" (according to the dispositions in our nature).[29]

However, Leibniz dissociates himself from some aspects of the Cartesian doctrine, and he seems to give a different orientation to the analysis. First, there is no reason why one should rely on metaphysics to ascertain the objectivity of particular hypotheses. The proofs act from the inside if hypotheses are used properly: the probability of the matter (in a continuous progression to approximate truth, which would not make sense for Descartes) is based on the simplicity of the *explicans* together with the scope of its conformity to facts: the unfolding of rationality in the *explicans* objectivates a virtual order in the data; if a fully adequate analysis were completed, the order of reality would be approached asymptotically. Moral certainty yields physical certainty in its own right, and thus expresses the "reasons" in nature provided the hypothesis matches all the experiential determinations. But the most significant feature consists in the heuristic role vested in hypotheses. This function, when adequately fulfilled, changes physical certainty into a kind of truth. It is so because the analytic unfolding of terms in the hypothesis gives form to the manifold of phenomena under the necessity of architectonic principles, such as the principle of continuity. *Stricto sensu*, God's veracity, which serves as a bail of truth in Descartes's system, is an unfruitful *deus ex machina* when compared with the analytic potential that a Leibnizian hypothesis can display in inventing well-framed explicata. And one must add that the adequacy criterion for complex notions does not consist anymore in the delusive clearness of a necessarily partial, limited, and truncated representation, but it will henceforth consist in spelling out logical implications of concepts framed to account for unexplored connections in the realm of phenomena and to concatenate the elements of a progressive revisable system. And this would be a quite special form of the 'More Geometrico' ideal as applied to natural science, if there were a general unequivocal expression of this synthetic notion for so-called "classical rationalism", which I doubt.

Université de Montréal

NOTES

[1] Cf. Yvon Belaval, "Premières Animadversions sur les 'Principes' de Descartes", *Etudes leibniziennes*, Gallimard, Paris, 1976, p. 58 n. 8.

[2] P. VIII, 2 verso, in Louis Couturat, *La logique de Leibniz*, (1901), Olms, Hildesheim, 1969, p. 199.

[3] Cf. Couturat, *op cit.*, p. 206: "Ainsi le fondement essentiel de la déduction est le principe suprême et unique de la Logique, et non pas le principe du syllogisme (le *Dictum de omni et nullo* d'Aristote), car celui-ci n'est pas, comme on le croit souvent, un axiome identique, mais un théorème qui se démontre au moyen du principe prédédent."

[4] L. J. Beck, *The Method of Descartes. A Study of the Regulae*, Clarendon Press, Oxford, 1952, pp. 78–79.

[5] *Die Philosophischen Schiften*, Gerhardt, (ed.) VII, p. 300.

[6] *Ibid.*, VII, pp. 295–296.

[7] *Ibid.*, IV, p. 438.

[8] As quoted in Weingartner's text.

[9] Yvon Belaval, *Leibniz critique de Descartes*, Gallimard, Paris, 1960, pp. 410–411.

[10] Cf. Conring an Leibniz, 16/26 Feb., 1671, *P.* I, pp. 171–172.
[11] From the classification of ideas, one could probably infer what the minimal conditions are for ascertaining the rational capacity of scientific concepts.
[12] *P.* I, p. 174.
[13] Cf. *P.* I, p. 174, and especially, p. 195 (Leibniz an Conring, 19 March, 1678).
[14] *P.* I, p. 174.
[15] Cf. *P.* I, p. 186.
[16] *P.* I, p. 202.
[17] *P.* I, p. 334.
[18] *P.* I, p. 336.
[19] *P.* I, p. 336.
[20] Yvon Belaval, *op. cit.*, "Premières Animadversions sur les 'Principes' de Descartes", pp. 57–85.
[21] Leibniz an Malebranche, 22 June, 1679, *P.* I, p. 332.
[22] Leibniz an Foucher, *P.* I, p. 372.
[23] Cf. *P.* I, p. 381: "En matière de connaissances humaines il faut tâcher d'avancer, et quand même ce ne serait qu'en établissant beaucoup de choses sur quelque peu de suppositions, cela ne laisserait pas d'être utile, car au moins nous saurons qu'il ne nous reste qu'à prouver ce peu de suppositions pour parvenir à une pleine démonstration, et en attendant, nous aurons au moins des vérités hypothétiques, et nous sortirons de la confusion des disputes. C'est la méthode des Géomètres . . . "
[24] Leibniz an Conring, 3 Jan., 1678, *P.* I, p. 186.
[25] Conring anLeibniz, 26 Feb., 1678, *P.* I, pp. 190–191.
[26] Leibniz an Conring, 19 March, 1678, *P.* I, p. 194.
[27] *Ibid.*, p. 194.
[28] *Ibid.*, pp. 194–195.
[29] *Principia,* § 206, *A. T*, VIII–1, pp. 328–329.

THOMAS M. LENNON

THE LEIBNIZEAN PICTURE OF DESCARTES

The picture Professor Weingartner gives of Leibniz seems to me clear in the Leibnizean sense of that term – it is certainly Leibniz whom one picks out by means of it. His picture of Descartes, however, I find rather less clear, so much so that with it I hardly recognize the philosopher whom I know as Descartes. The problem as I see it is that Weingartner gives us essentially one picture for both Leibniz and Descartes, and my primary aim here will be to indicate why this is problematic. My thesis will be that Leibniz's own appreciation of the differences between his own philosophy and Descartes' as significant, is correct.

At the end of 500 pages of detailed argument, Yvon Belaval begins the conclusion of his *Leibniz: Critique de Descartes*[1] as follows. "Méthode, sciences, savoir, en toutes ses parties et dans sa totalité elle-même, la philosophie de Descartes est donc attaquée par Leibniz."[2] This is no intramural quarrel. Nor is the thesis Belaval argues a reconstruction of what Leibniz said, still less a reconstruction of what he ought to have said against Descartes. It is rather a summary of what Leibniz in fact said.

Leibniz wrote to Foucher in 1675 or 76 that he had not yet read Descartes' works with the care he intended one day to devote to them;[3] but Belaval notes that in fact Leibniz here is speaking only about Descartes' geometrical work, and shows moreover that Leibniz early on was familiar not only with Descartes' published work but with the unpublished as well (the *Rules*, the *Search After Truth, the introduction to the Géometrie*, etc.).[4] In addition, Leibniz followed the Cartesian-anticartesian debates of the period and corresponded with many of the disputants. The upshot is that with his own main views in place Leibniz was from early on sufficiently informed to begin attacking what for the rest of his life never ceased to be his primary targets in Cartesianism: the rejection of final causes, the dissociation of religion and metaphysics, the proofs of the existence of God, the explanation of transubstantiation, the view of sin as mere privation, the failure to establish the immortality of the soul, the failure to define thought and the consequences of this for the ontological argument, the insufficiency of clarity and distinctness as criteria of truth, the frivolousness of methodical doubt, the conservation of quantity of motion measured by the product of volume and speed,

William R. Shea (ed.), Nature Mathematized, 215–226.

the essence of body as extension, motion defined as change of place, the lack of force beyond rest to explain cohesion, the rules for communication of motion, and so on.

The catalog of criticism is extensive, reaching as it does to most of the major issues with which Descartes concerned himself, and it is also systematic: the Leibnizean reaction to Cartesianism is no piecemeal opposition of isolated views, but springs rather from a perception of worldviews as irreconcilable due to a single issue. "L'accusation fondamentale qui commande toutes les autres se résumerait sans doute en disant: Descartes a péché contre la logique incréée."[5] I fully agree with Belaval both that there is a pivotal complaint and with his identification of it as the charge that if the logic reflecting the *ordo essendi* as well as the *ordo cognoscendi* is created, it is arbitrary, capricious and ultimately unintelligible. Here I shall attempt to elaborate a Leibnizean critique solely in terms of this issue as it appears most strikingly in Descartes *viz.* the doctrine of the created eternal verities, to which in fact Belaval properly reduced Leibniz's opposition to Descartes.

That the eternal verities are created is not the least known of Descartes' views; nonetheless, it will be useful to recall certain of its features. When he wrote to Mersenne on the 6th of May 1630 in one of his earliest statements of the view, Descartes had in mind Suarez (whose *Disputationes Metaphysicae* he in fact quotes):

> In regard to the eternal truths, I say once more that they are true and possible because God knows them to be true and possible, and not rather that they are known by God to be true as though they were true independently of Him. And if men understood the meaning of their words they would never say without blasphemy that the truth of something precedes God's knowledge of it, for in God, to will and to know are but one, that is to say, that because He wills something, so He knows it and thus only is such a thing true. One should therefore not say that if there were no God, nevertheless those truths would be true.[6]

The eternal verities said to be created include both moral and logico-mathematical principles. In his *Reply* to the sixth set of *Objections* Descartes says:

> [God] did not will to create the world in time because He saw it would be better than if He created it from all eternity; nor did He will the three angles of a triangle to be equal to two right angles because He knew they could not be otherwise. On the contrary, because He worked to create the world in time, it is for that reason better than if He had created it from all eternity; and it is because He willed the three angles of a triangle to be necessarily equal to two right angles that this is true and cannot be otherwise.[7]

And later on in the same *Reply* he tells us that "nothing can be which does not depend on Him, not only everything subsisting, but also all order, every law and every reason of truth and goodness" – otherwise God would not be indifferent and thus not free.

A second point of interest is that God is said to be the *efficient* cause of these truths. On 27 May, 1630 he wrote to Mersenne that God is in the same sense the author both of the essence of things as well as their existence – He is as free to make it not true that the lines drawn from the center of a circle to its circumference are not equal as not to create the world.

A final point concerns Descartes' use of modal terms. On 2 May, 1644 he wrote Mesland arguing that infinite omnipotence entails a freedom of indifference to make it not true that the three interior angles of a triangle are equal to two straight angles, or "generally that contradictories cannot be together". In fact, however, God wills these to be necessarily true – but in doing so He is in no way necessitated.[8] The upshot of this, I take it, is that Descartes has no analytic-synthetic distinction, that the only significant distinction for him is between those propositions which happen always to be true or always false, which he describes as necessary, and those which happen to be sometimes true and sometimes false.

But now what about the principle of non-contradiction just referred to; in what sense might it have been false? Or what of the doctrine of the created verities; could *it* have been false? The only move of interest that Descartes makes in this regard is the claim repeated in letters to Mersenne, Mesland and Arnauld that conceivability is sufficient but not necessary for possibility; we can be certain that God can do all that we can understand, but not that He cannot do what we do not understand. If some things are impossible, we cannot know that they are impossible. I don't think that in the end this move is of much help with regard to the questions just raised and I introduce it only to emphasize the question about to be pursued, *viz*. why it was that Descartes adopted such a non-rationalist, not to say irrationalist, posture.

I shall outline three hypotheses as to why Descartes held the doctrine of created verities; they are not necessarily competing and are here suggested as complementary. The first, which was initially proposed by Emile Bréhier,[9] might be called *the anti-holism hypothesis*. The argument is as follows: both for Descartes and for the scholastics in whose wake he worried about the eternal verities, it is theologically unacceptable for there to be anything uncreated other than God. To account for the eternal verities within this restriction, Aquinas, for example, appealed to the Neo-Platonic theme that the essence of individual things is the divine essence insofar as it is or can be participated in by individual things. The mutliplicity of essences is a multiplicity only on the side of the things instantiating them; on the side of the One there is no multiplicity or diversity. God's contemplation of the eternal verities is a self-contemplation of the ways in which this simple substance can be the principle of intelligibility of a multiplicity of things He can create. Details aside, one historical upshot of this solution was the view that knowledge of any individual thing (or proposition) was necessarily incomplete and deficient because its essence was incomplete, that in fact the knowledge of anything involved the knowledge of everything else. Bréhier has persuasively argued that the applicability of clarity and distinctness as criteria of truth, and thus of the whole Cartesian methodology, required absolutely discrete essences: what is clear and distinct is self-evident in the sense of being acceptable as true on the basis of nothing else. The only theologically acceptable way of securing this was to regard them as created.

The Leibnizean response to all this was threefold: (1) he rejected the way of clear and distinct ideas as hopelessly psychologistc; I shall return to this later; (2) he rejected the Cartesian logical, or epistemological atomism and, with his monads mirroring the whole of their universe, espoused internal relations with a vengeance; (3) he rejected the Cartesian strategy as theologically unacceptable. The being who creates with the blind omnipotence of the Cartesian deity is in no significant way different from the Spinozan deity who creates from blind necessity. Especially with regard to perfection is Leibniz concerned, for example, at *Discourse on Metaphysics*, 2: "In saying . . . that things are not good according to any standard of goodness, but simply by the will of God, it seems to me that one destroys, without realizing it, all the love of God and all His glory; for why praise Him for what He has done, if He would be equally praiseworthy in doing the contrary? Where will be His justice and His wisdom if He has only a certain despotic power, if arbitrary will takes the place of reasonableness, and if in accord with the definition of tyrants, justice consists in that which is pleasing to the most powerful?"[10]

A second hypothesis as to why Descartes held the doctrine of created verities might be called *the architectonic*. No one as far as I know has proposed it in the form about to be offered here, but Frankfurt's work has provided more of its components than any other.[11] The argument is that a primary aim of the Cartesian metaphysics is the refutation of skepticism, which especially in the *Meditations*, takes the form of a *reductio*. Rather than validating reason by proving its reliability – the enterprise that involves the notorious circle – Descartes in fact attempts to show what is just as useful to his program, *viz.* that "far from leading to the discovery of reasons for mistrusting reason, . . . the most conscientious use of reason leads to the discovery that such mistrust has no rational ground".[12] What Descartes does is to make the best case against reason and knowledge, and then show that the same grounds make a better case in their favour. The strongest case against clear and distinct perceptions, on this hypothesis, would be that God could have failed to create what appears to be their object and what would make them true, *viz.* the simple natures or the relevant eternal verities. In the case of geometry, extension either would be non-Euclidean or just not exist at all. In answer to Mersenne's question whether there could be real space had God not created anything, Descartes replied that "not only would there be no space, but that even those truths called eternal, such as that the whole is greater than its part, would not be truths unless God had established it [i.e. *space*] as such".[13] I shall not enter into the question of Descartes' mathematical realism; nor shall I consider the extent to which the doctrine of the created verities was actually motivated by methodological concerns suggested by the architectonic hypothesis or conversely. Of more importance is the Leibnizean response, which is, that the pretended Cartesian refutation of skepticism is worthless.

In his *Remarks on the Principles* Leibniz tells us that "the much praised rule: only what is clear and distinct ought to be accepted as true . . . has no value unless better criteria of clearness and distinctness than those given by Descartes are established".[14] The problem he raises is not the criterion of criteria problem posed by Gassendi in the *Objections*; rather, it is that the Cartesian criteria are psychological only, whereas what is required is a specification of *reasons* for accepting a proposition as true. He makes the same complaint against the Cartesian method of doubt: "What Descartes says here [*Principles* I, 1] on the necessity of doubting anything in which there is the slightest uncertainty would have been better expressed by the following more satisfactory and precise principle: The degree of assent or dissent which any proposition deserves must be considered; or still more simply, for every proposition the reasons must be examined. Then the cavilling about the Cartesian doubt would have ceased."[15] The upshot is that the difference between Weingartner's Cartesian principle

1.1.5. If *p* is known *clare et distincte* then *p* is true

and his Leibnizean principle

1.1.5.1. If *p* is known *clare et distincte* and *p* is known to be possible, then *p* is true is not just the additional conjunct in the antecedent of the Leibnizean principle, but a very important difference in the meanings of *clare et distincte*. In addition at least four of Weingartner's six characteristics of rationalism assume radically different significance for Descartes and Leibniz when viewed in this light. Take just the first two:

(1) There are first evident principles which are the sources of human knowledge.

Evidence for Descartes, however, means only indubitability, a notion carried over into the second:

(2) There are logical and mathematical principles of deduction by which theorems (further knowledge) can be derived from these first evident principles.

Weingartner notes later that although Descartes describes the operation of deduction, he doesn't state explicitly any logical rules of deduction. The reason for this in my view is that for Descartes there are no such rules since all inference is immediate. Deduction is "a continuous and uninterrupted movement of thought",[16] and it is this movement alone which distinguishes it from intuition, with the result that the mark of genuine and reliable deduction is precisely that of intuition, *viz.* indubitability or constraint of belief. Even the longest chain of deductions can at least in principle for Descartes be reduced to intuition by increasingly rapid repetitions of its separate links. There are no rules, then, because they would be superfluous; what is required is not an inference ticket legitimating the move from one proposition to another, but directions for placing the mind in such a position that it will of itself move, or be constrained to move, from one proposition to another – and these *regulae ad directionem mentis* Descartes does provide. Nor is this stance an unconscious prejudice on his part. Descartes' complaint against scholastic formal logic is its sterility; what he wanted was more like the natural logics of the sixteenth century, a material logic whose *ars inveniendi* followed from reasoning anchored, as Caton puts it, in the natural illation of ideas reflecting the natural articulation of things.[17]

The third and last hypothesis I shall consider as to why Descartes held the eternal verities to be created might be called *the anti-finalism hypothesis*. In short it is the contention that the doctrine is required in order definitively to expunge consideration of final causes from physics. It has been espoused by a number of people, most notably by Gilson and Gouhier, each of whom garnish it, however, with another consideration, that Descartes had his view of Omnipotence for other reasons as well, for example, because of his association with the Oratory, with Bérulle and especially Gibieuf. But the main point is that since in God there is no distinction between willing and understanding, talk of divine purpose is meaningless, and thus the search for purposes, whether in the whole of nature or its parts taken separately, is to no avail. I think this thesis needs to be modified in light of *Principles* III, 2: "that we ought to beware lest we presume too much in supposing ourselves to understand the ends which God set before Himself in creating the world".[18] But while there are purposes achieved by creation, they are arbitrary and not rooted *in rerum natura* and thus the search for them *in rebus* is no less futile. Only the mechanism of the universe is a proper object of physics. Another part of this hypothesis might be Kenny's thesis that Descartes's rejection of substantial forms entailed the rejection of essences and with them "the element of stability in the flux of phenomena which made it possible for these to be universally valid scientific knowledge."[19] To connect one moment of a thing's history to another Descartes appealed to the immutable will of God, which in creating the eternal truths provides the foundations of physics.

Having argued that force rather than the Cartesians' quantity of motion is preserved, Leibniz felt "obliged to restore certain beings or forms which they have banished". (*Discourse* 18)[20] With these forms of course, comes the utility of final causes, even in physics (*Discourse* 19) – which brings us to the outset of a long story with respect to which I shall limit myself to a few remarks that relate directly to the Leibnizean critique of Descartes. (i) In *Discourse* 22 Leibniz argues the compatibility, indeed the complementariness of mechanical and teleological explanations. "The method of efficient causes . . . goes much deeper and is in a measure more immediate and apriori", but often is lacking and less efficient with respect to details than the method of final causes, "which is easier and can be frequently employed to find out important and useful truths which we should have to seek a long time"[21] through efficient causes. Especially anatomy, but also optics provide examples of the heuristic value of final causes. Not only is Descartes' mechanical derivation of Snell's Law less satisfactory – presumably because less elegant – than Snell's and Fermat's version by final causes, but his method likely would not have yielded the law it did unless guided by Snell's previous discovery of it. (ii) Nor is consideration of final causes merely of heuristic value. In *Discourse* 21 Leibniz argues that without the metaphysical considerations which yield a system of final causes the laws of collision would be other than what they are. The case he has in mind is the one he had argued as early as his *Theoria motus abstracti* presented to the *Académie des sciences* in 1671, *viz.* if bodies were only extension, and motion change of place, then the least body in motion must carry with it the greatest body at rest without diminution of motion, i.e. only speed is relevant. (iii) Finally, the rejection of final causes entails the rejection of the *conspiration universelle* and with it the distinction Leibniz tried to draw between his own view, on the one hand, and the Spinozan rejection of contingency which in this context, as we have seen, hardly differed from Cartesian fatalism. I shall only quote a now famous passage which will indicate the long story I have in mind:

There is something which had long perplexed me – how it is possible for the predicate of a proposition to be in the subject without making the proposition necessary. But the knowledge of geometrical matters, and especially of infinitesimal analysis, lit the lamp for me, so that I came to see that notions too can be resolvable to infinity.[22]

Systematic pursuit of these themes throughout the whole of Professor Weingartner's paper would clearly take me beyond the bounds of my own. Instead, I shall briefly focus attention on the three principles Rescher takes to be at the basis of Leibniz's system. I realize that Weingartner does not regard them as fundamental in the same way, but they or a version of them emerge as sufficiently important in his paper so that a critique of the whole will at least be suggested.

The first of Rescher's fundamental principles to be encountered is what he calls the principle of Identity or of contradiction, the principle, namely, that all finitely analytic propositions are necessarily true. Weingartner reads it as a soundness principle: every proposition which is derivable in the axiomatic system is necessarily true. This is said to be "more a principle of Leibniz than of Descartes, but when the terms are explained one can see that it is also applicable to Descartes."[23] The axiomatic system must be that arrived at by his method, the exact observance of whose rules will ensure that no one ever takes anything false to be true. But for Descartes, while intuition and deduction guided by method yield only truth, the truth arrived at need not be analytic. Rescher's Leibnizean principle of identity can be interpreted as a soundness principle by Weingartner only because for him the axioms are analytic – all of them are identicals. This is an important difference since it means that the ground for the truth of a proposition whose truth analysis preserves is that proposition itself. For Descartes, meanwhile, that ground must be sought elsewhere, *viz*. in the arbitrary will of God.

Rescher's first fundamental principle, the principle of sufficient reason, is that every true proposition is analytic. Weingartner interprets this for Leibniz as a completeness claim for the comprehensive deductive system of scientific knowledge, where scientific knowledge must mean everything that is true. Does Descartes have such a principle? Weingartner believes that while Descartes has neither this nor the later version of the principle, and has no analogous principle of such importance, he nonetheless makes the completeness claim in *Rules*, 4 and 7. But rule 7 only gives directions to ensure the completeness of our science without specifying the sense of its completeness, and rule 4, while it speaks of the method by which we may arrive at the knowledge of all things, does so in a way that suggests our knowledge thus obtained is in fact limited to "all that does not surpass [our] powers".[24] That is, the aim of method is not to get us all that is true, but to keep us from error and get us all we can know, with the result that scientific knowledge is only as complete as it can be.

Rescher's third fundamental principle, that of perfection is that every infinitely analytic proposition is true, i.e. contingently true. This is either a version or a consequence of the principle under which Weingartner elaborates the mathematical order of the universe, the *conspiration universelle*: "in all states and processes of the universe, some property is at a maximum or a minimum." As a principle of the best, *viz*. that in all states and processes of the universe perfection is maximized, this principle is meaningless for Descartes. A universe *a* can meaningfully be said to be most perfect only if there is another, possible universe *b*, which, had it (*b*) been actual, would have been less perfect than *a*. But this condition cannot be satisfied for Descartes, for whom every possible universe, had it been actual, would have been no less perfect than any other actual one. This I take to be a consequence of the view that goodness depends on God in the sense that whatever universe He creates is *ipso facto* good. The Leibnizean corollary that in all states and processes of the universe there is a maximization of existence is similarly meaningless for Descartes. Though compossibility, unlike compatibility, is transitive, the compossibility of substances is in the end a matter of the logical compatibility of their states, so that maximization of existence means all that can logically exist. But for Descartes, what can logically exist depends no less than goodness on the arbitrary divine decree. And similar remarks can be made about the other corollaries of the minimax principle.

Postscript

It might be argued that much of the difference I have tried to draw between Descartes and Leibniz rests on metaphysical issues so ethereal as to be of historical interest only – all this talk of God, after all – and that once Descartes' pathological concern with skepticism, for example, is cured or ignored, much of my case begins to evaporate. That is, once clarity and distinctness are accepted as reliable or ignored as irrelevant, discourse is restricted to the realm of the logically possible and knowledge for Descartes might be regarded *à la Leibniz* as a deductive system which is sound, complete, finitely axiomatizable and mathematical. Put another way, there is no difference of interest between the Leibnizean world and any Cartesian world such as our own in which God is good. Two comments.

(1) For Leibniz "every truth which is not an absolutely first truth can be demonstrated from an absolutely first truth". (*Omnis veritas quae non est absolute prima, demonstrari potest ex absolute prima.*)[25] Not so for Descartes. While not every possible world is logically possible, every logically possible world is Euclidean; however there is more than one logically possible world, one being distinguished from another on the basis of the quantity and distribution of motion, the determination of which is logically arbitrary. While the principles of every logically possible world will be the same, *viz.* extension and the geometry it grounds plus the three laws of nature and the rules of collision derived from them, those principles will be insufficient for the deduction of all that is true of that world. The relevant text is *Principles* III, 46: "We have noted above . . . [he lists some general principles] ; but we could not determine in the same way [i.e. the same *a priori* way] how large the parts are into which this matter is divided, nor what the speed is with which they move, nor what circles they describe. For since these things were ordainable by God in an infinity of different ways, it is through experience alone, and not by the force of reasoning, that we can know which from all these things He chose." Thus the recourse in principle to hypothesis and experiment, the admissibility of assuming in explanations what is in fact false, the requirement only that the deductive consequences agree with the observed phenomena, and so on.

(2) The first half of Weingartner's third chapter, the ideal of mathematization deals with the mathematical language of science. For Leibniz this is the *chatacteristica universalis*, whose primitive terms can all be represented by numbers. Do we find such a language or one analogous to it in Descartes? We do not. All we find is that whatever can be known should have the same certainty and inter-connection as are found in mathematics, but not that it should be mathematically expressible. Extension is only one of the simple natures and there is no suggestion that the others are mappable onto it. The historical *dénouement* was twofold. On the one hand was the materialist attempt to provide just this mapping, but in a way which hardly mobilizes the metaphysics of the *characteristica universalis*. The other tack was taken

by Malebranche, who denied that we have a clear and distinct idea of the self or thought, claiming that the second *Meditation* establishes only the existence and not the nature of the self. To be sure, the self and its creation are intelligible – but not to us. There is a geometry of the mind – but we are wholly ignorant of it. If we had a clear and distinct idea of the mind then instead of having to rely on experience to determine the states of which it is capable, pains for example, we could deduce such states in the way in which the states of matter, of which we do have a clear and distinct idea, are actually deduced. This tack points in the equally non-Leibnizean direction of David Hume.

University of Western Ontario

NOTES

[1] Gallimard, Paris, 1960.
[2] *Ibid.*, p. 529.
[3] *G. W. L. Samtliche*, II, 1, p. 247; Gerhardt I, 368n.
[4] Belaval, p. 11 fn. 3.
[5] *Ibid.*, p. 532.
[6] *AT* I, 149–50.
[7] *AT* VII, 431; *HR* II, 248.
[8] *AT* IV, 118–19.
[9] *V.* 'The Creation of the Eternal Truths in Descartes's System', in Willis Doney (ed.), *Descartes: A Collection of Critical Essays*, Doubleday, Garden City, New York, 1967, pp. 192–208.
[10] *Leibniz: Basic Writings*, Trans., George R. Montgomery, Open Court, La Salle, Illinois, 1968, pp. 4–5.
[11] Frankfurt's actual view on the creation of the eternal verities is that Descartes was led to it because for him there is no difference between God's will and His understanding and thus the scholastic view above, the only alternative to viewing the verities as independent of God, is not open to him. Harry Frankfurt: 'Descartes on The Creation of The Eternal Truths', *The Philosophical Review* **LXXXVI**, 1 (1977), 36–57. I do not find convincing his very brief attempt (n. 7, p. 41) to explain *why* Descartes denied any difference between divine will and understanding, which denial can be regarded as a metaphorical assertion of the creation doctrine. On the other hand, his discussion provides good evidence for his well-known thesis on the Cartesian distinction between truth and rational assertibility.
[12] Harry G. Frankfurt; 'Descartes' Validation of Reason', Willis Doney, *op. cit.* p. 221.
[13] 27 May, 1638; *AT* II, 138.
[14] Re: *Principles* I, 43. *Monadology And Other Philosophical Essays*, Trans., P. & A. M. Schrecker, Bobbs Merrill, Indianapolis, 1965, p. 35.

[15] Re: *Principles* I, 1. *Ibid.*, p. 22.
[16] *Regulae* III, *HR* I, 8.
[17] Hiram Caton: 'Will and Reason in Descartes' Theory of Error', *Journal of Philosophy* 72 (27 February, 1975) 4.
[18] *HR* I, 270.
[19] Anthony Kenny: 'The Cartesian Circle And The Eternal Truths', *Journal of Philosophy* LXVII (1970), 698.
[20] *Loc. cit.*, p. 33.
[21] *Ibid.*
[22] L. Couturat, *Opuscules*, p. 18, cited by Nicholas Rescher, *The Philosophy of Leibniz*, Prentice-Hall, Englewood Cliffs, New Jersey, 1967, p. 43.
[23] *V.* P. [ ].
[24] *HR* I, 9.
[25] *Cf.* Weingartner [ ].

PART IV

P. M. HARMAN

FORCE AND INERTIA: EULER AND KANT'S *METAPHYSICAL FOUNDATIONS OF NATURAL SCIENCE*

The conceptual problems of natural philosophy remained a subject of enduring interest for Kant. Despite shifts in outlook, engendered by a deeper understanding of natural philosophy and the transformation in philosophical perspective which led to the development of the critical philosophy, the conceptual status of 'force' and the relationship between 'force' and 'matter' remained at the core of Kant's treatment of the problems of natural philosophy. These issues pervade the chapters on 'Dynamics' and 'Mechanics' in the *Metaphysical foundations of natural science* (1786), Kant's mature and most systematic discussion of natural philosophy, where he elaborated the ways in which the transcendental categories of the *Critique of pure reason* (1781) were to be applied to the concept of matter. It has become a commonplace to describe Kant's intentions as an attempt to demonstrate the validity of Newtonian physics, but as Buchdahl has emphasised Kant's 'metaphysics of nature' purports to demonstrate links between physical theory and the transcendental principles rather than to claim that the actual inductive validity of Newtonian physics can be derived from *a priori* premises.[1] Moreover, in the *Metaphysical foundations* Kant's examination of the metaphysical foundations of Newtonian physical concepts led him to a reappraisal of the conceptual status of 'force', 'matter' and inertia' in physical theory. The argument of the *Metaphysical foundations* thus has a complex relationship to the conceptual structure of Newtonian natural philosophy.

In this paper I propose to follow a single strand in the complex of problems associated with the status of 'Newtonian physics' in Kant's philosophy, tracing his changing attitude to the relationships between the concepts of 'force', 'matter' and 'inertia' from his early writings on natural philosophy to the mature formulation in the *Metaphysical foundations*. To do so necessarily imposes limitations, both historical and philosophical, but may serve to highlight some important features of Kant's treatment of natural philosophy. In focusing on the conceptual status of 'force' and 'inertia' in Kant's natural philosophy, I shall emphasise the impact of Johann Bernoulli's theory of 'living force' and of Euler's treatment of the conceptual status of 'force' and 'inertia' on Kant's discussions of physical theory.[2] While Kant's discussions of natural philosophy display a marked divergence from the theories of

William R. Shea (ed.), Nature Mathematized, 229–249.

nature formulated by Bernoulli and Euler, he drew important insights from their work which influenced his construal of the problems of natural philosophy. In the *Metaphysical foundations* Kant follows Euler in stressing the disjunction between the concepts of 'force' and 'inertia', but rejects Euler's view that 'inertia' was a defining property of 'matter', maintaining that 'forces' were the defining characteristics of material substances. Nevertheless, the arguments of Bernoulli and Euler helped to shape Kant's treatment of natural philosophy. For Euler, and for the Kant of the *Metaphysical foundations*, the problems posed by Newton's concepts of 'force' and 'inertia' were of crucial importance, and in particular Kant seeks to provide an analysis of the conceptual framework of natural philosophy, the assumptions about 'matter' and 'force' which underlie Newton's statement of the laws of motion and the concept of universal gravitation. Kant's examination of the metaphysical foundations of 'Newtonian' physical concepts led him to a reappraisal of the conceptual status of 'force' and 'inertia' in physical theory, and to reject the 'Newtonian' concept of 'inertia' as a defining property of 'matter' in favour of the ontological priority of 'force' over the concept of 'matter'. In the *Metaphysical foundations* this re-location of the conceptual relations between 'force', 'inertia' and 'impenetrability' was developed under the formal guidance of the categorial principles, and fostered by Euler's critique of Newtonian natural philosophy. Kant's account of the 'metaphysics of nature' yields a framework of principles which differs from the philosophical assumptions of Newton's theory of nature, providing a reappraisal of the conceptual structure of 'Newtonian' natural philosophy.

Before turning to an account of relevant aspects of Kant's early writings on natural philosophy and an examination of Euler's critique of Newton's discussion of the concept of 'force of inertia [*vis inertiae*]' and its impact on Kant's argument in the *Metaphysical foundations*, some preliminary remarks on Kant's notion of 'metaphysical' foundations' are necessary. Kant's treatment of the 'metaphysics of nature' sought to remedy a crucial deficiency in his early writings on natural philosophy, viz. the gap between a mathematical theory of nature and physical reality. In the 'Preface' to the *Metaphysical foundations* he makes a distinction between the general part of the metaphysics of nature (in the *Critique*), concerned with the general 'laws which make possible the concept of a nature in general', and a 'special metaphysics of nature' (the subject-matter of the *Metaphysical foundations*), in which the transcendental concepts and principles are applied to the empirical concept of matter. Kant maintains that the demonstration of the possibility of 'determinate natural things', as distinct from the demonstration of the possibility of the concept of a nature in general, requires the metaphysical 'construction of concepts', to show that mathematics can be applied to 'determinate natural things'. Kant emphasises that mathematical physics 'presupposes metaphysics of nature' because 'principles of the construction of concepts that belong to the possibility of matter in general must precede' and 'make possible the application of mathematics' to the concept of matter.[3] However Kant denies that 'forces' are entities that can be constructed, and while his 'special metaphysics of nature' sought to bridge the gap between physical reality and the mathematical representation of nature, Kant does not claim to establish that matter can be represented by 'mathematical construction', but only purports to show 'the possibility of a mathematical doctrine of nature'.[4]

I

In his *Thoughts on the true estimation of living forces* (1747) Kant was concerned to analyse the arguments about whether 'force' was to be measured by the 'quantity of motion' or by the 'living force [*vis viva*]', a controversy that had occupied leading natural philosophers in the 1720s and was still the subject of debate, Daniel Bernoulli, Euler, Boscovich and d'Alembert all making contributions to the argument in the 1740s. Kant believed that the 'Cartesian' measure of 'force' ('quantity of motion') could be established by mathematical argument, while the 'Leibnizian' measure of 'force' ('living force'), not being subject to the principle of the conservation of force (in Kant's view) was 'found to be false in mathematics'.[5] Nevertheless he maintained that the 'Leibnizian' measure of 'force' provided a true description of physical reality: 'I will maintain that living forces are really to be found in nature' he declared.[6] Kant emphasised the disjunction between mathematics and physical reality, arguing that 'the mathematical concepts of the properties of bodies and their forces are quite distinct from the concepts found in nature'.[7] There was therefore a gap between mathematics and nature, and Kant set himself the task of resolving this problem, maintaining that while 'living forces' would remain concealed from mathematical argument they would be revealed by a 'metaphysical investigation'.[8] By a 'metaphysical' investigation here Kant did not mean the demonstration of a 'metaphysics of nature' in the sense of the *Metaphysical foundations*. While Kant's concern in the *Metaphysical foundations* was to establish that mathematics could be applied to physical entities, in *Living forces* he stresses the disjunction between the 'mathematical' status of 'force' and its ontological status as a 'physical' entity existing in nature. By a 'metaphysical' investigation in the *Living forces* essay Kant means an analysis of the conceptual status of 'force' as a physical entity.

Kant was conversant with the extensive literature on 'living forces' published during the controversies of the 1720s, and his distinctive view of the conceptual status of 'living forces' cannot be understood without reference to these works, notably the authoritative essays by Johann Bernoulli to which Kant refers on several occasions, and which had been reprinted in the collected edition of Bernoulli's works published in 1742.[9] Kant's theory of 'living force' displays striking contrasts to the arguments of Leibniz and Bernoulli. In Leibniz's dynamics the conceptual status of 'living force' and its intelligibility as a dynamical principle was grounded on his theory of substance. He argued that 'living force' was a phenomenal manifestation of the 'primitive active force' which, as he put it in *Specimen dynamicum* (1695), unlike extension and motion was 'real' and substantially present in all corporeal entities. 'Derivative' forces such as 'living force' were the phenomenal manifestations of 'primitive' forces, and the science of dynamics was concerned with 'derivative' forces alone, the forces by which 'bodies actually act and are acted upon by each other'. A 'derivative' force such as 'living force' was thus considered to be a 'well-founded appearance' which was part of the order of nature and subject to the laws of nature but which did not possess the status of 'primitive' forces which characterised the nature of substances.[10]

Leibniz's mathematical argument was based on a distinction between 'living force' and 'dead force [*vis mortua*]', both being conceived as 'derivative' forces. Leibniz maintained that 'dead force' (mdv) arose in the tendency of a body to motion, whereas 'living force' (mv^2) arose in 'actual motion' such as the collision or fall of bodies, and 'arises from an infinite number of continuous impressions of dead force'. 'Living force' was thus to be considered as arising from an infinite number of infinitesimal impulses of 'dead force'. Leibniz emphasised that in referring to infinitesimal quantities such as 'dead force' he did 'not mean that these mathematical entities are really to be found in nature as such, but merely that they are the means of making accurate calculations'.[11]

These remarks in the *Specimen dynamicum* remained among Leibniz's few public statements of his concept of 'living force' and the existential status of infinitesimal entities; and his view of infinitesimals as 'well-founded fictions', as he expressed it in a letter to Varignon,[12] and of the relationship between 'dead force' and 'living force' as a purely mathematical model was not echoed by Johann Bernoulli in his writings on 'living forces'. Substantial portions of Bernoulli's mathematical and philosophical correspondence with Leibniz were published in 1745,[13] and in correspondence with Bernoulli Leibniz had maintained that the theory of 'living force' as arising from successive infinitesimal impulses of 'dead force' was in consonance with the 'law of continuity', which implied that 'motion does not change by a leap' and hence that a body would not change from one state of motion to another without passing through all the intermediate degrees of motion.[14] In formulating his theory of living forces Bernoulli drew directly on Leibniz's arguments, though he rejected Leibniz's view of 'living force' as a 'well-founded phenomenon'.

The conceptual kernel of Bernoulli's treatise on the *Laws of the communication of motion* (1727) was his adherence to the fundamental presupposition of Leibnizian dynamics, the law of continuity, and hence to the principle that the mathematical concept of infinitesimals had its counterpart in nature. The law of continuity established the harmony between nature and mathematics, the 'perfect conformity which prevails between the laws of nature and those of geometry'.[15] Bernoulli thus follows Leibniz in supposing that motion is produced in infinitely small degrees in accordance with the law of continuity, and hence that 'living force' is produced as a result of an infinite number of infinitesimal impulses of 'dead force' and was 'inherent in a body when it is in uniform motion'.[16] Bernoulli's concept of 'living force' was explained mathematically, and his theory of motion contrasts with Kant's disjunction (in *Living forces*) between the mathematical and physical representations of 'force'. Bernoulli did not however repeat Leibniz's claim that the infinitesimal entities posited in Leibniz's dynamics were merely mathematical quantities, and did not possess the ontological status of physical entities. In correspondence with Leibniz, Bernoulli had affirmed the view that infinitesimal quantities were physically existent entities, and given his assumption of the harmony between mathematics and nature, it is likely that he intended to imply the physical reality of 'dead force'.[17] In an essay *On the true concept of living forces* (1735) Bernoulli amplified his account of the theory of living forces, characterising 'living force' as a 'real and substantial entity',[18] a view that conflicts with Leibniz's statement of the ontological status of 'living force' as a 'well-founded phenomenon', an entity which did not possess the status of substance. Bernoulli's dynamical writings elaborate an interpretation of 'living force' and 'dead force' as real physical entities.

In his essay on *Living forces* Kant quoted with approval Bernoulli's formulation of 'living force' as a 'real and substantial entity' persisting in natural bodies. While Bernoulli had 'expressed his meaning as a mere geometer' his argument was nevertheless 'completely intelligible': 'this formulation expresses my view to no small advantage'. However Kant believed that Bernoulli was mistaken in seeking to establish the physical status of 'living force' by appeal to the Leibnizian law of continuity, the mathematical relation between infinitesimals and finite quantities.[19] Kant criticised the Leibnizian distinction between the 'beginnings of motion' and 'actual motion', arguing that 'whatever is valid when a body has been in motion for some time . . . must also apply at the commencement of motion'. Assuming that both the 'forces' possessed by the body in these two states of motion were real physical entities (as Bernoulli had done), Kant turned the 'law of continuity' against Leibniz and Bernoulli, asserting that 'if a body possesses living force when it has been in motion for some time (for however small a time increment), then it must also have that force when it begins to move'.[20] For Kant there was a continuity and hence identity between the 'forces' in these two states of motion; he rejected the conceptual basis of the Leibnizian distinction between 'dead force' and 'living force'. He concluded that the Leibnizian measure of 'force' was disproved by mathematical argument, and that the geometrical arguments of the Leibnizians – Bernoulli, Bilfinger and Hermann – which purported to demonstrate the law of the conservation of 'living force' were invalid.[21]

There are parallels between Kant's critique of Leibnizian theory and arguments advanced by Euler in a paper on *The force of impact and its true measure* (1745), a notable contribution to the debate on the conceptual status of 'force'. Euler notes that according to Leibniz impact occurred in an instant giving rise to 'dead force'; 'living force' then arose from an infinite number of impressions of 'dead force'. The two 'forces' were therefore 'heterogeneous', a term which had been used by Bernoulli.[22] Euler questions the 'heterogeneity' of 'living' and 'dead force' by turning the 'law of continuity' against the Leibnizian argument. According to this 'supremely constant law of nature, in virtue of which nothing occurs suddenly', impact would require a time interval; and so he argues that both 'living' and 'dead force' must arise in the same way, in a time interval, and hence have the same quantitative measure. Euler concludes that 'the fundamental principle, on which rests the measure of living force, is not only unsteady, but falls entirely in ruin; for it is demonstrated that the effect of impact of two or more bodies is not produced in an instant, but that it requires a certain interval of time'. Euler concludes that Newton's first law of motion, the law of inertia, defined the measure of 'force': 'there is no other force but inertia, by which [a body] conserves its state, and this force is always the same, whether a body remains at rest or whether it moves'.[23]

In a very similar fashion Kant concluded that 'mathematics does not permit a body to be endowed with a force . . . that is not the external cause of its motion . . . this is the fundamental law of mechanics . . . for which there is no other measure [of force] than the Cartesian'. Newton's first law of motion thus established the 'Cartesian' measure of 'force' by mathematical argument. However, he went on to claim that a 'natural body' possessed a 'quite different property', the capacity of 'increasing in itself' the 'force' which was the external cause of its motion, so as to attain 'a degree of force that does not arise from the external cause of motion'. This property of physical bodies, Kant maintained, could not be reconciled with the mathematical theory of 'force': 'the mathematical concepts of the properties of bodies and their forces are quite distinct from the concepts found in nature'.[24]

Kant therefore stresses the disjunction between the mathematical status of 'force' and its ontological status as an entity occurring in nature. He explicitly concurs with Bernoulli's characterisation of 'living force' as a 'real and substantial entity' and, by implication, rejects Leibniz's distinction between 'primitive' (substantial) and 'derivative' (physical) 'forces'; and he contests the 'Leibnizian' justification of the 'living force' concept by the appeal to the mathematical relation between infinitesimals and finite quantities. In stressing the disjunction between mathematics and physical reality, Kant faced the question: 'how is it that a law which is found to be false in mathematics can occur in nature?'[25] Thus, the central issue of the essay on *Living forces* became the relationship between mathematics and physical reality. Kant's answer to this question took the form of a 'metaphysical investigation', viz. a conceptual analysis of 'living force' that would demonstrate the intelligibility of the concept as denoting something physical. He declared that 'we must connect the metaphysical laws with the rules of mathematics in order to determine the true measure of force in nature'.[26] To do so he provides an 'estimation of living forces' which, though not purporting to yield a mathematical explanation in the sense of the Leibnizian theory of 'living force' as arising from an infinite number of infinitesimal impulses of 'dead force', nevertheless attempted to provide a mathematical illustration of 'living force' as a physical entity existing in nature.

Kant's 'estimation' of 'living force' is based on his introduction of a concept which he terms 'intension', the 'basis of activity' of physical bodies, representing their 'perpetual endeavour to action', the property of physical bodies which, Kant had maintained, could not be explained by the mathematical theory of 'force'. 'Force' was measured by the product of 'intension' and velocity. At the commencement of motion the striving towards motion ('intension') is indeterminate, so the 'force' is proportional to the velocity (corresponding to the Leibnizian 'dead force'). Once motion is established, the 'intension' is finite, so the 'living force' is proportional to the product of 'intension' and velocity. If velocity and 'intension' are each represented by a line, then the 'force' at the commencement of motion would be represented by a line, but the 'living force' would be represented by a square, corresponding to the square of the velocity and the representation of 'living force' as mv^2.[27] The comparison between 'dead force' and 'living force' as between a 'line and a surface' had been drawn by both Bernoulli and Euler.[28] The difference between the 'force' at the commencement of motion and when motion was established was thus due to the way in which the 'intension' of bodies was manifested, not (as in Leibnizian dynamics) as a result of a disjunction between the two states of motion, a heterogeneity which Kant believed was contrary to the law of continuity.

In Kant's 'estimation' of 'living force' the concept of 'living force' is explained as a physical property of bodies. He juxtaposes the mathematical and physical explanations of 'force', one yielding a 'Cartesian' and the other a 'Leibnizian' measure of 'force'. Kant did not succeed in resolving the controversy over the meaning of 'force', only in highlighting the tension in his own theory of nature, the disjunction between the mathematical definition of 'force' and its ontological status as a physical entity.

Kant's essay on *Physical monadology* (1756) provided an important elaboration and development of his view of the ontological status of 'force'. He now asserted that bodies were constituted of 'monads' or 'simple substances' which possessed inherent 'forces'. Kant's view of the physical reality of the 'force' possessed by the 'monads' constituting physical bodies resembles Wolff's theory of monads as indivisible points rather than Leibniz's theory of monads as 'metaphysical points' characterising the nature of 'substance'.[29] Kant's theory of monads as possessing inherent 'forces' which defined their physical reality is in consonance with his acceptance of Bernoulli's view of 'living force' as a 'real and substantial entity' in his *Living forces* essay; but the introduction of the concept of 'monads' as an explication of the physical status of 'living force' probably reflects his response to the controversy over the Wolffian 'monad' theory in the late 1740,[30] an emphasis that was foreshadowed in his essays on fire and cosmology of 1755, where he enunciated a theory of an 'inherent striving' of nature.[31]

In *Physical monadology* Kant argues that the monads constituting physical bodies possess an inherent 'force of inertia'; and that while the 'inertia' of bodies was due to an inherent 'force', the contact action of bodies was explained in terms of the 'forces of impenetrability'. Kant now introduces the 'Newtonian' notion of attractive and repulsive forces, which he interprets as being inherent in matter and which determined the properties of bodies; the conjunction and degree of these 'forces' determined the limit of the extension of bodies.[32] The use of attractive and repulsive 'forces' here is of fundamental importance in Kant's natural philosophy. In the *Metaphysical foundations* he attempts to clarify the relationship between the concepts of 'force' and 'inertia' under the formal guidance of the categorial principles. While he remained committed to one of the central assumptions of the *Physical monadology*, that of the primacy of 'force' over the concepts of solidity and impenetrability, in the *Metaphysical foundations* he denies that 'inertia' was due to an inherent 'force', in favour of a concept of 'inertia' as denoting the passivity of matter, an argument that was foreshadowed in his *New conceptual study of motion and rest* (1758) where he denied that 'inertia' was 'an inner force of nature'.[33] Now this reappraisal of the conceptual status of 'inertia' reflects Kant's indebtedness to Euler's critique of the Wolffian monadology and of the conceptual status of 'force', 'impenetrability' and 'inertia' in Newton's natural philosophy. Although Kant did not accept Euler's view that 'inertia' was a defining characteristic of matter, his restructuring of the relation between the concepts of 'force' and 'inertia' suggests the influence of Euler's critique of the expression 'force of inertia'. Before turning to an examination of Euler's arguments, I will first briefly review Newton's discussion of 'inertia', to provide a perspective for the analysis of Euler's examination of the intelligibility of Newton's concept of 'inertia'.

II

The main sources of Newton's concept of 'inertia' for eighteenth-century natural philosophers were Definitions III and IV of *Principia* (1687) and the third Rule of Philosophising that Newton appended to the second edition of *Principia* (1713). In Definition III Newton states that 'force innate [*vis insita*] in matter is the power of resisting whereby each individual body, inasmuch as it is in it to do so, perseveres in its state of resting or of moving uniformly straight on', adding that because the 'innate force' did not differ from the 'inertia' of the mass of the body other than in the manner of our conceiving it, the 'innate force' could be called the 'force of inertia [*vis inertiae*]'. In Definition IV Newton added that a body persevered in each state of motion by its 'force of inertia' alone.[34] In Rule III he listed 'inertia' together with extension, hardness, impenetrability and mobility as the 'universal' qualities of matter, the essential properties an entity would have to possess as a necessary and sufficient condition of its materiality.[35] As commentators have noted, Newton's use of the term *force* of 'inertia' and his association of 'inertia' and the 'innate force' of matter bristles with difficulties. Newton referred to 'inertia' both as a passive property of matter and as an inherent active 'force' of matter, an innate power of resisting, actualised when bodies interacted with one another.[36] If Newton considered matter to be totally passive, then the use of the expression 'force of inertia' was ambiguous; yet Newton stressed that a body perserved in each state of motion by its 'force of inertia'. The first law of motion of *Principia*, that 'every body perserves in its state of resting or moving uniformly straight on, except inasmuch it is compelled by impressed forces to change that state',[37] was thus grounded, it seemed, on a concept of 'inertia' that lacked adequate explication.

Whatever Newton's intentions, this is how the issue was construed by Euler, subjecting Newton's concepts of 'force' and 'inertia' to a sharp critique in an essay *On the origin of forces* (1750). Euler emphasises that strictly speaking the concepts of 'force' and 'inertia' were 'directly contrary to one another'; while 'force' was the cause of a change of motion, 'inertia' preserved a body in a state of motion or rest. Hence the expression 'force of inertia' was self-contradictory. Euler however follows Newton in emphasising the importance of impenetrability as a defining property of matter, arguing that impenetrability was as essential a property of bodies as extension and inertia. In his third Rule of Philosophising Newton had distinguished the 'universal' qualities of matter such as impenetrability and inertia from other properties of matter (including gravity) by arguing that the 'universal' qualities of matter were not subject to alteration of degrees of intensity, and Euler again follows Newton in arguing that 'impenetrability does not admit of degrees'. Euler was firmly committed to Newton's concept of matter and to Newton's philosophical criterion by which the doctrine of the 'universal' qualities of matter was established. He concluded that 'if a body is not completely impenetrable, it is penetrable'; impenetrability was an absolute quality of bodies, consisting in 'an absolute impossibility to admit penetration', and hence it was 'impossible for two bodies, or only two of their smallest particles, to exist at the same time in the same place'.[38] Euler claims that there is a close conceptual association between 'force' and 'impenetrability', that 'force' is 'inseparable from' impenetrability. Forces were necessary to maintain the impenetrability of bodies and 'derive their origin from the impenetrability of bodies'.[39] In the *Metaphysical foundations* Kant inverts Euler's assumption that the concept of 'force' was grounded on impenetrability, rejecting Euler's assumptions about the nature of matter, and arguing that 'forces', not extension and solidity, were the defining properties of material substance.

Euler developed his views on 'force' and 'inertia' at greater length in his *Letters to a German Princess* (1768–72), again emphasising that impenetrability was 'a necessary property of all bodies', and defining 'body' as an 'impenetrable extension'. Euler therefore agrees with Newton rather than Descartes in maintaining that while extension was a necessary characteristic of bodies, 'extension and mobility alone do not constitute the nature of bodies' for 'something may be extended without being a body'. Another property was required, that of 'impenetrability', without which 'extension cannot be body'. The concept of 'impenetrability' was fundamental to Euler's theory of motion. He argues that the motion of bodies was changed in collisions 'no more than is necessary to prevent penetration'; because collision was the source of 'all the changes which occur in the world' impenetrability was therefore 'the great spring by which nature works all her effects'.[40]

When Euler turns to a discussion of 'inertia' and its relation to 'force' he states that 'inertia' is 'as necessary to bodies as extension and impenetrability'. 'Inertia' was a defining characteristic of material bodies: 'it would be impossible for a body to exist without inertia'. He defines 'inertia' as 'a repugnance to everything that tends to change the state of bodies', to be contrasted with 'force' which was the 'external cause' which signifies 'everything that is capable of changing the state of bodies'. Unlike 'inertia' which was 'essential' to matter, 'force' was not '*inherent* in matter'. Euler thus rejects the Wolffian monad theory which had held that bodies make a continual effort to change their state by means of an inherent force. That doctrine clearly contradicted his account of 'inertia'. Moreover, the term 'force of inertia' was an abuse of language; as already noted, 'inertia' was 'the opposite of a force'.[41]

Euler then attempts to explain the origin of 'force', for despite the fact that 'force' was not inherent in matter it was apparent that 'forces' were associated with material entities. He finds the explanation in the property of 'impenetrability'; the impenetrability of bodies 'always exerts the force necessary to change their state . . . [so as] to prevent penetration'. Hence the 'impenetrability of bodies' was the source of the 'true origin of the forces which continually change the state of bodies in the world'. Collision, contact action, was the basic mode of change in nature. The 'force' arising from 'impenetrability' was only manifested in resisting penetration: 'it is only to prevent penetration, that impenetrability becomes active', that is, 'provides a force sufficient for the required effect'. While 'impenetrability' was an essential property of body, the 'force' to which it gave rise was dispositional: 'though impenetrability provides these forces' it was not 'endowed with a determinate force; it is rather in a condition to manifest force' necessary in order to prevent penetration. The 'impenetrability' of bodies, a quality necessary to their materiality, would therefore manifest the 'forces' sufficient to prevent the mutual penetration of bodies. The concept of 'force' was thus explicated in terms of the more fundamental property of 'impenetrability': the 'origin of forces is based on the impenetrability of bodies'.[42]

In the *Metaphysical foundations* Kant follows Euler in rejecting the theory that bodies make a continual effort to change their state by means of an inherent 'force', accepting Euler's strictures against the term 'force of inertia' as a linguistic and conceptual confusion.[43] In his essay on *Living forces* he expounded his disjunction between mathematical and physical theories of nature by contrasting the concepts of 'inertia' and 'living force', the former concept expressing a mathematical law while the latter was conceived as an inherent active 'force', a physical entity persisting in natural bodies. Euler had criticised the doctrine that bodies possessed 'internal forces' as contradicting the concept of 'inertia', and in seeking to establish the intelligibility of a unified natural philosophy, bridging the gap between the mathematical and the physical representations of nature proposed in *Living forces*, in the *Metaphysical foundations* Kant rejects the concept of 'living force' in favour of 'inertia'. In the *Metaphysical foundations* he seeks, under the guidance of the categories, to clarify the distinction between the concepts of 'force' and 'inertia', rejecting the theory of 'force' propounded in *Living forces*, that bodies possessed inherent 'activity', a 'perpetual endeavour to action'.[44]

Kant does not however accept the Newton-Euler doctrine that 'inertia' and 'impenetrability' were 'essential' properties of matter. In the *Metaphysical foundations* he conceives 'inertia' as a law of mechanics which implied the passivity of matter, not as an 'essential' or defining property of material substance. As in his essay on *Physical monadology* he conceives the concept of 'force' as ontologically prior to 'impenetrability', and emphasises that 'forces' were the defining properties of matter. While Kant did not accept Euler's views uncritically, Euler's writings on the conceptual foundations of mechanics were clearly instrumental in deepening his understanding of the problems of natural philosophy, bringing about a re-structuring of the relations between the concepts of 'force', 'inertia' and 'impenetrability' in the *Metaphysical foundations*.

III

In the *Metaphysical foundations* Kant elaborates a 'special metaphysics of nature' which considered 'matter' as 'it must be represented in accordance with the universal laws of thought'. The four chapters of the book, 'Phoronomy', 'Dynamics', 'Mechanics' and 'Phenomenology' correspond to the categories of quantity, quality, relation and modality, which subsume 'all determinations of the universal concept of matter in general'.[45] While an early commentator maintained that the terms 'Phoronomy', 'Dynamics' and 'Mechanics' had little significance, merely deriving from Hermann's *Phoronomia* (1716), d'Alembert's *Dynamique* (1743) and Euler's *Mechanica* (1736),[46] this assertion is clearly mistaken. The term *phoronomia* had been used by Leibniz to denote motion as considered merely geometrically, precisely the sense of Kant's 'Phoronomy'.[47] The term *dynamica* has clear Leibnizian echoes, denoting Leibniz's theory of 'forces'; and in the chapter on 'Dynamics' Kant is concerned with the concept of 'force'. The term *mechanica*, however, probably does relate to Euler's treatise, though the connection is a substantive one, for in 'Mechanics' Kant is concerned to discuss the laws of motion, corresponding to Euler's attempt to construct an axiomatic science of mechanics based on the concepts of mass and force.

The discussion of the concepts of 'force', 'impenetrability' and 'inertia' which Kant gives in the chapters on 'Dynamics' and 'Mechanics' in the *Metaphysical foundations*, indicates in a very interesting fashion some of the distinctive features of Kant's 'special metaphysics of nature', in the context of his reconstruction of the relationships between these physical concepts, and his attempt to avoid the disjunction between mathematical and physical representation that had dominated his essay on *Living forces*. The chapter on 'Dynamics' begins with the fundamental 'dynamical explication' of, as Kant emphasises, the 'empirical concept' of 'matter' as 'the movable insofar as it fills a space'. This 'dynamical' treatment of matter has been preceded, and presupposes, its 'phoronomic' explication as 'the movable in space' now supplemented by having 'the capacity of resisting a motion within a certain space'.[48] As I have already observed, Kant rejects the 'Newtonian' doctrine (also espoused by Euler) of 'essential' qualities, that 'solidity' and 'impenetrability' were defining characteristics of material substances. He maintains that 'forces' define the essence of matter, that 'force' is 'inherent in matter'.[49]

The 'dynamical' account of 'matter' supposes that 'matter fills a space, not by its mere existence, but by a special moving force'. Matter fills space by 'repulsive' or 'extensive' forces which resist the penetration of other forces and which thus form the 'ground of its impenetrability'. The property of 'impenetrability' thus 'rests on a physical basis', for the 'extensive force makes matter itself, as something extended filling its space, first of all possible'.[50] Moreover Kant claims that 'the possibility of matter requires a force of attraction, as the second essential fundamental force of matter', a force limiting the dispersive effect of the repulsive force. Kant argues that the 'possibility of matter' requires the assumption of 'an original force of matter acting in the opposite direction to the repulsive [force]', an 'attractive force' which 'belongs to all matter as a fundamental force appertaining to its essence'. Without the attractice force matter would 'disperse itself to infinity'; and without repulsive force 'all parts of matter would approach one another without hindrance and diminish the space that matter occupies'. Hence both attractive force and repulsive force belong to the 'essence of matter', and 'one cannot be separated from the other in the concept of matter'.[51]

This 'dynamical' explication of matter is carried out with an eye to the categories of 'quality' ('reality, negation and limitation').[52] These categories are clearly meant to correspond analogically to the account of 'matter' in terms of attractive and repulsive 'forces', and the limiting or balance between these 'forces' which determined the 'degree of the filling of space' by matter. There is thus the suggestion of a correspondence between these categories and explication of matter in terms of attractive and repulsive 'forces': 'attractive force as a penetrative force . . . in combination with its counteracting one, namely, repulsive force [and] the limitation of this latter'. This 'dynamical' account thus demonstrates the 'possibility of a space filled in a determinate degree' by matter.[53]

Kant rejects the concept of 'absolute impenetrability' (as expounded by Euler), the doctrine that 'matter, insofar as it is matter, resists all penetration unconditionally and with absolute necessity'. In place of this Newtonian 'mathematico-mechanical' concept of matter Kant proposes his 'metaphysico-dynamical' mode of explanation, which supposes that 'impenetrability rests on a physical basis', being grounded on the 'extensive' or repulsive force. He argues that this force 'has a degree which can be overcome, and hence matter's space of extension can be diminished' by a 'compressive force' counteracting the repulsive force. Hence the filling of space by matter must be dependent upon the 'degree of compression' or 'relative impenetrability' of matter. Thus the 'extension' and 'impenetrability' of matter could change by degrees of intensity, both being accounted for in terms of the concept of 'force'. In contrast with Newton's doctrine, according to which 'impenetrability' could not change by degrees, and impenetrability, like extension and solidity, was an absolute quality defining the materiality of matter, Kant maintains that impenetrability and extension are derivative, being dependent on the fundamental 'forces' which define the concept of 'matter'. As I have already pointed out, Kant thus inverts Euler's treatment of the relationship between 'force' and 'impenetrability'.[54]

In emphasising that 'the possibility of matter requires a force of attraction' Kant arrives at a conclusion of fundamental importance, the justification of the possibility of the action of matter at a distance, a concept which (as Kant remarks) had given such offence to Newton's contemporaries and even to Newton himself. Kant sought to justify the notion of 'action at a distance' as being 'possible without the mediation of matter lying in between', thus demonstrating the intelligibility of Newton's theory of gravitational attractive force. The justification of the possibility of the concept of gravity is given in terms of the 'metaphysico-dynamical' concept of matter, which supposes that 'matter occupies a space without filling it' by means of an attractive force, by which 'matter acts through empty space upon other distant matters'.[55]

It should be noted in particular that Kant's 'metaphysico-dynamical' approach is not in any way intended to yield any specific empirical laws of 'forces' by which nature is regulated, nor is it claimed that the physical actuality of the 'forces' follows deductively from the general categorial scheme. The metaphysical argument establishes only the possibility of matter in general as composed of fundamental 'forces' of attraction and repulsion, not that one can 'presume to assume either of them as actual'. Kant emphasises that 'one must guard against going beyond what makes the universal concept of matter in general possible and against wanting to explain a priori the particular or even specific determination and variety of matter'. He declares that 'no law whatever of attractive or of repulsive force may be risked on a priori conjectures; but everything, even universal attraction as the cause of gravity, must, together with the laws of such attraction, be concluded from data of experience'.[56]

Kant rejects the view of 'Lambert and others', presumably including Euler, who considered 'solidity' to be a defining property of matter, an ultimate concept which entailed the resistance of matter to penetration. He argues that the proponents of this 'mathematico-mechanical' theory suggested that 'solidity' was a concept which could not be 'further constructed' because it was 'an initial datum of the construction of the concept of matter', and that they implied that 'solidity' was 'incapable of any mathematical construction in order thereby to prevent a return to the first principles of natural science'.[57] Kant maintains that his 'metaphysico-dynamical' explication of 'matter' has exposed the confused nature of this argument, by demonstrating that 'solidity' and 'impenetrability' were not ultimate concepts but were derivative, being dependent on the 'fundamental forces'. While he admits that the 'mathematico-mechanical mode of explication' had the advantage of explaining a 'great specific multiplicity of matters', he regards the conceptual foundations of this schema as being unsatisfactory. Kant maintains that 'force' is the 'initial datum' of construction, and he argues that the 'metaphysico-dynamical' mode of representation contrasted favourably with the 'mathematico-mechanical' theory, for the latter was a 'merely mathematical physics' based on the 'empty concept' of 'absolute impenetrability', a conception of nature in which 'forces were philosophised away'. In the 'mathematico-mechanical' theory 'forces' were arbitrarily superadded to matter whereas the 'metaphysico-dynamical' mode of representation was based on the 'proper forces of matter' which were conceived as being essential to matter.[58]

Nevertheless Kant was careful not to claim too much for his 'metaphysico-dynamical' argument. He denies that his 'dynamical' explication established the 'possibility of fundamental forces' because 'all means are wanting for the construction of this concept [of force]'. Kant argues that 'forces' cannot be constructed because they are 'fundamental' and 'cannot be further derived from any source', and also because they are empirical and are known from the 'data of experience' and cannot be exhibited *a priori.*[59] 'Forces' can therefore only be 'assumed', and Kant does not claim to have established a mathematical physics based on the supposition of 'fundamental forces'. In seeking to establish the intelligibility of a unified natural philosophy based on the 'fundamental concept' of 'force', bridging the gap between physical reality and the mathematical representation of nature proposed in the essay on *Living forces*, Kant denies that 'forces' could be constructed, and thus emphasises that the metaphysical argument only establishes the 'possibility of a mathematical doctrine of nature'.[60]

In the next chapter of the *Metaphysical foundations*, on 'Mechanics', Kant gives an account of the relationship between the concepts of 'force' and 'inertia', demonstrating the application of the categories of 'relation' ('subsistence', 'causality' and 'community') to the concept of 'matter'.[61] The principle of 'causality', as enunciated in the 'second analogy' of the *Critique*, states (as Kant puts it in the *Metaphysical foundations*) that 'every change has a cause'. Applied to the concept of 'matter' it yields Kant's 'second law of mechanics: every change of matter has an external cause'. Kant places brackets round his appended statement of Newton's first law of motion to which this metaphysical 'law of mechanics' analogically corresponds, implying that the metaphysical argument merely suggests the possibility of Newton's first law of motion, not its physical actuality. Nor does his 'proof' of the 'second law of mechanics' purport to validate Newton's first law of motion, but to establish that all changes of matter are 'based on an external cause' and that matter 'undergoes no changes except by motion', postulating a link between the category of 'causality' and Newton's first law of motion. Newton's first law of motion is therefore construed as the empirical embodiment of the principle of 'causality'.[62] The metaphysical argument thus only establishes the 'possibility of a mathematical doctrine of nature'.

Kant describes his 'second law of mechanics' as the 'law of inertia', because the 'inertia of matter is and signifies nothing but its lifelessness', its inability to 'determine itself to motion or rest as change of its state'. He refers to the capacity of a substance to change its state by an 'internal principle' as 'hylozoism', the capacity of living things to determine to act by means of thought, and he argues that only living things could manifest a 'positive effort' to change or maintain their state. He concludes that the doctrine that matter possesses an 'internal principle' of 'inertia', the concept of a 'special force of matter under the name of the force of inertia', would entail 'hylozoism' and would therefore contradict the 'law of inertia'. Kant's rejection of the supposition of an 'internal principle of a substance to change its state' was fundamental to his proof of the 'law of inertia': because 'matter has no absolutely internal determinations' hence 'all change of matter is based upon an external cause'.[63]

Kant therefore is an agreement with Euler in rejecting the construal of 'inertia' as a 'special force of matter'. He maintains that the 'designation force of inertia' was self-contradictory: 'a special and entirely peculiar force merely to resist, but without being able to move a body, would under the name force of inertia be a word without meaning'. The expression 'force of inertia', employed in his essay on *Physical monadology*, was therefore abandoned. To suppose that matter possesses an 'internal principle' of 'inertia', which he considers to be implied by the term 'force of inertia', would contradict the 'law of inertia' which entails the denial that matter possesses an inherent power enabling it to change its state. Kant's interpretation of the connotations of the term 'force of inertia' in the *Metaphysical foundations* does however give a misleading impression of his use of the concept of 'force of inertia' in his essay on *Physical monadology*, where 'hylozoism' (as defined in the *Metaphysical foundations*) was not implied.[64]

Kant's argument illustrates his reconstruction of the concepts of his earlier period, following his re-interpretation of the relationship between 'force' and 'inertia' under the guidance of the categorial principles. In the *Metaphysical foundations* his 'metaphysico-dynamical' explication of 'force' as the ontological basis of 'matter' replaces his 'physical monadology' in which bodies were supposed to be constituted of 'monads' endowed with inherent 'forces'. The notion of inherent activity came to be associated strictly with living things; and the concept of an 'internal principle' or 'force of inertia' was abandoned together with the concept of 'living force' as an inherent active 'force' persisting in physical bodies. The 'law of inertia' thus entailed the denial of the principle that the cause of any change of matter could be 'internal'.[65]

Nevertheless this did not imply an acceptance of the Newton-Euler doctrine that 'inertia' was a defining property of 'matter'. Kant regards the concept of 'inertia' as a 'law of mechanics' which implied the passivity of matter, not as a defining property of matter. He retains his commitment to the ontological priority of 'force' over the concept of 'matter', though in the *Metaphysical foundations* he expounds this theory in terms of his 'metaphysico-dynamical' explication of 'fundamental forces'. Kant's shift from the assertion of his 'physical monadology' to his formulation of his 'metaphysico-dynamical' theory of matter thus demonstrates his reconstruction of the relationship between the concepts of 'force' and 'inertia' under the guidance of the categorial principles; whilst his reinterpretation of these concepts manifests his changed approach to natural philosophy under the influence of Euler.

ACKNOWLEDGEMENTS

I wish to express my gratitude to Gerd Buchdahl for several helpful discussions during the preparation of this paper, for his detailed comments on the draft, and for suggesting some valuable improvements.

I also wish to thank the Royal Society for financial support in obtaining source materials during the preparation of this paper, and the British Academy for an overseas conference grant enabling me to attend the Third International Conference on the History and Philosophy of Science in Montreal in August 1980, where this paper was presented.

University of Lancaster

NOTES

1 G. Buchdahl, *Metaphysics and the philosophy of science: the classical origins. Descartes to Kant*, Blackwell, Oxford, 1969, p. 678.
2 H. E. Timerding: 'Kant und Euler', *Kant-Studien* 23 (1919), 18–64.
3 I. Kant, *Metaphysische Anfangsgründe der Naturwissenschaft* [1786], in *Kant's gesammelte Schriften herausgegeben von der Königlichen Preussischen Akademie der Wissenschaften*, Vol. 4, Reimer, Berlin, 1911, pp. 469–72. I have used the recent translation by J. Ellington of Kant's *Metaphysical foundations of natural science*, Bobbs-Merrill, Indianapolis and New York, 1970. Quotations from the *Metaphysische Anfangsgründe* in this paper are taken, with occasional modifications, from this translation; and all page references are to the *Akademie* edition, page numbers in that edition being given as marginal numbers in Ellington's translation.
4 Kant, *Schriften*, Vol. 4, pp. 473, 476. See G. Buchdahl: 'The conception of lawlikeness in Kant's philosophy of science', *Synthese* 23 (1971), 24–46.
5 I. Kant, *Gedanken von den wahren Schätzung der lebendigen Kräfte* [1747], in *Kant's gesammelte Schriften*, Vol. 1, Berlin, 1902, p. 139. For especially relevant background on the *vis viva* controversy, see: T. L. Hankins: 'Eighteenth-century attempts to resolve the *vis viva* controversy', *Isis* 56 (1965), 281–97; R. Calinger: 'The Newtonian – Wolffian confrontation in the St. Petersburg Academy of Sciences (1725–1746), *Journal of World History* 11 (1968), 417–35; and C. Iltis: 'Madame du Châtelet's metaphysics and mechanics', *Studies in History and Philosophy of Science* 8 (1977), 29–48.
6 Kant, *Schriften*, Vol. 1, p. 59.
7 *Ibid.*, p. 107.
8 *Ibid.*, p. 60.
9 *Ibid.*, pp. 57, 91, 150. Bernoulli's writings on 'living force' were collected in the third volume of his *Opera Omnia*, 4 vols., Bousquet, Lausanne/Geneva, 1742.
10 G. W. Leibniz: 'Specimen Dynamicum', *Acta Eruditorum* 14 (1695), 145–57; in Leibniz, *Mathematische Schriften*, 7 vols. C. I. Gerhardt (ed.), reprinted: Olms, Hildesheim, 1960–61, Vol. 6, pp. 235–7; trans. by L. E. Loemker in Leibniz, *Philosophical papers and letters*, D. Reidel Publ. Co., Dordrecht, Holland, 1969, pp. 435–444.
11 Leibniz, *Mathematische Schriften*, Vol. 6, p. 238.
12 Leibniz to Varignon, 20 June 1702, *Mathematische Schriften*, Vol. 4, p. 110. *Cf.* J. Mittelstrass, *Neuzeit und Aufklärung: Studien zur Enstehung der neuzeitlichen Wissenschaft und Philosophie*, de Gruyter, Berlin, New York, 1970, pp. 489–501.
13 *Got. Gul. Leibnitij et Johan Bernoullij commercium philosophicum et mathematicum*, 2 vols., Lausanne/Geneva, 1745. A more complete collection of the Leibniz-Bernoulli correspondence is in Leibniz, *Mathematische Schriften*, Vol. 3, parts 1 and 2.
14 Leibniz to Bernoulli, 20/30 September, 1698, *Mathematische Schriften*, Vol. 3, p. 544.
15 J. Bernoulli: 'Discours sur les loix de la communication du mouvement' [1727], *Opera omnia*, Vol. 3, p. 58.
16 Bernoulli, *Opera omnia*, Vol. 3, p. 23.
17 Bernoulli to Leibniz, 6 December, 1698 and 7 January, 1699, in Leibniz, *Mathematische Schriften*, Vol. 3, pp. 555 and 563. On Bernoulli's relation to Leibniz see

P. M. Heimann: ''Geometry and nature': Leibniz and Johann Bernoulli's theory of motion', *Centaurus* **21** (1977), 1–26.

18 J. Bernoulli: 'De vera notione virium vivarum earumque usu in dynamicis' [*Acta Eruditorum* (1735), 210–30], *Opera omnia*, Vol. 3, pp. 239–41. *Cf.* Heimann, *op. cit.*, for a full account of Bernoulli's theory of 'living force'.

19 Kant, *Schriften*, Vol. 1, pp. 150–1. *Cf.* E. Adickes, *Kant als Naturforscher*, 2 vols., Berlin, 1924–25, Vol. 1, p. 98. While Adickes correctly contrasts Kant's and Bernoulli's theories of 'living force', he fails to note Kant's reference to Bernoulli, which has a significant bearing on the relationship between them. The account of Kant's 'living forces' paper by Buchdahl, *op. cit.*, (note 1), pp. 553–6 provides a valuable analysis, but ignores Kant's relation to Bernoulli, and hence gives a slightly misleading account of the status of Kant's 'living force' concept.

20 Kant, *Schriften*, Vol. 1, pp. 34, 139.

21 J. Hermann: 'De mensura virium corporum' and G. Bilfinger: 'De viribus corporum moto insitis', *Commentarii Academiae Scientiarum Imperialis Petropolitanae* **1** (1728), 1–42 and 43–121. See Kant, *Schriften*, Vol. 1, pp. 43–5, 79–84.

22 Bernoulli: 'Discours sur le mouvement', *Opera omnia*, Vol. 3, p. 37.

23 L. Euler: 'De la force de percussion et de sa véritable measure', *Mémoires de l'Académie Royale des Sciences de Berlin* **1** (1745, [published 1746]), 21–53 (on pp. 29–31).

24 Kant, *Schriften*, Vol. 1, pp. 107, 140.

25 *Ibid.*, p. 139.

26 *Ibid.*, p. 107. *Cf.* Buchdahl, *op. cit.* (note 1), pp. 553f.

27 Kant, *Schriften*, Vol. 1, pp. 26, 141–8.

28 Bernoulli: 'Discours sur le mouvement', *Opera omnia*, Vol. 3, p. 37; Euler: 'Force de percussion', *op. cit.* (note 23), p. 29. *Cf.* Leibniz to de Volder (undated, late 1698), in *Die Philosophischen Schriften von G. W. Leibniz*, 7 vols., C. I. Gerhardt (ed.), reprinted: Olms, Hildesheim, 1960–61, Vol. 2, p. 154, comparing 'living force' to 'dead force' as a 'line' to its 'elements'. Leibniz had compared 'living force' to 'dead force' as a 'surface to a line' in a manuscript, *Mathematische Schriften*, Vol. 6, p. 121.

29 J. Ecole: 'Cosmologie wolffienne et dynamique leibnizienne', *Les études philosophiques* **19** (1964), 3–10.

30 I. Polonoff, *Force, cosmos, monads and other themes in Kant's early thought*, Bouvier Verlag Herbert Grundmann, Bonn, 1973, pp. 77–89; R. Calinger: "The Newtonian – Wolffian controversy (1741–1759)', *Journal of the History of Ideas* **30** (1969), 319–30, and also *idem*: 'Kant and Newtonian science: the pre-critical period', *Isis* **70** (1979), 349–62. Euler had taken a strongly anti-Wolffian stance in these controversies, in his *Gedanken von die Elementen der Cörper*, Berlin, 1746, so in the mid-1750s Kant's position was anti-Eulerian, a view he was ultimately to invert.

31 Kant, *Schriften*, Vol. 1, p. 226. On these issues see S. Schaffer: 'The phoenix of nature: fire and evolutionary cosmology in Wright and Kant', *Journal for the History of Astronomy* **9** (1978), 180–200.

32 I. Kant: 'Monadologia Physica', *Schriften*, Vol. I, pp. 473–87.

33 I. Kant: 'Neuer Lehrbegriff der Bewegung und Ruhe' [1758], *Schriften*, Vol. 2, p. 20.

34 D. T. Whiteside, ed., *The mathematical papers of Isaac Newton*, Vol. 6, Cambridge University Press, Cambridge, 1974, p. 93.

[35] I. Newton, ***Mathematical principles of natural philosophy***, ed. and trans. A. Motte and F. Cajori, University of California Press, Berkeley, 1934, p. 399. *Cf.* J. E. McGuire: 'The origin of Newton's doctrine of essential qualities', *Centaurus* **12** (1968), 233–60.

[36] *Cf.* A. Gabbey: 'Force and inertia in seventeenth century dynamics', *Studies in History and Philosophy of Science* **2** (1971), 1–67; R. S. Westfall, *Force in Newton's physics: the science of dynamics in the seventeenth century* Macdonald, London, 1971, pp. 448–56; E. McMullin, *Newton on matter and activity*, University of Notre Dame Press, Notre Dame/London, 1978, pp. 33–43.

[37] Whiteside, *op. cit.* (note 34), p. 97.

[38] L. Euler: 'Recherches sur l'origine des forces', *Mémoires de l'Académie des Sciences de Berlin* 6 (1750 [published 1752]), 419–47; in *Leonhardi Euleri opera omnia*, series II, Vol. 5, J. O. Fleckenstein (ed.), Orell Füssli, Lausanne, 1957, pp. 112–5.

[39] *Ibid.*, pp. 116, 131.

[40] L. Euler, *Lettres à une princesse d'Allemagne sur divers sujets de physique et de philosophie*, 3 vols., St. Petersburg, 1768–72; *in Leonhardi Euleri opera omnia*, Series III, Vol. II ed. A. Speiser, Orell Füssli, Zürich, 1960, pp. 150–3.

[41] *Ibid.*, pp. 161–6.

[42] *Ibid.*, pp. 167–71.

[43] *Cf.* Timerding; *op. cit.* (note 2), p. 50.

[44] Euler, *op. cit.* (note 40), p. 172; Kant, *Schriften*, Vol. 1, pp. 26, 141.

[45] Kant, *Schriften*, Vol. 4, pp. 473, 476, 477.

[46] *Gehler's physikalishes Wörterbuch neu bearbeitet*, 10 vols., Leipzig, 1825–44, Vol. 2, p. 715.

[47] G. W. Leibniz, *Theoria motus abstracti*, Mainz, 1671; in Leibniz, *Mathematische Schriften*, Vol. 6, p. 71. *Cf.* R. Palter: 'Kant's formulation of the laws of motion', *Synthese* **24** (1972), p. 111.

[48] Kant, *Schriften*, Vol. 4, pp. 472, 480, 496.

[49] *Ibid.*, p. 502.

[50] *Ibid.*, pp. 497, 502, 508.

[51] *Ibid.*, pp. 508, 509, 511.

[52] *Ibid.*, p. 523.

[53] *Ibid.*, pp. 518, 519.

[54] *Ibid.*, pp. 501, 502, 525.

[55] *Ibid.*, pp. 508, 511, 512, 516.

[56] *Ibid.*, pp. 524, 534, *Cf.* Buchdahl, *op. cit.* (note 4).

[57] Kant, *Schriften*, Vol. 4, pp. 497, 498.

[58] *Ibid.*, pp. 498, 524, 525.

[59] *Ibid.*, pp. 513, 525, 534.

[60] *Ibid.*, pp. 473, 524.

[61] *Ibid.*, p. 551.

[62] *Ibid.*, p. 543. *Cf.* Buchdahl, *op. cit.* (note 1), pp. 676–8.

[63] Kant, *Schriften*, Vol. 4, pp. 543, 544, 549.

[64] *Ibid.*, pp. 549, 550, 551.

[65] *Cf.* Buchdahl, *op. cit.* (note 1), p. 556.

KATHLEEN OKRUHLIK

KANT ON THE FOUNDATIONS OF SCIENCE

The most common view of Kant's *Metaphysical Foundations of Natural Science* (*MFNS*)[1] is that is represents an explicit effort by Kant to provide an *a priori* justification for the Newtonian world view. It is usually argued that Kant attempted to provide this justification by deducing the main tenets of Newtonian mechanics directly from his critical philosophy.

In fact, it is not surprising that this view of the relationship between Kant and Newton should be the dominant one. Both the quasi-geometric mode of presentation which Kant affects in *MFNS* and certain striking similarities of phrasing between Newton's three laws of motion and Kant's three laws of mechanics tend to promote and support this interpretation. I shall argue, however, that these formal and verbal similarities conceal rather than reveal Kant's true intentions. In particular, they disguise the extent to which Kant rejects major aspects of the Newtonian natural philosophy. Consequently they make it difficult for the reader to see *MFNS* for what it really is: one stage in a lifelong struggle by Kant to combine in a single unified system the conceptual advantages of a force ontology with the empirical successes of mechanism.

I shall explicate and defend this claim in the early parts of this essay. Subsequently, I shall examine Peter Harman's claims[2] regarding Kant's debt to Euler and shall argue that caution is also required when attempting to determine the extent to which Kant has appropriated Euler's insights. Here again it will be shown that even the most strikingly similar verbal formulations may mask a radical divergence in meaning.

I. PRELIMINARY CONSIDERATIONS

In the following sections of this essay we shall examine some very interesting and rather sophisticated reasons for denying the claim that Kant's *MFNS* was an attempt to deductively establish the laws of Newtonian mechanics. The arguments there will concern the basic ontological and methodological commitments of Newton and Kant, respectively, as well as their different conceptions of lawlikeness. Before undertaking this more subtle task, however,

William R. Shea (ed.), Nature Mathematized, 251–268.

it would be well to dispense at the outset with some of the grosser misunder–standings of Kant's natural philosophy.

Suppose we set aside temporarily considerations of underlying methodological and ontological commitments and deal with Newton's theory at face value – simply as a collection of laws which 'describe the phenomena'. Suppose we strip our account of Newtonian matter theory to its positivistic essentials and consider only the three laws of motion and the inverse square law for universal gravitiation. Can we then argue that, in this very modest sense of Newtonian theory, Kant was attempting to provide an *a priori* deduction of its validity? Even a cursory reading of *MFNS* reveals that the answer to this question is 'No', and that there are several good reasons for rejecting the common view of the Kantian project.

Consider first the inverse square law for universal gravitation. Kant nowhere argues that this law expresses a necessary truth, but instead is quite specific abount maintaining the contrary position. Necessity characterizes only conceptual truths, and the truth of the inverse square law can be confirmed only by experience. Kant indicates this by discussing the attractive force of matter, not in the chapter called 'Mechanics' where the proper modality is that of necessity, but in the 'Dynamics' where *actuality* is the appropriate modality.

Though it is clear that Kant sometimes yearns after an *a priori* deduction of the inverse square law, he reluctantly admits that no such derivation is possible. The proper measure of the attractive force can be determined only empirically. Froces must be encountered in nature; they cannot be constructed *a priori*.

Consider, secondly, Kant's three laws of mechanics. Though these *are* meant to express necessary truths (unlike the inverse square law), it is important to note that they are not identical with Newton's three laws of motion. The three propositions discussed by Kant are as follows:[3]

(1) With regard to all changes of corporeal nature, the quantity of matter taken as a whole remains the same, unincreased and undiminished.

(2) Every change of matter has an external cause.

(3) In all communication of motion, action and reaction are always equal to one another.

Note first that nothing here corresponds to Newton's second law in either its modern ($F = ma$) or original ($F = \Delta mv$) form. In his discussions, Kant treats Δmv as the true measure of force but nowhere gives this identity the status of a law. In its place, so to speak, he substitutes the conservation of matter. It is true that Kant's second and third laws bear some resemblance to Newton's first and third laws, respectively. They are not, however, identical to them. Notice, in particular, what a huge gap there is between saying that every change of matter has an external cause, on the one hand, and stating on the other that every body perseveres in its state of rest or rectilinear uniform motion unless acted upon by an external force. There is nothing in Kant's second law of mechanics which tells us what counts as a 'change of matter'. It was exactly the task of coming to see that rectilinear uniform motion does not constitute a change of matter that had cost the seventeenth century so much pain and that had ultimately become its greatest triumph. Although Kant immediately glosses his second law in terms of Newton's first, he offers no derivation of the latter from the former.

This is also the point, of course, at which Newton's very famous definition of the "force of inertia" becomes important:[4]

> The *vis insita* or innate force of matter is a power of resisting, by which every body, as much as in it lies, continues in its present state, whether it be of rest or of moving uniformly ahead in a straight line. This force is always proportional to the body whose force it is, and differs nothing from the inactivity of mass, but in our manner of conceiving it. A body, from the inert nature of matter, is only with difficulty changed in its state of rest or motion. Upon which account, this *vis insita* may by a most significant name be called the force of inertia. But a body only exerts this force when a body impressed on it endeavors to change its condition. And the exercise of this force may be considered as both resistance and impulse: *resistance*, insofar as the body opposes the impressed force in order to maintain its present state; *impulse*, insofar as the body endeavors to change the state of another body by not easily giving way to the impressed force of the other . . .

This is one of the most discussed passages in the *Principia*,[5] in large measure because it reveals so poignantly the tensions inherent in Newtonian mechanics and in seventeenth-century matter theory generally.

These tensions (which will be discussed more fully in the next section) revolve chiefly around the problem of how to reconcile the allegedly passive character of matter with with some of its apparently active properties. In order to guarantee the applicability of mathematics to bodies in motion, it seemed necessary to construe matter simply in terms of geometric extension. Such a treatment, however, could not account for other properties of matter – such as impenetrability and inertia. In the definition just quoted, we find Newton struggling to portray inertia as "nothing [different] from the inactivity of mass" as conceived of by the reigning mechanical philosophy. But, of course, by the end of the definition, it becomes very clear that *vis inertiae* is something quite different, as indicated by R. S. Westfall:[6]

If matter is endowed with an inherent force by which it resists efforts to change its state, it cannot be wholly inactive, or in the classic phrase of the seventeenth century wholly indifferent to motion. Indeed, Newton's formulation of the laws of mechanics implied a conception of matter at once inert and active, unable to initiate any action itself, passively dominated by external forces, but endowed with a power to resist their actions.

In other words, *vis inertiae* is not just the power to persevere in inertial motion, but is also a *resisting force*. This resisting force does not operate during inertial motion, but is exerted only when a force acts to change the motion of a body. It appears to function both as a source of reaction in its own body and as an agent responsible for motion in any body which collides with it. Newton sometimes treats *vis inertiae* as if it were commensurable with impressed force and employs a parallelogram of forces to work out the resultant force.

Kant, of course, denies that the persistence of inertial motion requires a cause at all; only changes of state must be causally explained. Unlike Newton, Kant has no backdrop of absolute space against which states of rest and inertial motion could be distinguished in principle. Remaining in a state of rest does not qualify as a causal sequence; so neither does persevering in a state of rectilinear uniform motion. Hence Kant has no need for *vis inertiae* in its conserving role. His law of inertia is nothing more than a denial that matter has any power to determine its own state. Newton's law, on the other hand, seems to depend crucially on just such a power.

The situation is even more striking when we consider Newton's and Kant's third laws, respectively. Verbally these two laws are very nearly identical; but once again quite different things are going on beneath the surface. In Newton's case, the third law (like the second law) is underwritten by *vis inertiae* – this time in its guise as resisting force. The equality of action and reaction refers to a sort of fictive equilibrium between any impressed force and the force of resistance.

For Kant, the third law can refer to no such equilibrium of forces because he denies that there could be a special power to resist that could not also initiate change. Rather, for Kant, the third law is simply a straightforward consequence of the relativity of space. When two bodies interact, change of relation is entirely reciprocal. The motion of a body A twoard another body B at rest must be considered as if both bodies participate equally in the motion – although in common appearence the motion is attributed entirely to A. So we divide the velocity which we might normally attribute to A between the two bodies in inverse proportion to their masses. Therefore, the momenta of the two bodies are equal in magnitude but opposite in direction.

On Kant's understanding, the equality of action and reaction depends entirely on the relativity of the spatial framework within which interaction occurs and not at all on any innate force of resistance possessed by individual particles of matter. The verbal similarity between Kant's third law of mechanics and Newton's third law of motion disguises two very different conceptions of the nature of bodies and their interactions. This will be discussed in somewhat more detail when we examine Euler's understanding of the properties of matter in general and of inertia in particular.

We have seen so far that Kant nowhere in *MFNS* attempts to provide an *a priori* deduction of Newton's laws of motion or the inverse square law of universal gravitation. There is still, however, a more fundamental reason for rejecting the view that Kant was attempting to provide an *a priori* validation for Newtonian mechanics. This is because he was not attempting to provide an *a priori* proof for any physics – Newton's, his own, for anyone else's. Science, according to Kant, stands in no need of *a priori* proof for its own safety or certainty; for it "rests . . . upon experience and its thorough confirmation."[7] Scientific theories are established inductively upon the empirical evidence, not deductively by reference to metaphysical categories.

The success of scientific practice is not something that Kant must guarantee or defend; rather, it is something he must *explain*. It is his starting point, his explanandum. The question to which he addresses himself is *not*: How can we guarantee the unassailable status of Newtonian mechanics? Rather, Kant must answer questions of the following sort: How is science possible? Why should nature be amenable to mathematization? How is it possible that we are able to anticipate nature through laws?

If the critical philosophy should prove to be unable to answer these questions, Newtonian mechanics would not be undermined. It would be no more and no less well-confirmed than if the enterprise were to succeed. Rather the explanatory failure would reflect badly on the critical philosophy itself. Kant was perfectly well aware of this order of dependence, even if some of his readers are not.

Kant proposes to answer the question about the possibility of science by appealing to the categories and to his analysis in the first *Critique* of the conditions which make knowledge possible. In order, however, to appreciate the argument in *MFNS*, we must not view it simply as a (somewhat unsubtle) application of the architectonic machinery to the empirical concept of matter, but also as a continuation of Kant's lifelong struggle to create an adequate theory of matter. In particular, we must view it (at least in part) as an attempt to determine the proper relationship between two competing approaches to matter theory: the metaphysical/dynamical on the one hand, and the mathematical/mechanical on the other.

II. DYNAMISM VS MECHANISM: LEIBNIZIAN ASPECTS OF KANT'S MATTER THEORY

The most important consideration leading Kant to reject crucial aspects of the Newtonian world picture was his dissatisfaction with the "merely mechanical" approach to matter theory and his subsequent preference for dynamism. Throughout his long academic career Kant maintained that the mathematical/mechanical approach fails to do justice to the actual concrete existence of physical bodies and that it therefore must be supplemented with a metaphysical analysis. The distinction between the two forms of explanation comes from Leibniz. And in some very fundamental respects, Kant's matter theory is more Leibnizian than it is Newtonian.

Recall that Leibniz had strenuously and effectively criticized the mechanistic theory of matter. He had attacked it on many fronts; but only two (closely related) criticisms will concern us here. First, Leibniz appreciated more keenly than any of his contemporaries the chief difficulty afflicting seventeenth century analyses of matter. It was in fact, the same problem which plagued Newton's definition of *vis inertiae*: In order that matter might be mathematized, it was conceived of as purely passive extended stuff. So long as matter had no properties except those derived from its geometric extension, one could be sure that its behavior was describable in terms of geometric laws. But the problem then was how to explain the *activity* of matter: its apparent ability to cause and resist change of motion. Clearly, activity could not be injected directly into passive extension without thereby violating its very nature and definition. On the other hand, it was not clear that one could simply superadd forces from outside and still account for all the properties of matter, especially its impenetrability and resistance to change. Leibniz maintained that the whole effort was confused – the result of a conceptual muddle.

So for the mechanistic ontology of passive extension he substituted a radically new force ontology. He maintained that force is prior to extension and that it is in fact *constitutive* of matter, not something superadded to it from outside. Physical bodies and the forces which govern them are merely phenomenal manifestations of the substantial forces which are real and metaphysical in character. These forces ground and make possible the so-called primary properties of the mechanical philosophy which no longer have to function as unexplained explainers.

This brings us directly to Leibniz's second and more fundamental criticism of the reigning mechanical philosophy: that it does not go far enough in its explanatory endeavors. Leibniz was, in a certain sense, the most mechanistic of philosophers. He believed that there is a mechanical explanation for *every* change in matter, including human and other animal motions; but he also believed that the mechanical laws themselves stand in need of explanation:[8]

> ... [T]here is in matter something else than the purely Geometrical, that is, than just extension and bare change. And in considering the matter closely, we perceive that we must add to them some *higher or metaphysical notion, namely, that of substance, action, and force*; and these notions imply that anything that *is acted on* must act reciprocally, and *anything which acts must receive some reaction* ... For although I an convinced that everything in corporeal nature is done mechanically, I also continue to maintain that the very principles of Mechanics, that is, the first laws of motion, have a more sublime origin that those furnished by pure Mathematics.

Leibniz believed that the dynamic approach to matter theory provided part of the required explanation of mechanical laws; but the force ontology alone was not enough. It was part of a larger picture which included the Principle of Perfection and certain architectonic principles in accordance with which God ordered and created the centers of metaphysical force (i.e. the monads).

Now these two Leibnizian criticisms of the mechanical philosophy are obviously very closely related. I have separated them simply to emphasize one general and one particular aspect of Leibniz's proposal for an adequate theory of matter: in general, it is necessary that mechanical laws be grounded in something 'more sublime' then mere mathematics. And, in particular, one aspect of this 'grounding' must be a force ontology. That is, a dynamical theory of matter must be substituted for (on at least must provide the metaphysical underpinning for) the mechanistic theory of matter.

When Kant distinguished the metaphysical/dynamical from the mathematical/mechanical approach and gave priority to the former, he was carrying on the Leibnizian research tradition. We shall see that the metaphysical/ dynamical approach was Kant's preferred mode of explanation almost from the beginning of his career and that it remained so until the very end. It is this preference that led him to reject the basic ontology of Newtonian mechanics and to replace it with something very different. Kant's dynamism survived even the Copernican Revolution, and the critical philosophy was used in *MFNS* to legitimate his pre-existing preference for dynamism. His presentations of dynamics and mechanics were both transformed when they took their place within the critical philosophy; but their basic relationship to one another remained unchanged, as did Kant's affinity for the former.

From his youth Kant argued that mathematical principles are insufficient to give us a full understanding of the properties of matter. In the *Living Forces*[9] essay of 1747, he set out to demonstrate that *vis viva*, although it is not a mathematical property of matter, is nevertheless perfectly real. *Vis viva* is not determinate and is not, therefore, mathematizable. It is neither conserved nor an essential property of bodies. However, *vis viva* is *actual*, and any adequate theory of matter must come to grips with it. Kant announced at the outset that it was his purpose in this essay "to establish the doctrine of living forces once and for all".[10]

He went on to criticize the Leibnizians, including Jean Bernoulli, for having become too mathematical and for having neglected metaphysics. He argued that their accounts of the nature of bodies had become distorted because they overemphasized the mathematical, abstract properties and had neglected their concrete nature as active entities in the world. In doing so, they had lost sight of Leibniz's great discovery: that mathematics is not enough and that it must be supplemented with metaphysics.

In particular, Kant argued that *vis viva* cannot be accounted for mathematically and that, therefore, all purported geometric 'proofs' of its existence or conservation are illegitimate. Only a metaphysical argument for the existence of *vis viva* is possible. Harman has already given a brief summary of Kant's proposed solution to the *vis viva* problem; so I shall not repeat it here. We should, however, draw attention to some very important characteristics of Kant's position in the *Living Forces* essay.[11]

At this stage in his intellectual development, Kant thought of mathematical and metaphysical properties as co-existing side by side. Mathematization fails to capture all real properties of matter, and its explication can be rendered complete only by adding a metaphysical analysis. It is not the case in this essay that the metaphysics *grounds* the mathematics; rather it *complements* it. And in some ways Kant even seemed to indicate that the mathematical properties are the more fundamental properties of matter. He said, for example, that *vis viva* is only accidental, whereas the mathematical properties are essential.

Kant never in late life gave up the belief that both mathematics and metaphysics are indispensable for an adequate understanding of material substance. One important change, however, did occur *within* the dualistic framework. At the time of the *Physical Monadology*[12] (1756), Kant altered in a fundamental way the relationship between the metaphysical and mathematical approaches. He made the former *prior to and foundational for* the latter rather than complementary to it.

This transformation was part and parcel of Kant's thoroughgoing acceptance of dynamism. As Barbara Jill Buroker has pointed out,[13] Kant – although he did lip service to dynamism in the early days – cannot be called a full-fledged dynamist prior to the *Physical Monadology*. This is because prior to that essay Kant dealt almost entirely with *derived* forces – i.e., with those which affect already existing bodies and cause them to intereact with one another. Until 1756, he had not made forces to be actually *constitutive* of the bodies themselves. This is exactly what he did in the *Physical Monadology*. Here he did not simply start with fully constituted bodies and then discuss the mathematical laws governing the interaction among them. Rather he explained the mechanical properties of the bodies themselves in terms of the original constitutive forces. He explained impenetrability, cohesion, density, and elasticity in terms of two original forces – one repulsive and the other attractive. This same sort of explanatory scheme was still present thirty years later in *MFNS*.

Chapter two of that work is entitled "Dynamics", and Kant describes there an ontology of forces similar in most major respects to the force ontology of the *Physical Monodology*. The forces described there are the forces responsible for the very existence of matter; they are neither derived from other bodies nor impressed from outside. Chapter three is called "Mechanics", and presupposes the dynamical properties of matter as described in the previous chapter. It is only here that Kant deals with impressed forces; i.e., with forces which govern the interactions of bodies in physical space. He maintains that the original forces which constitute matter make possible these interactions.

The scientist who proceeds in the merely mechanical way does not bother to discover the real nature of matter as constituted by attractive and repulsive forces. Instead he simply *posits* the existence of entirely empty spaces and absolutely solid particles of whatever geometric shapes the phenomena seem to require. His mode of procedure is rather like that of the mathematician who can proceed synthetically because he is not bound by the concrete natures of physical existents, only by the demands of mathematical consistency.[14] For this reason the mechanical mode of explanation "has, under the name of the atomistic or corpuscular philosophy, always maintained its authority and influence on the principles of natural science."[15] Kant freely admits that:[16]

. . . the mathematico-mechanical mode of explication has over the metaphysico-dynamical mode an advantage that the latter cannot provide, namely, from a completely homogeneous material, by means of the manifold shape of the parts, with empty intermediate spaces interspersed, this mathematico-mechanical mode can accomplish a great specific multiplicity of matters, according to their density as well as their mode of action (if foreign forces be superadded). For the possibility of the shapes as well as of the empty intermediate spaces can be proved with mathematical evidence.

This appears to have been Newton's methodology when he simply assumed that God created perfectly hard atoms possessed of various geometric properties and placed them in the void of absolute space. The advantage of the mathematico-mechanical mode is that one can move ahead quickly on a hypothetical basis to do physics with perfect confidence that the properties which one is attributing to matter are mathematically constructible.

The price which a merely mathematical physics must pay for this advantage is, however, very high:

(1) "It must lay at its foundation an empty concept (that of absolute impenetrability)",[17] and it must presuppose the existence of completely empty spaces which, of course, can never be the objects of experience.

The best (and only) defense of empty space is that it must be presupposed as a necessary condition for the explication of the different degrees of the filling of space. But Kant's dynamical approach provides an alternative explication of varying density by showing how all spaces can be thought of as full yet as filled in varying measure, thereby stripping empty space of its alleged necessity. We ought, therefore, to give up our belief in the existence of the absolute void and should embrace instead the ontology of original attractive and repulsive forces. Similar considerations apply to the concept of absolute impenetrability.

(2) The person who embraces the mathematico-mechanical mode of explication "must give up all the proper forces of matter."[18]

We have seen already the problems one encounters when one strips matter of all but geometric properties and then tries to reintroduce active properties later on when "complete indifference to motion" fails to do justice to the phenomena. Dynamism avoids these difficulties by building the active properties of matter in at the ground floor.

(3) Finally a merely mathematical science gives too much rein to the imagination because it does not pursue rigorous explanation far enough.

Proper explanations in physics are explanations in terms of forces. The mechanical philosopher explains changes of motion among a group of bodies by appealing to impressed forces. But when called upon to explain how these impressed forces themselves can operate, he must give up the explanatory ghost far too quickly. Instead of explaining the operation of impressed forces in terms of original constitutive forces, as the dynamist does, the mechanical philosopher builds a world "merely according to fancy out of fullness and emptiness."[19] Kant rails against this approach:[20]

Everything that relieves us of the necessity of having recourse to empty spaces is an actual gain for natural science. For these give far too much freedom to the imagination to supply by fiction the lack of intrinsic knowledge of nature. Absolute emptiness and absolute density are in the doctrine of nature approximately what blind chance and blind fate are in metaphysical science, namely, a barrier for the investigating reason, with the result that either fiction occupies the place of reason or else reason is lulled to sleep on the pillow of occult qualities.

Though dynamism can never provide an *a priori* construction of the original forces of nature, it can explain given forces in terms of a smaller number of more fundamental forces; and that is all we can legitimately ask of it.

III. SYSTEMATICITY AND LAWLIKENESS: THE IMPACT OF THE CRITICAL PHILOSOPHY ON KANT'S THEORY OF MATTER[21]

The preceding section shows that the *MFNS* represents a continuation of Kant's persistent effort to provide a metaphysical foundation for the laws of mechanics by grounding them in an ontology of forces. This part of the Kantian project carries over largely intact from the pre-critical period. We have already seen that much of the analysis had been worked out 30 years previously in the *Physical Monadology*.

In the meantime, however, Kant had written the first *Critique*, and 'metaphysics' had acquired a whole new meaning. To provide a metaphysical foundation for mechanics now meant to explain the conditions which make a science of mechanics possible. This involves reference to the conditions of knowledge, not simply to the micro-structure of independently existing matter.

In *MFNS* the old and new notions of metaphysics come together in a very complex way, and both kinds of "metaphysical foundation" are being provided for mechanics. It is impossible to appreciate the complexity and structure of the argument in *MFNS* without bearing this fact always in mind. The old founding of mechanics in dynamics is maintained, but is subsumed under, and re-interpreted in terms of, the new metaphysical project. The 'Dynamics' underwrites the 'Mechanics', and the categorial framework underwrites the entire scientific enterprise so construed. The contribution of the critical philosophy to a re-interpretation of Kant's matter theory occurs at a number of levels.

First, the architectonic machinery provides a (somewhat artificial) organizing device for Kant's presentation of his matter theory. Secondly, it allows Kant to defend the dynamical approach by arguing that although the fundamental forces cannot be constructed *a priori*, they are subsumable under the categories and *are* possible objects of experience (unlike perfectly hard, impenetrable atoms floating in the void of absolute space). Finally, and most importantly, the critical philosophy provides a new conception of lawlikeness.

Lawlikeness is no longer discovered in nature, but is *ascribed to* nature.[22] It does not emerge from below (so to speak) from the properties of indivdual atoms or corpuscles, but is imposed from above by the scientific activity of human beings. The necessary connections among events arise from the conceptual structure we impose upon them in our efforts to organize experience. Nomological necessity applies only to *relations* among bodies, not to individual particles of matter considered in total isolation from the community of matter.

Kant's three laws of mechanics in the third chapter of *MFNS* are meant to explain the possibility of a science of mechanics by showing that it is possible for the human knower to impose a certain sort of system on the interactions of corporeal bodies. Each law puts a different constraint on the systematizing activity of the human scientist. The first law tells us that we cannot allow interactions in which matter either comes into or goes out of existence. The second law forbids us to attribute the cause of any body's change of state to a power within that same body. And the third law rules out one-way interactions among bodies. Kant refers to these as the laws of the subsistence, lifelessness, and reaction of matter, respectively.[23]

The first two laws guarantee that we can successfully treat matter as a closed system of non-self-determining stuff. Matter has only external determinations and it is conserved; therefore, mathematics is applicable. The third law guarantees that we have *one* system, subject to a unified treatment.

It is precisely because matter *can* be thought in this way that the science of mechanics is possible. The principles of the subsistence, lifelessness, and reaction of matter correspond exactly to the categories of substance, causality, and community. So the categories applied to the empirical concept of matter (as given in the 'Dynamics') yield the possibility of mechanics. This is not to say that the categories dictate any unique set of particular mechanical laws; they simply explain how a science of mechanics is possible, why mathematization works.

The foregoing considerations will allow us to better understand the difference between Kant's and Newton's "laws of inertia". Newton's law depends crucially on a power attributed to individual particles of matter. Kant's law amounts to a simple restraint on the system-making activity of the scientist when considering the *relations* among moving bodies.

It is at this point that Kant thinks he is finally able to reconcile the advantages of dynamism and mechanism, respectively. We have discussed the advantages of dynamism at great length. Its chief drawback in Kant's eyes was that it seemed to promote hylozoism, thus precluding the applicability of mathematics to matter theory. If one were going to make matter essentially active, then the temptation was always there to attribute to matter some power to change its own state (in the form of "endeavor", *conatus*, or even "intension"). The results were disastrous, and the mechanical approach to matter theory enjoyed much greater empirical success despite the conceptual difficulties attending it.

What Kant is saying in *MFNS* is that we can consistently employ a force ontology without falling into this trap. We can endow matter with original attractive and repulsive forces without thereby giving it any power whatsoever to change its own state. Inertia is not a dynamical property as the younger Kant and many others had thought. It does not derive from the repulsive force of matter and does not refer to a property of the individual particles of matter considered separately. Instead it refers to the lifelessness which we ascribe to matter in order to make it a system. This analysis makes it possible for Kant to maintain the advantages of dynamism without endangering the laws of mechanics.

The critical philosophy has thus allowed Kant to justify and re-interpret his pre-existing preference for dynamism, while at the same time explaining the possibility of a science of mechanics. He has completed his original project of accounting for the success of Newtonian mechanics. In the process of doing so, however, he has thrown out much of Newton's ontology and has introduced a radical new conception of lawlikeness.

IV. KANT AND EULER: ON BANISHING THE 'FORCE OF INERTIA' FROM PHYSICS

Let us turn now to the relationship between Kant and Euler in order to see whether it is any more straightforward than the relationship between Kant and Newton. Harman has argued persuasively that Kant was greatly influenced by Euler, in particular, that Kant's treatment of the so-called force of inertia owes a great deal to him. This is undoubtedly true and will not be in dispute here (though I would not go so far as some commentators have in stressing Kant's debt to Euler).[24] What I should like to argue, however, is that once again verbal agreement may be misleading and may cause us to neglect significant differences between the two positions.

Euler and Kant are agreed that the phrase "force of inertia" is linguistically unacceptable. It contains within itself an explicit contradiction. (Interestingly enough, the Wolffians had criticized Leibniz's *vis passiva* along exactly the same lines.) They are also agreed that matter is not innately sluggish and that matter does not strive always to change its own state.

Beyond this, however, they part company. For Euler, inertia is a sort of *internal cause* operating on bodies. The word 'force', however, is reserved for *external causes*; and that is why the expression 'force of inertia' is inappropriate.[25]

Despite his rejection of this misleading phrase, however, Euler's concept of inertia seems to be much like Newton's. It is an essential property possessed by every particle of matter, however small. Inertia is also quantifiable and proportional to the quantity of matter possessed by the body. The inertial behavior of macro-bodies is a direct result of the inertia of their smallest parts. The law of inertia and the law of action and reaction are both founded upon this essential property. Recall that for Newton *vis inertiae* as a conserving force was central to the first law and as a resisting force was central to the third law.

Though Euler is no longer willing to call it a "force", inertia still seems to function in both capacities for him as well. It maintains uniform rectilinear uniform motion[26] and it actively resists change of state by opposing itself to any impressed force.[27] Euler does call inertia a *cause*[28] and a *power.*[29]

Kant differs from Euler on all these counts:

(1) Inertia is not for Kant an essential property of bodies. The law of inertia does not arise from the collective inertia of all micro-particles considered jointly. Rather, the law of inertia is a sort of prescription which directs us in our search for physical causes. It tells us always to look outside a body for the cause of any change in its state.

(2) Inertia does not signify a cause or a power for Kant; it only signifies lifelessness.

(3) Kant does not say that inertia is proportional to quantity of matter. He has no '*quantum in se est*' clause. Indeed it is hard to see how he *could*

say any such thing. After all, lifelessness appears to be an absolute property which does not admit of degrees. One body cannot properly be said to be more lifeless than another. So it would follow that a more massive body cannot properly be said to have more Kantian inertia than a less massive body.

(4) Kantian inertia does not cause uniform rectilinear motion because such motion does not require a cause, whether internal or external. Neither does it signify any sort of resistance to impressed force: "Nothing but the opposite motion of another body can resist a motion, but this other's rest can in no way resist a motion. Here, then, inertia of matter, i.e., mere incapacity to move of itself, is not the cause of a resistance. A special and entirely peculiar force merely to resist, but without being able to move a body, would under the name of a force of inertia be a word without any meaning."[30] It is very significant here that Kant completely severs the law of inertia from the law of action and reaction. For Newton and Euler, the single property of inertia had been at the heart of both laws. For Kant, on the other hand, the law of action and reaction was based entirely on the relativity of spatial frameworks.

(5) This same spatial relativity points to another profound difference of opinion between Kant and Euler on the subject of inertia. For Kant there is no actual backdrop of absolute space against which the motion of bodies occurs. We can only refer motion to larger and larger *relative* frameworks. Therefore, it is impossible to distinguish even *in principle* between rest and rectilinear uniform motion. Thus, we are anything but arbitrary when we employ the same rubric ("remaining in the same state") to describe resting bodies, on the one hand, and bodies which persevere in a state of rectilinear uniform motion, on the other. Yet Euler argues that the idea of a state is "of itself arbitrary".[31] He deduces two distinct laws of inertia, one for bodies at rest and one for bodies in uniform motion. This appears to be entirely appropriate for him because Eulerian space is absolute. Therefore, there is a real ontological difference between rest and uniform motion in a straight line; and the idea of a state must refer to a conjunction of the two distinct axioms.

(6) There is a final consideration which strikingly illustrates the divergence between the roles played by inertia in Kant's and in Euler's natural philosophy. Recall that Kant's law of inertia was intended to preclude from mechanics the existence of indeterminate forces (such as his own youthful 'intension'). Such indeterminate quantities would not be amenable to mathematization, and a science of mechanics would not be possible. Contrast this with Euler's treatment of impenetrability. Though he maintains that

extension, impenetrability, and inertia are all essential properties of matter, he suggests that impenetrability is the *most* fundamental property and that inertia can (in some very loose sense) be deduced from impenetrability which is the mainspring by which nature accomplishes all her works. Yet impenetrability is not a determinate quality: "[I] t is only to prevent penetration that impenetrability becomes active, and supplies a force sufficient to oppose it. When, therefore, a small force suffices to prevent penetration, impenetrability exerts that and no more; but when a great force is necessary for this purpose, impenetrability is ever in a condition to supply it. Thus, though impenetrability supplies these powers, it is impossible to say that it is endowed with a determinate force . . . "[32]

The contrast could not be more startling. For Kant inertia is the guarantor of determinateness; for Euler it is dependent upon the exercise of a power which is itself indeterminate.

The foregoing should be sufficient to suggest that Kant's and Euler's concepts of inertia are very different although the language they use to banish the 'force of inertia' is similar. We ought, therefore, to be cautious in evaluating the role that Euler played in the development of Kant's thought. There is no doubt that he was extremely influential, but it would be just as mistaken to infer that Kant was attempting to prove that *a priori* truth of Eulerian mechanics as it is to assume that he was attempting to establish the truth of Newtonian physics on *a priori* grounds. (Harman does not make this mistake; my difference with him is only one of emphasis.)

When Kant borrows an idea from someone else, he almost invariably transforms it to suit his own purposes. Everything is grist for his own special mill. The misleading feature of this whole procedure lies in the extent to which he maintains verbal agreement with his original sources, while beneath the surface everything has changed. This insight makes it clear that *MFNS* is a richer and more complicated document than it might first appear. It is not just a somewhat hokey, faintly disreputable deduction of Newton's laws from a heavily theory-laden *Critique of Pure Reason*. It is instead Kant's attempt to locate his long-cherished beliefs regarding the proper relationship between dynamics and mechanics (metaphysics and mathematics) within the critical philosophy. Viewed in this light *MFNS* becomes a much more creative and interesting endeavor than many would have guessed.[33]

University of Western Ontario

NOTES

[1] I. Kant, *Metaphysical Foundations of Natural Science*, trans. by J. Ellington, Bobbs-Merrill, Indianapolis, 1970. (This translation is based on volume four of the Königliche Preussische Akademie der Wissenschaften edition of Kant's works. Page numbers in that edition will be given in parentheses.)

[2] Elsewhere in this volume.

[3] *MFNS*, pp. 102–117 (541–553).

[4] I. Newton, *Mathematical Principles of Natural Philosophy*, trans. by Motte and Cajori, Definition III.

[5] Two of the most helpful discussions are to be found in: R. S. Westfall, *Force in Newton's Physics: The Science of Dynamics in the Seventeenth Century*, American Elsevier, New York, 1971, and E. McMullin, *Newton on Matter and Activity*, University of Notre Dame Press, 1978.

[6] Westfall, *ibid.*, p. 450.

[7] I. Kant, *Prolegomena to any Future Metaphysics*, ed. by Beck, Bobbs-Merrill, Indianapolis, 1950, p. 75.

[8] G. W. Leibniz: 'Whether the Essence of a Body Consists in Extension', *Journal de Savans*, June 18, 1691. Reproduced in *Leibniz Selections*, P. Wiener (ed.), Scribner's, New York, 1951, pp. 100–102.

[9] I. Kant, *Gedanken von der wahren Schätzung der lebendigen Kräfte und Beurtheilung der Beweise* (etc.), Königsberg, 1747.

[10] Cited in I. Polonoff, *Force, Cosmos, Monads, and Other Themes of Kant's Early Thougt* (sic), Bonn, Bouvier Verlag Herbert Grundmann, 1973, p. 43.

[11] The ensuing discussion is based largely on Polonoff, *ibid.*, especially Chapter 1.

[12] I. Kant, *Metaphysicae cum geometria iunctae usus in philosophis naturali, cuius specimen I. continet monadologiam physicam*, Königsberg, 1757.

[13] Barbara Jill Buroker: 'Kant and the Dynamical Tradition: A Case Study of the Role of Matter in Explanation', *PSA 1972*, Schaffner and Cohen (eds.), D. Reidel Publ. Co., Dordrecht, Holland.

[14] Cf. Kant's discussion of the difference between the methods of mathematics and philosophy toward the end of the first *Critique* (A 713 / B 741 and following).

[15] *MFNS*, p. 91 (533).

[16] *MFNS*, p. 79 (525).

[17] *MFNS*, p. 80 (525).

[18] *MFNS*, p. 80 (525).

[19] *MFNS*, p. 78 (524).

[20] *MFNS*, p. 90 (532).

[21] This section is primarily a response to a question asked by Robert Butts at the conference where this paper was read.

[22] The best discussion of this and related topics is to be found in G. Buchdahl, *Metaphysics and the Philosophy of Science*, Blackwell, Oxford, 1969, Chapter VIII.

[23] *MFNS*, p. 114 (551).

[24] Cf. Y. Elkana: 'Scientific and Metaphysical Problems: Euler and Kant', *Methodological and Historical Essays in the Natural and Social Sciences*, Cohen and Wartofsky (eds.), D. Reidel Publ. Co., Dordrecht, Holland, 1974, pp. 277–305.

[25] L. Euler, *Letters of Euler on Different Subjects in Natural Philosophy Addressed to a German Princess*, Harper and Brothers, New York, 1843, vol. I, p. 252. (Hereafter this work will be referred to as *LGP*.)
[26] *LGP*, vol. I, p. 251.
[27] *LGP*, vol. I, p. 251.
[28] *LGP*, vol. I, p. 252.
[29] *LGP*, vol. I, p. 260.
[30] *MFNS*, p. 114 (551).
[31] L. Euler, *Theoria Motus Corporum Solidorum*, Chapter II, §§78–84. Reproduced in *The Concepts of Space and Time*, M. Čapek (ed.), D. Reidel Publ. Co., Dordrecht, Holland, 1976, pp. 111–119.
[32] *LGP*, vol. I, p. 263.
[33] I wish to thank J. R. Brown for his assistance in preparing this essay.

VLADIMIR KIRSANOV

NON-MECHANISTIC IDEAS IN PHYSICS AND PHILOSOPHY: FROM NEWTON TO KANT

The evolution of the world picture after the appearance of Newton's *Mathematical Principles of Natural Philosophy* is remarkable for many reasons. Without any doubt this work of Newton became the source for all further investigations in mechanics, physics and astronomy which created modern science. However, the process of mastering the heritage of Newton was so closely connected with social, political and cultural factors that some important features of Newton's understanding were distorted in order to satisfy them. As Leon Rosenfeld aptly remarked, not before the middle of the XVIIIth century there "emerged a somewhat streamlined newtonianism, from which all the deeply significant hesitations of the creator's thought, arising from his intimate insight into the difficulties of the subject, had been ironed out by less subtle epigons".[1] These "deeply significant hesitations" concerned, as a rule, fundamental notions in physics as space, time, matter and force as well as concepts and ideas closely connected with them.

Two directions as a way for the search of initial cause are essential for Newton's mechanics – reducing mechanics to geometry and reducing action at a distance to the interaction between matter and aether. I should like to note that both of these directions remain fundamental nowadays. The general relativity with its equivalence between gravity and characteristics of space is a direct result of investigations in this first direction. If we substitute the word 'aether' by the word 'field', we shall come to the second set of problems characterizes science today. I want to emphasize that Newton's aether is not the kind of aether of Michelson or Lorentz, but it is a vague idea about presence in science and universe, something that cannot be described within mechanical frames and cannot be discovered by the experimental technique of a given time. We called field by name of matter and did not get so much from the epistemological point of view. It is clear that such a matter should have nothing to do with matter in common sense. In Newton's opinion we called matter something absolutely immaterial and made progress in the description of phenomena only. Our time differs from Newton's only by our ability to create, to record and to describe fields. Here, I should not like to go far 'into the problem of difference beween explanation and description, so that I should only press the point that Newtonian tendency to come back

William R. Shea (ed.), Nature Mathematized, 269–276.

again and again to the question about 'immaterial' agents of material interactions had led-at least-to the creation of the modern field theory.

Let us remember the basic characteristics and functions of Newton's aether. First of all, it was essentially non-mechanistic. In a draft letter to Leibniz, Newton even wrote that aether might be "a substance in which bodies move and float without resistance and which therefore has no *vis inertiae*, but acts by other laws than those that are mechanical".[2] The concept of aether was an important hypothesis by means of which Newton tried to explain a rather broad scope of problems – from the transmission from the sense organ to brain to the problem of gravity. His use of aether for the explanation of action at a distance was especially important.

"It is inconceivable", Newton wrote to Bentley, "that inanimate brute Matter should, without the Mediation of somthing else, which is not material, operate on, and effect other Matter without mutual Contact (. . .) that one Body can act upon another at a Distance thro' a Vacuum, without the Mediation of anything else, by and through which their Action and Force may be conveyed from one to another, is to me so great Absurdity, that I believe no Man who has in philosophical Matters a competent Faculty of Thinking, can ever fall into it".[3] This expression, cited so often, is by no means an accidental delusion of Newton. Newton occupied himself with the idea of aether for more than 30 years (from his first memoir of 1675 and his letter to Boyle (1679) to the second English edition of his *Optics* of 1717). Unpublished papers of Newton edited by Halls and McGuire also approved this opinion.

Newton understood his aether as a substance which fills space elsewhere – both in the space surrounding bodies and within bodies, in their pores. The structure of this aether allows the existence of other forces than attractive ones. Newon wrote (Query 31): the "attractive Force (of aether particles) can reach but to a small distance from them (. . .) where Attraction ceases, there a repulsive Virtue ought to succeed".[4] On the other hand, the size of these particles is also very important. The smaller these particles are, the greater forces appear between them. Newton states (Query 21) that these "Particles are exceedingly smaller than those of Air, or even than those of Light: The exceeding smallness of its Particles may contribute to the greatness of force by which those Particles may recede from one another, and thereby make that Medium exceedingly more rare and elastic than Air".[5] Thus, aether fills all space, and there are repulsive forces between its particles. These forces are quite different from attractive ones which act between material bodies. They are not proportional to the mass, but become greater when the size of a particle decreases.

This concept of Newton might probably be the consequence of his early views on the nature of space which he understood as the possibility of existence of material bodies. In a paper entitled "On the Gravity and Equilibrium of Fluids" he defined a body or a substance by its action upon other bodies, and considered it as a part of space which was endowed with certain qualities. Extention as a part of space is the potentiality of figure in space: "We firmly believe that the space was spherical before the sphere occupied it, so that it could contain the sphere; and hence there are everywhere spaces that can adequately contain any material sphere, it is clear that space is everywhere spherical. And so of other figures".[6] It is quite possible that further evolution of Newton's concept of space as the possibility of existence of bodies led him to the idea that space is the possibility of existence of forces, since he divided later the notions of body and aether.

Rosenfeld's expression cited above properly describes the situation arising in the process of mastering the heritage of Newton. I dare say that the works of the founders of modern science already contained all the initial ideas and concepts which characterize science – I mean physics – today. Nonclassical science began at the moment of birth of classical science, from the time of Descartes and Newton. I should like to emphasize not the modernistic interpretation of many vague and therefore capacious expressions and assertions, but the tendency to speculate in nonclassical terms.

I think that the existence of such a tendency was partly stipulated by deep religious feelings of the founders of modern science. These feelings which Einstein later called "cosmic religion", were identical for them with faith in God. Scholars of the XVIIthe century tried to separate the system of mechanical laws from God. Nevertheless, His existence was a guarantee of existence of reasons and laws which were quite different from mechanical ones. It guaranteed the possibility of other explanations than those that were accepted in mechanical frames.

Creation of a fundamental system is always connected with the going beyond that science within which the newly invented system begins to operate. Nowadays, a creation of such a system is connected with an encroachment into ethics, morals and other domains which are not subjected to logical analysis. At the time of Newton religion overlapped with all these fields. And Newton's deep religiousness was a ground for his freedom of creative power.

It is interesting to note that in the second half of the XVIIIthe century there is a strengthening of the tendency to exclude from the world picture those traits which so essentially complemented it in the works of Newton.

Central force physics became an all-embracing system which excluded the possibility of speculations concerning the essence of things and the source of natural laws. In working out and spreading this opinion French scholars played the leading role and it is easy to see that at the same time French social and philosophical thought acquired clear anticlerical and antireligious character. Voltaire, an atheist, became a propagandist of Newtonian ideas on the continent, and mechanics, restricted by hard frames of logic and experiment, became a doctrine opposing the religious concept of the world. The system of classical mechanics acquired the status of final truth, which closed to some extent the way of studying the fundamental problems of science. One of the causes of the inadequate interpretation of Newton's ideas and concepts lies in this self-restriction of science, which was partly caused by its opposition to religion. For many scientists of the XVIIIth and XIXthe centuries this parting with religion meant the loss of freedom of creative power.

XVIIIthe century showed also the final partition of science and philosophy. It turned out that many problems, being purely scientific ones but too difficult for immediate solving, were treated as purely philosophical. At the same time, Laplace announced his ban on arbitrary speculations about the structure of atoms, and philosophy turned to the investigation of an *a priori* structure of matter. The attitude to this problem determined to some extent that 'line of demarcation' which divided natural science and philosophy. Post-Newtonian tradition, best of all expressed by the slogan "not to frame hypotheses", had even led to such a curious case: a mathematician and physicist, Lesage, who invented a purely mechanical theory of gravitation, was treated by his colleagues as philosopher only. As Ferdinand Rosenberger said, "Readily treating him as a talented philosopher, they stopped to pay attention to him for the future, cited him from time to time but did not consider his system".[7] Thus, it was philosophers who were destined to speculate about the structure of matter, and the remarkable investigation of this problem belongs to Immanuel Kant. His *Metaphysical Principles of Natural Philosophy* were published in 1786, and Kant tried to solve the problem following from his previous studies in the *Critique of Pure Reason*, how is it possible to represent the structure of matter? I'd like to note that this book was published almost simultaneously with *Système du Monde* by Laplace, and was to some extent a response to Newton's *Principia*. Not for nothing Kant chose an analogous title for his book.

I shall not dwell upon the study of Kant's constructions in detail, but point out those among them which are characterized by Newton's ideas and concepts, i.e. by conceptions of the same Newton who did "frame hypotheses". Let's begin with the following remarkable extract:

" . . . würde man es nicht unmöglich finden, sich eine Materie zu denken, (wie man sich etwa den Aether vorstellt) die ihren Raum ohne alles Leere ganz erfüllete und doch mit, ohne Vergleichung minderer Quantität der Materie unter gleihem Volumen, als alle Körper, die wir unseren Versuchen unterwerfen können. Die repulsive Kraft muss am Aether, in Verhältniss auf die eigene Anziehungskraft desselben, ohne Vergleichung grösser gedacht werden, als an allen andern uns bekannten Materien. Und das ist denn auch das Einzige, was wir bloss darum annehmen, weil es sich denken lässt, nur zum Wiederspiel einer Hypothese (der leeren Räume), die sich allein auf das Vorgeben stützt, dass sich, dergleichen ohne leere Räume nicht denken lasse. Den ausser diesem [dem Aether – V.K.] darf weder irgend ein Gesetz der anziehenden, noch zurückstossenden Kraft, auf Muthmassungen a priori gewagt, sondern alles, selbst die allgemeine Attraction, als Ursache der Schweren, muss sammt ihrem Gesetze aus Datis der Erfahrung geschlossen werden".[8]

From this extract it is clear that all the main features of Newton's constructions constituted the basis of Kant's natural philosophy: the real space is conceived as a space filled with the forces operating in it. Therefore, space is not empty. These forces are repulsive ones, and Kant says that they are proportional to the third power of a distance, but not to a squared distance as it was the case for attractive forces. Repulsive forces appear at infinitely small distances between particles. And finally, it is aether which carries these forces. Kant thought that force is inherent in matter. It is its essence which does not need to be reduced to something else. Kant apparently was the first who introduced the notion of dynamical philosophy into science, and determined "mechanical" and 'dynamical' approaches to the explanation of the phenomena of Nature: "Die Erklärungsart der specifischen Verschiedenheit der Materien durch die Beschaffenheit und Zusammensetzung ihrer kleinsten Theile, als Maschienen, ist die mechanische Naturphilosophie; derjenige aber, welche aus Materien, nicht als Maschienen, d.i. blossen Werkzeugen äusserer bewegender Kräfte, sonderm ihren ursprünglich eigenen bewegender Kräften der Anziehung und Zurückstossung die specifische Verschiedenheit der Materie ableitet, kann die dynamische Naturphilosophie gennant werden".[9] Further, he pointed out that the dynamical way of explanation is much more useful – " . . . die [dynamische Erklärungsart] der Exprimentalphilosophie weit angemessener und befördlicher ist, indem sie geradezu darauf leitet, die den Materien eigene bewegende Kräfte und deren Gesetze auszufinden".[10]

The correctness of these words was proved by the further development of physics.

In his *Metaphysical Principles of Natural Science* Kant considers force as the possibility of the fact that matter can be conceivable by experience.

However those forces which he studies in his book, attraction and repulsion, give the possibility of the experimental knowledge, and cannot be discovered in experiments. Therefore, studying these forces belongs to the field of metaphysics, while the real forces putting matter in motion is the subject of physics. The problem of physical approach was studied in Kant's last work published only 84 years after his death, under the title *Von dem auf Principien a priori gegründeten Ubergange von den metaphysischen Anfangsgründe der Naturwissenschaft zur Physik.*[11] In his opinion the main problem arisen by the examination of such a transition is a question of existence of aether (which Kant calls sometimes 'caloric'):

"Es gibt aber in dem Übergänge von den metaph. Anf. Gr. der NW. zur Physik eine dergleichen unvermeidliche Aufgabe: ob nämlich ein in Weltraum durchgängig (mithin auch durch alle Körper durchdringend verbreiteter Stoff den man etwa Wärmestoff nennen könnte (ohne doch dabei ein gewisses Gefühl des Erwärmens, weil es bloss das Subjektive in einer Vorstellung, als Warnehmung, Betrifft in Anschlag zu bringen), – ob sage ich ein solcher Stoff, als die Basis aller bewegender Kräfte der Materie vorhanden sein oder nicht sein: oder ob seine Existenz nur zweifelhaftsein; mit anderen Worten: ob er, als bloss hypothetischer Stoff, von den Physikern nur zur Erklärung gewisser Erscheinungen angenommen werde, oder kategorisch als Postulat zu statuiren sein – diese Frage ist für Naturwissenschaft als System, von der grössten Wichtigkeit; zumal sie vom Elementarsystem derselben zum Weltsystem die Leitung gibt".[12] The logical scheme of Kant's proof of the real existence of aether is the following: space is conceivable only by experience, empty space cannot be an object of experiment, and therefore it must be filled with something. The substance filling the space must be in a state of permanent motion, otherwise it cannot influence our sense organs. Therefore, Kant concludes, aether does exist:[13] "der Wärmestoff ist wirklich und kein bloss zum Behuf der Erklärung gewisser Phänomene gedichteter, sondern aus einem allgemeinen Erfahrungs-prinzip".[14] Kant's aether has the same functions as Newton's one – it is the field of forces filling the space. Like Newton, Kant considered aether as a kind of "active principle" which was necessary for constructing the system. The same things which had been vague and uncertain in Newton's works acquired an attribute of necessity, and the concept of aether gained a solid reason.

Kant's last work is usually regarded as the result of senility, as it is said, for example, in the commentary for the Russian translation of this work. However, its obvious logical connection with his previous writings, *Metaphysical Foundations of Natural Science* and the *Critique of Pure Reason*, proves the soundness of the alternative point of view expressed by such outstanding philosophers and historians as Kuno Fischer and Ferdinand Rosenberger. "Aether is an absolutely inevitable and necessary construction in Kant's philosophical system, and it is by no means an empty invention of the weakened old man".[15]

Let us mark, finally, one more feature of the succession. For Newton, the concept of the "Analogy of Nature" was an essential element of the entire and harmonious world picture. He wrote in the draft of the third Rule of Reasoning that "The laws (and properties) of all bodies on which it is possible to institute experiments, are laws (and properties) of all bodies whatsoever".[16] Apparently, Kant wanted to claim the same thing, talking about the unity of experience.

Kant's natural philosophy is also characterized by the concept of hierarchy of Nature which is followed from integrity of experience, and therefore from integrity of the world, from the fact that its different parts are subjected to the same laws. The contradiction of Newton's constructions, i.e. the attempt to isolate aether into a category of substances subjected to "other laws than those that are mechanical", is eliminated through exclusion of aether from the number of substances on which it is possible to make experiments. Kant introduces aether "between" physics and metaphysics as the necessary condition for the possibility of experiments. However complex and artificial these constructions were for recent scientists, the ideas of unity of the scientific world-view (*Weltbild*) where aether was conceived as a main component, led science beyond mechanics.

Institute for History of Science and Technology, Moscow

NOTES

[1] Leon Rosenfeld: 'The Velocity of Light and the Evolution of Electrodynamics', *Suppl. NUOVO CIMENTO* 5, (1957), p. 1639.

[2] See: J. E. McGuire: 'Force, Active Principles and Newton's Invisible Realm', *Ambix* 15, (1968), p. 203.

[3] *Four Letters from Sir Isaac Newton to Doctor Bentley*, London 1756, pp. 25–26.

[4] I. Newton, *Optics*, Dover (ed.), based on the 4th edition of 1730, N. Y., 1952, p. 395.

[5] I. Newton. *Optics* p. 351.

[6] A. Rupert Hall and Marie Boas Hall (eds.), *Unpublished Scientific Papers of Isaac Newton*, Cambridge University Press, Cambridge, 1978, p. 133.

[7] F. Rosenberger, *Istorija fiziki* (History of Physics) Russian translation, v. 3, part 1, Moscow-Leningrad, 1935, p. 40.

[8] I. Kant. Metaphys. *Metaphysische Anfangsgründe der Naturwissenschaften* . Königsberg, 1800, S. 81.

[9] *Ibid.*, S. 79.

[10] *Ibid.*, S. 90.

[11] *Kant's gesammelte Schriften*. Bd XXII, Berlin, Leipzig, 1938, SS. 543–615.

[12] Kant's gesammelte Schriften, Band XXII, S. 550.

[13] Kant gives the following definition of aether: "Unter dem Begriffe des Wärmestoffs verstehe ich eine allerver breitete alldurchdringende, innerlich in allen seinen Theilen gleichförmig bewegende und in dieser Bewegung beharrlich begriffende Materie, welche ein den Weltraum als Elementarstoff einnehmendes (occupants) und zugleich erfüllendes (replens) absolutes, für sich bestehendes Ganze ausmacht dessen Theile in ihrem Platze (. . .) continuirlich einander und andere Körper unablässig agitirend das System in beständiger Bewegung erhalten und als äusseres Sinnenobject die bewegenden Kräfte enthalten". (*Ibid.* S. 610).

[14] Kant's gesammelte Schriften, S. 551.

[15] F. Rosenberger, *Op. cit.* pp. 55.

[16] See: J. E. McGuire, 'The Origin of Newton's Doctrine of Essential Qualities', *Centaurus* 12, (1968), p. 236.

PART V

V. P. KARTSEV

V. V. PETROV'S HYPOTHETICAL EXPERIMENT AND ELECTRICAL EXPERIMENTS OF THE 18TH CENTURY

It is only in our time that the name of Russian academician Vasily Vladimirovich Petrov (1761–1834) – the discoverer of electrical arcs (1802) – became widely known among foreign historians of physics. It should be noted though, that in Russia itself academician Petrov was forgotten soon after his death and that his works fell into oblivion.[1] No documents elucidating his private life were preserved, no portrait of him was left, and his burial-place was lost. V. V. Petrov's now so well known book "News of the galvani-voltaic experiments which professor of physics Vasily Petrov had conducted by means of a particularly huge battery consisting at times of 4200 copper and zinc disks and installed at St. Petersburg Medicine and Surgery Academy", published in 1803 in St. Petersburg, was by chance discovered in a library in the town of Vilno at the end of the last century. Familiarization with the book showed that for the first time in world literature it described a series of physical phenomena of paramount importance related to electricity.

In this connection a broad discussion which took place in the 1950s is of special interest. The discussion started when academician S. I. Vavilov stated that descendants of the scientist for a long time had had in their possession a small water-colour portrait of V. V. Petrov. Soon after that a 7 × 10 cm picture in water-colours was discovered in the collection of the State Hermitage which many researches tended to regard as a "portrait of V. V. Petrov".[2]

Currently, proponents of the authenticity of V. V. Petrov's portrait produce as arguments proofs of scientific and historical nature apart from self-evident proofs connected with ethnographic, anthropological and biographical considerations.

A. I. Leushin, for instance, writes in his article[3] devoted to the portrait of Petrov: "What sort of a physical experiment is shown in the picture^ Portraits of scientists of that time often depicted objects connected with the most essential contribution by the respective scientist to science. Petrov, as it is known, studied electrical phenomena, the electrical arc in particular. He defined an electrical arc as "alight-conveying phenomenon by means of Galvani-Voltaic fluid" and maintained that the arc is a phenomenon of transference of light by means of a special fluid with electrical properties. By

William R. Shea (ed.), Nature Mathematized, 279–289.

the galvani-voltaic fluid he meant 'galvanism', what corresponds to 'electrial current' in modern terminology. Petrov tried to find the cause of this phenomenon by heating, using magnet, etc. Although it is unlikely that Petrov had any notion of the concept of 'direction of current' he, apparently, nevertheless supposed that elements of different poles participated in the creation of the galvani-voltaic fluid. He could as well set out to 'divide' galvani-voltaic fluid. It was probably for this purpose that he used the U-shaped lever. To obtain charges from the lever, he applied a magnet (in the picture it is positioned incorrectly which evidently is the painter's device to show the magnet better). In order to provide the current of galvani-voltaic fluid, Petrov heated one of the lever's ends on a burner. No voltaic pile-source is seen in the picture. Was it the painter's fault or Petrov's innovation^ . . . One can imagine several variants of Petrov's hypothetical experiment: search for the electrical phenomena which occur under the influence of temperature; attempts 'to separate charges' by means of a magnet and so on. Such a delicate physical experiment could be prepared only by a scientist who profoundly perceived the essence of physical processes in electrical phenomena. Only Petrov, who over many years had been engaged in research of this kind, could be such a physicist in Russia at the turn of the 19th century. So it must be him who is portrayed in the picture".[4]

This is suggested to the historian of science to determine retrospectively the possibility or authenticity of a historical scientific event of great importance on the ground of some pictorical material.

In what way can the historian of science approach this matter? First, in principle, by analysing the situation in physics as it had developed up to the potential moment of the historical scientific event (the end of the 18th century) while following the logic of scientific research of the period. Secondly, the historian inevitably should turn to the social situation which existed in Russia at that time and determine what is now usually called the "social demand" of society on science and technology. Thirdly, he inevitably should turn to studying Petrov's views and concepts and to what one can call his 'style of thinking'[5] as well as to his research program.

Finally, one should consider in detail what precisely is lying on the table in the supposed picture of Petrov and in what combination, and draw conclusions with respect to eventual experiments with the scientific equipment given.

The logic of research in the field of electrical investigations is in a comprehensive way reflected in works of Y. G. Dorfman[6] and it is difficult to add anything new to them.

The historical roots of research in the field of study of electrical phenomea, which had eventually led to Faraday's discovery, were exhaustively analysed by B. Kuznetsov.[7] The latter minutely considered major historical facts of civil and economic history and substantiated the emergence of various scientific theories in the field of electricity by analyzing society's productive forces and its needs (manufacturing, its energy base – mechanical limits of natural science – Cartesian teaching; technological revolution in England – machine industry, textile industry – development of chemistry – Davy, Volta; bourgeoisie in pre-revolutionary France – rationalism and exclusion of Cartesian techning – military technology and peculiarities of scientific and technological development in France – quantative methods – coulomb – mathematical analysis – Ørsted, Biot, Savart, Laplace, Arago, Ampère, etc.; steam technology and the unity of motion forms – application of the unity of motion forms to electricity – Faraday, etc.).

Despite the keen interest aroused by these constructions and their demonstrativeness, in our opinion, such an approach cannot – because of the large structure of the analysis – give an univocal answer to the question of possibility or impossibility for a certain specific individual to make one or another specific discovery since this approach cannot take into account the specific characteristics of the personality and those of the social environment in which the personality has to realize his creative potentialities.[8] Socio-economic factors influence a scientist first and foremost through the medium, i.e. the environment in which he works. This intermediate environment, this 'microsortium' can, precisely like the very personality of the scientist in which vicissitudes of his individual fate, his genetical peculiarities, breeding and education have been imprinted, in no smaller degree than society's global socio-economic characteristics determine the course of scientific work and even in a number of cases act in opposition to those characteristics.

To begin with, a very brief – in order to fit the terms of the paper – description of Vasily Vladimirovich Petrov's personality. He was born on July 19, 1761 in the small town of Oboyan in a clergyman's family. He studied at a church school, then at a "college" – institution of higher learning in Kharkov, and later at the teachers' seminary in St. Petersburg. He took special interest in mathematics and physics which he taught after graduation from the seminary in the Altai and later at the Medicine and Surgery Academy in the Capital. There he set up a physical labortory, the largest in Russia at that time, which had unique scientific equipment received from abroad, produced by Petrov himself and by craftsmen after his design, bought from rich dignitaries, etc. anatical passion for collecting modern instruments, particularly those related to electricity, was Petrov's distinctive trait – in his lifetime he had procured nearly 700 unique instruments and apparatus for the laboratory.[9]

Petrov created an original method of teaching physics by illustrating the process with demonstrative experiments and conducting well-grounded research. He was elected member of the Russian Academy and honorary member of the Erlangen physical and medical society.

Petrov's investigations expounded in the book "News of the Galvani-Voltaic experiments . . . " constituted the summit of his scientific technical activities. No sooner did Petrov learn of the discovery of the voltaic pile that he decided to build a gigantic battery which could consist of up to 4200 copper and zinc disks while being 10 feet in height. Petrov made an original improvement which overcame the earlier insurmountable obstacle for the construction of such big piles – he placed the pile horizontally in three separate boxes. The horizontal position of the pile permitted to avoid pressing-out of the electrolyte from between the plates and removed limits for the pile's height and capacity. The pile constructed by Petrov was undoubtedly one of the biggest piles of that time (to give an idea of Petrov's personality may I note that the equipment collected and valued by him was always distinguished by its unique characteristics, large size, etc.).

The pile's special size and capacity allowed Petrov to conduct a series of experiments which were impossible with smaller and less powerful piles.[10] In that way he discovered a steady electrical arc by which he illuminated a "dark chamber'; he also wrote about metal melting by means of the arc. "In Chapters V and VIII (of the book – V.K.) all experiments without exception are ones of which I have never read anywhere about their being conducted", Petrov wrote in the foreword to his book "News of the Galvani-Voltaic experiments . . . ".

To be noted is also the great variety of scientific subjects which Petrov dealt with: here chemistry, mechanics, optics, luminescence, strength, electrolysis, meteorology. Among his works there are very specific ones devoted, for example, to the cleaning of inkpots, installation of lightning rods. Such universalism was determined by the fact that numerous problems posed by practice were being solved at the Russian Academy and since the scientists were still few, each of them had to cope with a heavy burden. The elaboration by Petrov of a special pattern of thinking, which firmly linked scientific research to its immediate practical benefit and defined as necessary the correlation of nature's forces, should be considered a positive consequence of that fact. His following words can be regarded as a response to speculative 'philosophizing' and abstract reasoning which do not lead to any immediate benefit of science: " . . . it is much more reliable to seek for the real source of electrical phenomena in philosophizing to which well-nigh all of the scientists have so far constantly resorted but in the immediate consequences of experiments themselves".[11]

This side of Petrov's style of thinking is very definitely characterized by his references to Karazin's project entitled "The possibility of applying electrial power to human needs" (a project for using atmospheric electricity with the help of balloons). In the references he wrote directly "about the necessity for conducting by means of very powerful atmospheric electricity new remarkable physical, chemical and technological experiments, not only curious but also essentially useful in certain respects."[12]

Professor and academician Petrov's scientific credo can be defined as that of a scientist who confidently set practical tasks for science by contrast to the science which indulged in "philosophizing" and which virtually became obsolete at that stage of electrical research. In regard to electro-technical experiments Petrov definitely appeared one of the first electricians in the world, i.e. one of those scientists who consciously tried to utilize electricity for mankind's technological needs.

One more characteristic of V. V. Petrov and electrical investigations of that epoch should be pointed out.

After works by Franklin, Richman, Lomonosov and Aepinus it was necessary to go over to a new method of thinking, to find connections between heterogeneous electrical phenomena. The search for quantative methods was particularly peculiar to Aepinus who came close to the law of inverse squares for electrical charges; but the absence of appropriate measuring instruments prevented him from checking his supposition. The principles of Aepinus were caught up by many researchers of electricity, H. Cavendish among them.

The 'gap' which occurred in electrical research just after 1750 was caused first of all by the absence of appropriate instruments for electrical measurements (Coulomb, the 1890s).

Did academician V. V. Petrov perceive that? The archives of the Academy of Sciences contain many of his letters to the Academy's top officials with request to manufacture and purchase measuring instruments of high precision. In this respect he shared the view expressed by L. Galvani who wrote: "What good can be done with things, degree and measure of which have not been established, and particularly so in physics? How can the cause be determined without determining not only the quality but also the intensity of phenomena?"[13]

What instrument did V. V. Petrov use for measuring electrical phenomena? Luigi Galvani applied frog legs for the purpose, Henry Cavendish availed himself of his domestic's services, V. V. Petrov peeled skin from his fingers and estimated the 'electrical strength' (or, speaking in modern terms, tension of the battery) by the degree of pain sensations.

Thus, Petrov's style of thinking completely met the requirements of the new stage of electrical research.

Let's turn to the so-called "prtrait of V. V. Petrov".

A. I. Leushin's main argument for the authenticity of the portrait is the "delicateness of the physical experiment" which could be accomplished "only by V. V. Petrov". Actually, if it were proved that the man portrayed in the picture of the end of the 18th century conducted experiments pertaining to the search for connections between electricity, magnetism and heat, it would doubtless constitute a manifestation of a "claim" by the portrayed man to the major discoveries made many years later by Ørsted, Faraday, Joule and Lenz and it could become an argument in favour of it really being a portrait of V. V. Petrov himself.

Consequently, it is necessary first of all to specify what sort of objects are lying are lying on the table and in what combination. The first of these objects is called by Leushin a "magnet". Were there any magnets in Petrov's laboratory? The following fact mentioned in the "General Journal of Medical Science"[14] is worthy of consideration in this connection.

In 1795 V. V. Petrov selected a great number of physical instruments from the Anatomical laboratory of the State Medical Board and among those were "many an optical instrument, a copper steam-and-fire machine which cost no less than 1000 roubles alone as well as anear all the magnets with several devices". Those magnets (or magnets of the type) are being preserved in the State M. V. Lomonosov Museum. As it becomes clear from collating the 'magnet' in the picture with the magnets, they differ considerably and first of all by that the 'real' magnets are much more massive. The shape of a thin staple which the 'magnet' in the picture had is not – owing to its little magnetic mass – rational for natural magnets. Those should be as massive as possible.

Besides, it should be taken into account that the collection of physical instruments was gathered by Petrov in 1795 while the portrait, as it was believed by V. Chenakal[15] and by A. I. Leushin who supported the former's point of view, was – judging by the dress and other indications (furniture in the flat, etc.) – painted after 1791, that is after Petrov's return from the Altai, i.e. prior to the acquisition of the magnets. Let's concentrate on some other attributes of the portrait which lead the author (Chenakal) to the thought of a "unique character" of the experiment.

These are – first of all – a 'U-shaped lever', a "spirit burner – icon-lamp", a 'bracket', 'small pedestals'. To give our own interpretation of the objects shown in the picture let's "proceed from the opposite" and assume that portrayed in the picture is not academician Petrov conducting a 'unique experiment', but somebody absolutely unknown, a 'common gentleman' of that time in some way connected with the physical instruments – a scientist, a mechanic or at least, simply a dilettante interested in science.

It should be said that the passion for physical instruments was a firmly established fashion in the 18th and 19th centuries. Since O. Guericke and P. Musschenbroek made their discoveries, electrostatic machines and Leyden jars had become indispensible attributes of worldly pastimes. Successes in the study of electricity in the second half of the 18th century had contributed to that. As F. Rosenberger pointed out, "amazing phenomena of electrical light, electrical stroke, explanation of lightnings, direct extraction of electricity from the atmosphere evokes general enthusiasm even greater than the enthusiasm aroused one hundred years ago by Guericke's experiments. The desire to become acquainted with the new electrical phenomena and to experimence their wonderful effects upon oneself is penetrating the broadest sections of public; those who have failed to visit physical laboratories let themselves be electrified – with the view to improving their health – at least in fair show-booths during folk festivals, even though at the expense of their own purse in every case."[16]

In the above quotation of special interest is the reference to the broad popularity of electrical experiments in those years. The main elements of the experiments were the Leyden jar, discovered in 1745–1746, and the electrical machine with flat disk modernized in 1766 by the English mechanic Jesse Ramsden. Owing to its simplicity and efficiency the Ramsden machine quickly supplanted machines with spheres, the machines of M. Planquet, Sinot de la Phône, J. Ingenhousz and other inventors, and became practically the only popular source of static electricity at the end of the 18th century. This machine, described in detail in Guyot's book,[17] consists of a glass disk revolved by means of a crank between leather cushions. The charge is carried away from the disk by means of brushes supported by a traverse which rests on a stand while being insulated from it. The leather cushions seen in the figure, and fixed on a U-shaped brace 'H'.

Let's compare the description of this machine with the attributes of the so-called "portrait of V. V. Petrov". It's beyond all doubt that shown in the picture is just the Ramsden electrostatic machine. What A. I. Leushin takes for a 'U-shaped lever' is, apparently, a brush-traverse, the 'magnet" is a brace on which the cushions are fixed, "brackets" and 'small pedestals' are insulating brushes of the Ramsden machine, the 'handle'is a crank by means of which the glass disk is revolved. The disk itself is not to be seen in the picture – the difficult task of depicting a flat transparent object was, obviously, beyond the painter's ability. It might be, though, that the water-colours have faded with time and the painted signs of the glass disk in the picture have been lost. It is worthwhile also to remember that the portrait is a small one: 10 by 7 cm.

Thus the physical experiment shown in the picture is not unique. It consists in charging the Leyden jar by means of the Ramsden machine. Such an experiment could be conducted not only by a scientist but also by any dignitary who possessed the appropriate equipment. In particular, a rich collection of physical instruments was in the possession of Count Buturlin who sold it in 1802 to V. V. Petrov himself. Will this very circumstance, perhaps, help us solve the mystery of the portrait? Let's try this way, too.

The inventory of the apparatus purchased from Buturlin for the Academy's physical laboratory includes, inter alia, an electrical machine. One can visualize the machine through the following succinet description: "Mr. Petrov, professor of mathematics and of the Academy, had reported to the conference that among the instruments bought from Count Buturlin and belonging to St. Petersburg Medicine and Surgery Academy there is an electrical machine with its glass disk measuring 40 English inches in diameter and copper conductor being 5 feet in length and 5 inches in diameter."[18]

The dimensions of the machine bought from Buturlin considerably exceed those of the machine shown in the picture. So, while Buturlin's machine is at least by its size unique and can be relatively easily identified, the machine in the picture undoubtedly is one of those manufactured in large quantities 'for common use'. A unique electrostatic machine of a big size, designed by Petrov and made for him in 1804, has no resemblance to that standard machine either. Thus, this version is not confirmed either, unfortunately. The "portrait of academician V. V. Petrov" remains one of interesting historical and scientific enigmas.

But can we raise the question whether V. V. Petrov, with his views and the equipment available to him, was able to discover interdependence between electricity and magnetism, i.e. to accomplish the experiment of Ørsted?

As we already know, Petrov had a voltaic pile, unique by its power, at his disposal.

Were there any magnetric needles in Petrov's laboratory? One can answer this question by proceeding from archival data on the fact that in 181 Petrov engaged in rechecking the sensational experiments of Morrichini who in 1812 allegedly discovered magnetization of iron and steel needles by means of violet rays. Petrov repeated those experiments and tested scores of iron and steel needles and strips but he failed to obtain confirmation of any connection between magnetism and light. Nevertheless, he considered that subject important insomuch that he made it a competitive problem which was offered at St. Petersburg Academy.[19]

It was his principle to write only about what he had investigated personally and with his own eyes, and thus Petrov denied himself an opportunity to identify the 'galvani-voltaic fluid' with the "electric fluid" produced with the help of electrostatic machines: he saw with his own eyes that while the galvani-voltaic fluid provided a little spark and a strong arc, the electrical fluid created a long spark and no arc at all. Therefore, Petrov didn't conceive the two electricities as a united whole. This is understandable if we take into account the fact that Faraday either didn't identify the two kinds of electricity with each other until 1833.

Thus Petrov had both the necessary components in his style of thinking and all the "technical" prerequisites for discovering the law of electro-magnetic induction: voltaic picles, wires for which he had found a method of insulation (by means of sealing-wax), magnetic needles. He believed in the unity of nature's forces, even though he demanded that experiments in that field should be more demonstrative. Nonetheless, the discovery fell to the lot of Ørsted who effectively used ideas borrowed from Schelling's philosophy and applied the principles of correlation and interdependence and, at last, took advantage of his youth – at the time of the great electrical discoveries Petrov was already 60 years old and practically blind.

After all, didn't V. V. Petrov's life and activity, lost even in the memory of his nearest descendants, play their role in the development of science, culture and society? Generally speaking, isn't this really the destiny of all those whom circumstances make play the historical part of "unnoticed genius"?

Out of its scientific legacy society used only what was necessary to it; electricity had not yet become an object for practical application and all electrical investigations, which society still felt no need of, were therefore forgotten; the hostile scientific environment, in which there was no successor or continuer of Petrov's experiments to be found, quickly forgot him, too.

This was, of course, facilitated by the fact that his famous "News . . . " was published in Russian by the Medicine and Surgery Academy in a limited number of copies.

But was all what Petrov did for Russia and science futile? Although we cannot find in the archives of the period any clear evidence of practical application of Petrov's works we continue to hope that his labours were not in vain.

Commenting on V. V. Petrov's role in the development of Russian science academician V. I. Vernadsky wrote: " . . . Aepinus F. U. (1724–1802) continued the tradition of experimental work of St. Petersburg Academy launched by Kraft, Lomonosov, Richman and Brown. His assistant, V. V. Petrov – one of the original physicists of the 18th century – was an even greater experimentalist. A considerable part of his work was done within the Medicine and Surgery Academy before he was elected member of the Academy (1807–1837); there Petrov published, inter alia, the first description of the so-called Voltaic arc, 20 years prior to the actual recognized discovery. His experimental work still awaits evaluation. And still true are his words by which he defined his own activity: "I am a native Russian who has had no opportunity to avail himself of verbal instructions of foreign professors of physics and to engage in the studies of that science which so far has remained in complete obscurity among those with us contemporary amateurs." But for us it is already clearer than for the 'contemporary amateurs' that his experimental thinking was an independent creative work, original in design and implementation, and the existence of such a work in the Russian environment was a major cultural fact which doubtless had an impact on the development of the general structure of the Russian cultural life, even though we are not able to see it from the 'historical documents' and "miscellaneous excerpts from the past".[20]

'I hope', Petrov wrote once, "that educated and impartial physicists will at least some day in the future agree to do justice to my experimental work, the importance of which is well worthy of it".

Now, the time has come to do that justice. So we hail V. V. Petrov, and not only his immediate scientific achievement, but also the role he played in furthering Russia's cultural development.

Institute for History of Science, Moscow

NOTES

[1] S. I. Vavilov (ed.). *Academician V. V. Petrov*, Academy of Sciences of the U.S.S.R., Moscow-Leningrad, 1940.
[2] V. V. Chenakal. 'A valuable find', *Literaturnaya Gazeta*, June 6, 1952, N 75 (2948), p. 2.
[3] A. I. Leushin. On the authenticity of academician V. V. Petrov's portrait. *Problems of history of natural sciences and technogy*. N 1, 1980, 129–130.
[4] *Ibid.*
[5] This term has not been fully established in modern science and attempts are made to refer it to entire communities and scientific periods (see e.g. Novic I. B. *Style of Thinking in Natural Sciences*, Moscow, Progress, 1972.) and even reject it completely (see: Kedrov B. M. About scientific revolutions. Problems of History of Natural Science and Technology, N 3, 1980).
[6] Ja. G. Dorfman. History of Physics, v. I, Nauka, 1974, pp. 280–309.
[7] B. Kuznetsov. The historical roots of Faraday's works). in: '*History of Technology*'. II, ONTI, Moscow-Leningrad, 1934, pp. 22–56.
[8] V. P. Kartzev. *Scientists' intermediate social environment and their biography*, in: Men of science. 'Nauka', Moscow, 1974.
[9] L. D. Belkind, *V. V. Petrov*, in: 'People of the Russian science', GITTL. Moscow-Leningrad, 1948.
[10] This situation resembles the situation which arose later for J. Henry who, after having built an extremely powerful electromagnet, discovered very subtle effects, imperceptible with less powerful equipment. We can also note some other cammon features which characterized Petrov's and Henry's lives and activities, e.g. the 'provincialism' of their scientific environments, etc.
[11] V. V. Petrov. New electrical experiments. S. Petersb., 1884, p. 70.
[12] The archives of the U.S.S.R. Academy of Sciences, also see in the book 'Academician V. V. Petrov', p. 207.
[13] L. Galvani, A. Volta. *Selected works on animal electricity*, Academy of Sciences of the U.S.S.R., Moscow-Leningrad, 1937.
[14] See in the book 'Academician V. V. Petrov', 1940, p. 66.
[15] V. Chenakal. A valuable find – in '*Literaturnaya Gazeta*' . . .
[16] F. Rosenberger. *History of physics*, Vol. 2, ONTI, Moscow-Leningrad, 1937.
[17] M. Guyot. *Nouvelles récréations physiques et mathématiques*, Paris, 1800.
[18] Quoted from the book 'Academician V. V. Petrov', p. 62.
[19] Quoted from the book 'Academician V. V. Petrov', pp. 212–213.
[20] The archives of the U.S.S.R. Academy of Sciences, see also in the book 'Academician V. V. Petrov', p. 247.

KAREL BERKA

THE IDEAL OF MATHEMATIZATION IN B. BOLZANO

In my contribution I would like to draw attention to the views on the ideal of mathematization held by B. Bolzano, a later follower of Leibnizian rationalism. This analysis will show the evolution of conceptions elaborated in the epoch of *mathesis universalis* on this topic in a period basically influenced by the philosophy of Kant and other representatives of German classical philosophy.

Bolzano's revival of ideas, originally developed in the epoch of Enlightenment, on the nature and application of mathematics, especially its methodology, is determined by his view that there is a very close link between mathematics and philosophy. It is further influenced by his conviction that mathematics, in order to attain a still greater degree of precision, necessarily requires a logical revision of its foundations. His endeavours to base mathematics, as well as all other sciences, on objective grounds and to present its parts in appropriate scholarly treatises, follows from his criticism of the contemporary methodology of defining and also proving.

The discussions concerning the acceptability of the fifth postulate of Euclids *Elements* and the various attempts to prove it, seem to him to be clear evidence that the problem in question does not lie in the demonstration of the certainty of this postulate, but in finding the objective ground of its validity. In his work *Die drey Probleme der Rectification, der Complanation und der Cubirung* (1817), he claims that we cannot accept as a basic truth any proposition which admits a further ground of its truth. Bolzano does not doubt that this postulate is true, requiring only to have its validity grounded in an objective way, independently of our subjective feeling of certainty. He criticizes proofs of mathematical propositions using terms which are not necessary or sufficient, e.g., proofs of geometrical propositions based on the concept of movement, or proofs in pure mathematics (arithmetics, algebra, analysis) utilizing concepts of applied mathematics (geometry).[1] His objections to the usual exposition of mathematicians are summarized in his *Wissenschaftslehre* (*Theory of Science*) as follows: "so sind sie doch bis zur Erklärung ihrer ersten und allgemeinsten Begriffe noch nicht vorgedrugen, und ihre ersten Sätze stellen sie entweder ohne Beweise oder mit solchen Beweisen, die nicht echt wissenschaftlich sind, auf".[2]

William R. Shea (ed.), Nature Mathematized, 291–298.

Bolzano's conception, which extends and modifies the Leibnizian project of mathematization is explicitly proclaimed in part II, "On the mathematical method", of his *Beyträge zu einer begründeteren Darstellung der Mathematik* (1810) and further elaborated in other mathematical works, especially in his *Einleitung zur Grössenlehre*, and in the *Wissenschaftslehre*, where the logical aspects of mathematics and its methodology are taken into consideration.

In order to explain what he really means when speaking about the mathematical method, let me first of all make some remarks on his philosophy of mathematics in general. He defends against Kant the older standpoint of Leibniz, with new insights into the nature of mathematics, its relationship to philosophy and the admissible extent of mathematization in science. Bolzano vigorously denies the Kantian point of view according to which mathematics is a purely synthetic science based *a priori* on intuitions of time and space. In his argument against Kant, he relies on the analyticity of mathematical truth in the sense of Leibniz. Mathematical theorems are valid without exception and not only for objects existing in time and space. These intuitions appear neither in the fundamental arithmetical concepts of magnitude and number, nor in the formulation of arithmetical theorems. This also holds for "mathematical geometry".[3]

Bolzano accepts Kant's classification of judgments, but with a substantially different interpretation. The distinction of judgments *a priori* and *a posteriori* is transformed into the differentiation of conceptual propositions, e.g., "The square root of the number 2 is irrational", which consist merely of pure concepts containing no intuition, and intuitive (perceptual, empirical) propositions, e.g., "This is a flower", or "Socrates was an Athenian by birth", including at least one intuitive idea (*WL* II §133). Taking into account his logic of variation, he considered as analytic every proposition containing "just a *single* idea [in it] which can be arbitrarily varied without disturbing its truth or falsity", i.e., in our terminology, every propositional form which will yield for all admissible substitutions of its free variables either tautologies or contradictions. All other propositions, in which there is not a single idea "that can be arbitrarily varied without affecting their truth or falsity", are then synthetic propositions (*WL* II §148). Analytic and synthetic propositions appear in both of the above mentioned basic kinds, e.g., the proposition "Every triangle is a geometrical figure" is an analytic conceptual truth, whereas the proposition "This triangle is a geometrical figure" is an analytic perceptual truth (*WL* II §197).

Bolzano tries to refute Kant's view that mathematics is based on intuition, whereas philosophy is based on conceptual knowledge by pointing out that pure logic – as a part of philosophy – does not consist only of analytic propositions. He does not deny that the propositions of logic themselves are analytic, e.g., the proposition "If all men are mortal and Gajus is a man, then Gajus is mortal", but considers the corresponding syllogistic rule "that from two propositions of the form *A* is *B* and *B* is *C* follows a third proposition of the form *A* is *C*" as a synthetic truth. Similarly all existential propositions in logic, e.g., "There are simple and compound ideas", and many other propositions such as "Every sentence contains three or more parts" are synthetic propositions. From this argument, which anticipates the modern distinction of logic and metalogic, Bolzano concludes that it is mathematics rather than any other science, even logic, that has the right to be called an analytic science (*WL* III §315).

His basic argument against Kant is contained in his discussion of the concept of sum.[4] Bolzano argues that from the definition of this concept as "a collection in which the order of the parts can be regarded as something indifferent and the parts of the parts can be regarded as part of the whole" it follows that the associative law for addition is an analytic proposition. By applying this law and defining that $7 + 1 = 8$, $8 + 1 = 9$ etc. he proves – in a manner similar to Leibniz – that the proposition $7 + 5 = 12$ is a purely analytic truth and not a synthetic judgment *a priori*.

For Bolzano there is, for the most part, no difference between mathematics and philosophy in respect to proofs since both are purely conceptual sciences. There is, of course, a difference between the practical activities of mathematicians and the conceptual structure of mathematics itself, between intuitive constructions of geometrical figures and the formation of idealized geometrical concepts. The possible discrepancy between the objective nature of mathematics and its subjective exposition, which is not free from error,[5] can be overcome only by means of appropriate definitions, exact proofs and exposure of the objective ground-consequence relationship between their truths. The certainty of mathematical knowledge is therefore not guaranteed by the analyticity of mathematical propositions.

His endeavour to secure mathematical knowledge against all possible error, and to supress the disturbing factors of subjectivity so far as it is at all possible determines his views on the nature and function of the mathematical method. He is firmly convinced, in full agreement with the assessment of the Leibniz-Wolffs' school, that the basic aspects of the methods used by mathematicians in scholarly treatises can be applied in all other sciences. The so-called *methodus mathematica* is in its essence by no means different from scientific exposition. No human work is so accomplished that it could not be improved, even in the case of the mathematical method which attained already in Euclid such a very high degree of perfection. This is the goal of his foundational studies in the domain of mathematics, more exactly, in its speculative, i.e., philosophical, exposition.

Bolzano's concept of the mathematical method is not restricted to operations with magnitudes or numbers, nor to measurement procedures and other quantitative methods. The specificity of this method in his interpretation consists of conceptual and proof-theoretical methods concerned with concept formation, theory construction and deductive systematization of sciences, together with the underlying logic and practical instructions for writing scholarly treatises. This rather metatheoretical point of view, expressed clearly in the various subparts of his survey in *Beyträge* dealing with descriptions, definitions, and classifications, principles and postulates, theorems, corollaries, consequences and proofs, problems and their solution etc., explains the applicability of this method to any other branch of science. Mathematization of sciences in this specification serves, hence, to clarify their conceptual framework, deductive structure and philosophical foundations.

From this point of view Bolzano criticizes the usual definitions of axioms as obvious propositions which do not need any proof or as truths that we grasp intuitively when we understand their sense. This criticism follows from his assumption that intuition is not a reliable criterion for separating axioms or postulates from all other statements. *Anschauung* is for Bolzano a comparative term, differing from person to person. Further, to adopt a proposition as an axiom only because we do not know how to prove it is neither acceptable for deciding what has been already for objective reasons proven and what has not. We have, therefore, to acknowledge as axioms in their objective sense only such true propositions which are for themselves undemonstrable.

These initial considerations are clarified in the distinction of two kinds of proofs:[6] *objective proofs* or *demonstrations* (Begründungen), i.e., proofs where the truth of the proposition in question is demonstrated from its objective ground, and *subjective proofs* or *confirmations* (Gewissmachungen), which are aimed only at assurance. Demonstrations refer – according to Bolzano – even to basic propositions, if one can detect some objective ground of their truth. The application of confirmation is significant only in the case of propositions which are not obvious for subjective reasons, but not for propositions which everybody would conceive as true.

This distinction, as Bolzano explicitly stresses, is based on Aristotle who introduces demonstrations in which it is shown *that* (*hoti*) something is so and so and demonstrations which exhibit *why* (*dioti*) it is so and so.[7] It is also supported by Leibniz who in his *Nouveaux essais* (I c. 3 §24) adopted as one of his "grandes maximes, qu'il est bon de chercher les démonstrations des Axiomes même". In respect to Bolzano's 'objectivism', it expresses his doctrine of truths, which states "something to be as it is", leaving it undetermined whether such a "proposition has or has not been thought by anyone", in distinction to known truths, i.e., true propositions in the subjective sense.

Similarly, there are for Bolzano two kinds of definitions:[8] *explications* (Verständigungen) conveying terminological conventions or explaining the meaning of linguistic expressions, and *definitions* (Erklärungen) or *conceptual determinations* (Begriffsbestimmungen), which define or determine the content of some concept by determining whether a certain idea or proposition is simple or compound and, in the latter case, what parts it contains and how they are connected.

The peculiarity of Bolzano's theory of definition is rooted in his doctrine of simple and complex ideas (concepts) and their clarity (*Klarheit*) and distinctness (*Deutlichkeit*).[9] The relationship holding between the properties of clarity and distinctness and the classification of simple and complex ideas complicates his views but enables him to achieve consistency with Pascal's methodological postulates,[10] which he tries to respect in a modified way.

For Bolzano, it is not decisive whether a certain concept is primitive or not in a deductive system. His requirement to define all concepts, even the primitive ones, is significantly realizable for complex ideas only. Simple ideas cannot be defined, because they do not contain any parts. Unclear simple ideas can be, of course, explicated. Clarity itself is not a sufficient property for a scientifically exact concept formation since this property depends on intuition. In full agreement with Pascal's postulates, clear ideas need not be defined. Mathematical knowledge requires more than clarity of mathematical concepts. It is necessary to show what parts the complex ideas contain and how they are composed. Only in this case can their distinctness be exhibited and their content precisely determined. Bolzano's requirement of defining all concepts – whether primitive or not – relates, therefore, to their distinctness, but not to their clarity.

His approach results from his justified objections to the subjective acceptance of concepts and the use of sufficiently ungrounded assumptions. In a relevant sense, it stems from his conviction that mathematics can be founded on a safe base only when the objective relations holding between basic and derived propositions, primitive and defined concepts, are unveiled. It is, in a less obvious sense, influenced by his intentions to clarify the nature of logical concepts assumed in the conceptual net of mathematics and the logical structure of mathematical proofs.

When analyzing the mathematical method he is primarily interested in questions of concept formation because mathematical proofs depend, in his conception, on the nature of demonstrated propositions with respect to their composition. The special feature of concepts in mathematics, constructed from simple basic ideas by means of synthetic definitions, justifies for him the universal validity of mathematical propositions, providing the objective reason for their analyticity. He stresses, therefore, (and similarly for any other science) the precise determination of the corresponding universe of discourse reflected in its subset of simple ideas.

This ontologically grounded concept of simplicity, taken over from the rationalistic tradition of G. W. Leibniz, is in itself questionable. Bolzano has several difficulties when trying to make his viewpoint plausible, even with respect to his own principle of finding for every proposition, so far as possible, its objective ground. He does not make an attempt to enumerate those ideas which he holds to be simple. Neither does he indicate some practical procedure on how to decide for a given idea whether it is simple or not.

His thoughts on the nature of basic propositions and proofs do not result solely from this "ontological atomism" which assumes in every science the existence of special simple ideas. They follow as well from another philosophical assumption, namely that the proof of any proposition has to expose the objective ground of its truth. Some purely conceptual proposition, let us say A, can therefore only be an objective ground for some other proposition, B, if it does not contain simpler parts (*WL* II §221). From this point of view, Bolzano admits as axioms or postulates precisely those propositions which are grounds, but never consequences, of other propositions.

This interpretation of proofs in deductive systems entails a rather unacceptable evaluation of the role of proofs by reduction to absurdity, of apagogic or indirect proofs. Bolzano admits that these proofs are very frequently adopted in mathematics, nonetheless he claims that their utilization in a scholarly treatise is justified only if we "look to the objective of persuasion". For the sake of scientific proofs (demonstrations) this kind is, however, indequate, because "the proposition by which an apagogic proof supports the proposition it is supposed to prove can never represent its objective ground" (*WL* IV §530). Bolzano rejects such proofs for other reasons as well. He wants to avoid the adoption of false propositions as a premise in accordance with his doctrine of derivability (*Ableitbarkeit*).

The relation of derivability is a special case of compatibility (*Verträglichkeit*). The premises and the conclusion – according to Bolzano's less general theory of derivability – are in respect to the ideas i, j, . . . (selected as variables) *compatible, concuring* or *consistent* if and only if there are some ideas substituted for i, j, . . . which are so constituted that all propositions of this deductive chain become true at the same time (*WL* II §154). For this reason he does not admit deductions from an empty set of assumptions: neither does he allow derivations from an inconsistent set of premises according to Duns Scot's rule "*ex falso quodlibet*".

His approach to the indirect proof is quite understandable: Bolzano cannot be inconsistent with one of the most valuable results of his logical investigations. What is, however, less clear, is his obviously unwarranted statement[11] that he succeeded in transforming every indirect proof used in Euclid's *Elements* into a direct, ostensive one, a claim conceived in general (but with reservations) by Leibniz in his *Nouveaux essais* (IV c. 8 §2).

Bolzano's investigations concerning the mathematical method and its function in science, depending on the ideas evolved in the epoch of *mathesis universalis*, concentrate, therefore, clearly on questions of formal methods necessary for deductive systematization of scientific theories. In this sense he modifies the rationalistic ideal of universal mathematization in a more reasonable way. His standpoint – in spite of its ontological background and his neglecting to pay more attention to syntactical problems – is more acceptable than the present-day revival of this ideal by many theoreticians who attempt to utilize mathematical methods, especially methods of quantification and measurement in various branches of science, without sufficiently justified reasons. If we, however, understand the *methodus mathematica* in the same wasy as Bolzano, we can even defend the appropriateness of this ideal.

Československá Akademie Věd, Ústav pro Filosofii a Sociologii, Praha

NOTES

1 *B. Bolzano, Rein analytischer Beweis des Lehrsatzes, dass zwischen je zwey Werthen, die ein entgegengesetzes Result gewähren, wenigstens eine reelle Wurzel der Gleichung liege*, Praha, 1817, Introduction.

2 *Dr. Bolzanos Wissenschaftslehre, Versuch einer ausführlichen und grösstentheils neuen Darstellung der Logik mit steter Rücksicht auf deren bisherige Bearbeiter, herausgegeben von mehren seiner Freunde mit einer Vorrede des Dr. J. Ch. A. Heinroth*, Sulzbach 1837 (*WL*) III §305.

[3] Comp. *WL* III §§305, 315; *Einleitung zur Grössenlehre und Erste Begriffe der Allgemeinen Grössenlehre*, B. Bolzano Gesamtausgabe II A. 7, Stuttgart-Bad Cannstatt, Fr. Frommann Verlag 1975 (*EG*) I §3.
[4] Comp. *WL* I §84; III §305 and *Beyträge*, Addendum §8.
[5] Comp. *WL* I §42; III §307, 309, 317.
[6] Comp. *WL* IV §525; *EG* II §14.
[7] Arist. *Anal. pr.* II c. 2 p. 53b 4ff.; *Anal. post.* I c. 6 p. 75a 3ff; c. 13 p. 78a 22ff.
[8] Comp. *EG* II §9; *WL* IV §§554, 668.
[9] Comp. *WL* I §§56, 61; II §§280, 281.
[10] Comp. B. Pascal, De l'Esprit Géométrique, Section II, *De l'Art de Persuader*, Ouevres Complètes, vol. I, Paris 1956–57, p. 425ff.
[11] J. Berg (*Bolzano's logic*, Almquist & Wikselk, Stockholm-Göteborg-Uppsala 1962, p. 160) rightly points out that "the argument is based on very special assumptions and yet is inconclusive".

V. VIZGIN

"DIE SCHÖNSTE LEISTUNG DER ALLGEMEINEN RELATIVITÄTSTHEORIE": THE GENESIS OF THE TENSOR-GEOMETRICAL CONCEPTION OF GRAVITATION

One of the most effective ways in which physics and mathematics can influence each other in modern science, especially in developing fundamental theories, is such a use of abstract mathematical theories where all the relevant mathematical quantities and relations are identified with all the basic physical quantities and relations describing a certain class of physical phenomena. Two most fundamental physical theories which appeared in the first quarter of the XX century – the general theory of relativity (GTR) and quantum mechanics – have precisely this mathematical structure. From the mathematical viewpoint, the GTR is the Riemannian geometry and quantum mechanics is a theory of linear self-conjugated operators in the Hilbert space.

In actual history, of course, the identification of the gravitational theory, characteristic features of gravitation and the corresponding laws with the Riemannian geometry, metric of the Riemannian space, connection, curvature and theorems of this mathematical theory turns out to be a complex, multistage process associated with a reinterpretation of the theoretical principles of the previous theory of gravitation. The nucleus of this identification is a tensor-geometrical conception of gravitation (TGCG), according to which the basic tensor characteristics of the 4-dimensional Riemannian space (metric, curvature, etc.) coincide with the main characteristics of gravitational field (potentials, intensities, etc.), and the geometrical structure of the space coincides with the structure of this field. " . . . Diese Verschmelzung zweier vorher vollständig getrennter Gebiete – Metrik und Gravitation – muss als die schönste Leistung der allgemeinen Relativitätstheorie bezechnet werden", – Pauli wrote in his famous encyclopedic article concerning the theory of relativity [1]. When constructing the relativistic theory of gravitation, Einstein (partly together with his friend, mathematician M. Grossmann) effected this mergence, thereby converting the most vague (but genius!) prophecies of Lobachevsky, Riemann and Clifford into a physical conception: the tensor-geometrical conception of gravitation (TGCG), thoroughly developed mathematically and grounded experimentally. The development of this conception in 1912–1913, apparently, proved to be a cornerstone in the history of formation of the general theory of relativity (GTR) after the invention of the principle of equivalence (PE). The last step to the GTR,

William R. Shea (ed.), Nature Mathematized, 299–317.

namely, the derivation of the general-covariant equations of gravitational field on the basis of the TGCG, was made only in November 1915, which is now considered to be the data of the birth of the general theory of relativity.

An attempt has been made in this paper to review the process of formation of the TGCG, the basis of the GTR, relying not only on the documents and original works available to the author, but also on relatively recent historical and scientific works in the field [2–13].

1. THE PRINCIPLE OF EQUIVALENCE

Einstein's principle of equivalence, formulated by him for the first time in 1907 and stating the physical undistinguishibility between a homogeneous gravitational field and a uniformly accelerated frame of reference [14], proved to be a prototype of the TGCG and, in fact, the short from which it had grown. Actually, this principle meant the kinematization of homogeneous gravitational fields or, under the consistent 4-dimensional approach, their geometrization.

The radical reformation of physics in 1907–1910 caused by discoveries in the electron theory and by the development of the special theory of relativity (STR) resulted in several basic strategies for the construction of the theory of graviatation, called upon to replace the Newtonian theory and agreeing with new ideas in physics. These strategies, however, were of global character and served as original research programmes in many fields of physics, close in their sense to the research programmes of Lakatos [15].

The classical-mechanical programme, based on the Newtonian classical mechanics, atomistic conception, and the corresponding ideas of space, time and causality, dominated in the XIX century. At the end of the XIX and at the beginning of the XX century different versions of this programme entered the "regressive" stage, concerning, in particular, the problem of gravitation. In 1890-s the electromagnetic-field programme came to life, whose main ideas were electromagnetic field and the Maxwell field theory. A number of electromagnetic theories of gravitation appeared at the turn of the century, but they were unsatisfactory in many respects and did not lead to the 'progress' in the field programme. The STR gave birth to a relativistic programme and to several different trends in the development of the gravitational theory, based on that programme: theories of Poincaré-Minkowski which gave the Lorentz-covariant generalization of the Newtonian law of gravitation; scalar Lorentz-covariant theories of gravitational field, with which Einstein seemed to have started and which were further developed in 1912–1913 by

Nordström, Mie and others; theories based on the relativistic interpretation of the equality between inertial and gravitational masses, that is on the equivalence principle. The relativistic programme made rapid progress in many fields of physics and this gave hopes that it would also be successful in solving the problem of gravitation, despite a variety of approaches and considerable difficulties on this way.

Einstein started with the Lorentz-covariant generalization of the Poisson equation and came to the conclusion that such theories contradict to the equality between inertial and gravitational masses, the fact having a reliable empirical basis. Having taken this fact as an original postulate, he transformed it by methodological techniques of the relativistic programme into a peculiar extension of the special principle of relativity, the PE. But this principle undermined the bases of the programme itself and was far beyond its framework. The inclusion into consideration of uniformly-accelerated coordinate systems and the violation of the constancy of the velocity of light in the presence of even homogeneous gravitational field required the extension of the class of permissible coordinate transformations. This led to the necessity to refuse from the physical interpretation of differencies in coordinates (characteristic of the STR and relativistic programme, as a whole) as being the results of operational-measuring procedures performed with the help of scales and clocks. Therefore, the PE led as if to a peculiar extension of the relativisitic programme, though the contours of this extension had not been clearly outlined. The strategy based on the PE may be called, with some provisoes, the extended relativistic programme, or the project of such a programme, to be exact. The progressive character of this programme is illustrated by the prediction, on the basis of the PE, to two new effects observed under propagation of light in a gravitational field (bending of a light ray and shift of spectral lines to the red end of the spectrum).

Yet, the progress made in the PE in 1907–1908 was rather limited and difficulties associated with its transformation into a consistent theory of arbitrary gravitational fields were so considerable that even Einstein himself postponed for some time the development of the gravitational problem. Having directed his efforts to the field of the theory of quanta, he also attempted for a number of years at constructing the unified field theory, trying to guess the mathematical structure of differential equations of this theory [16]. The solution of this global problem could also ease the construction of a correct relativistic theory of gravitation. Einstein's attempts, however, were not a success, and in 1911 he again turned to the PE as the main initial point on the way to the relativistic theory of gravitational field [17].

2. SCALAR THEORIES

According to the PE, the velocity of light lost its property of being the universal physical constant and became a measure of the gravitational potential. Using this idea, a famous German theoretician M. Abraham, who had been working at that time in Milan, developed at the end of 1911 – at the beginning of 1912 a scalar theory of gravitational field based on the 4-dimensional (though non-relativistic!) generalization of the Poisson equation [18–21]. One of the leaders of the electromagnetic-field programme M. Abraham, known for anti-relativistic views, considered the refusal from the constancy of the velocity of light as the failure of the STR and the possibility of developing a theory of gravitational field permitting the construction of a privileged frame of reference related to the static gravitational field.

Since the beginning of 1912 Einstein starts the correspondence, and simultaneously polemics, with Abraham. At the same time, he develops, after Abraham, a scalar theory of static field based on some modification of the Poisson equation, which, in his opinion, more naturally accounted for the requirements of the PE [22, 23]. Einstein, however, confined himself only to static fields, since the contours of the extended principle of relativity were not clear and the application, under these circumstances, of the STR 4-dimensional formalism, giving reasonable relativistic extension of statics to non-stationary processes, seemed to him incorrect.

But even at the level of the static fields, the scalar approach results in difficulties associated with its consistence not only with the PE, but also with the laws of energy-momentum conservation. The problem, in what way the principle of special relativity should be extended, still remained unsolved. That is why Einstein is more and more disappointed in the scalar theories. In the autumn of 1912 fairly good versions of the Lorentz-covariant scalar theories (first theory of Nordström [24] and theory of Mie [25]) appeared as a response to the difficulties in the approach not covariant with respect to the Lorentz transformations. The 4-dimensional relativistic generalization of the Poisson equation corresponded, within fairly high accuracy, to the equality between inertial and gravitational masses (the exponential dependence of the mass of a particle on the gravitational potential being admitted). Einstein in his letter to Nordström pointed out, after making acquaintance with the proof of his article, that for a rotating system in the Nordström theory the equality between the masses is violated. That is why Einstein gave up a similar approach which he had tried to realize originally [24].

At that period Einstein together with M. Grossmann had been developing intensively the tensor-geometrical conception of gravitation. The scalar direction, both in the Abraham version and in the version of Nordstrom-Mie, seemed to him (in the autumn of 1912) a past stage).[1]

3. EINSTEIN IN PRAGUE (SCIENTIFIC CONTACTS AND PUBLICATIONS) [12]

The study of Einstein's publications, as well as his correspondence and some other materials of 1912 referring to the Prague period, shows the decisive role of that period in the genesis of the TGCG. Since the beginning of 1912, Einstein stimulated by works of Abraham on the scalar theory of gravitation develops two his own versions of the scalar theory of static field. These works were finished in February [22] and March [23], and both were published in the second half of May, 1912. In May of the same year Einstein finished another article on gravitation dealing with the discussion of the Mach conception of inertia from the viewpoint of his scalar theory of the static field [27]. Finally, at the beginning of July Einstein sent to the editorial board of "Annalen der Physik" a small polemic note "Relativity and Gravitation. Reply to Abraham's Remarks" that was published in August [28].

The analysis of these four papers clearly shows, what was the basis that Einstein used to develop the relativistic theory of inhomogeneous gravitational fields – the relativistic programme and equivalence principle, and what were the main difficulties (see p. 300 of this paper).

Those works also clearly indicate, how Einstein was approaching the TGCG. The main prerequisties of this conception which found their manifestation in the papers of 1912 were: the possibility of using non-Eucidean geometry in accelerated frames of reference; the infinitesimal approach to geometry; the Mach ideas of inertia and general relativity; serious employment of the 4-dimensional interpretation of the STR and attempts of using it, in particular, the use of 4-dimensional variational formulation of the inertia principle in the theory of gravitation and associated advanced role of the metric; recognition of unsatisfactory character of the scalar and vector potentials. Let us note immidiately, that each of these prerequisites was surprisingly related to certain scientific influences and contacts (physisists Ehrenfest, Abraham, Nordström, Planck, Laue and others, astronomer Freundlich, mathematicians Pick, Grossmann and others, and, finally, Mach).

4. "EHRENFEST DISK", "POCKET INSTRUMENTS" – AN IDEA OF NON-EUCLIDEAN AND INFINITESIMAL APPROACH TO GEOMETRY [12, 13]

In April of 1911 Einstein began to correspond with the famous at that time physisist Ehrenfest. One of the subjects of their discussion was, apparently, a hypothetic experiment suggested by Ehrenfest as early as in 1909 which proved that an absolutely rigid body could not be set in uniform rotation about its central axis [29]. Having admitted the possibility of such a rotation, Ehrenfest obtained the violation of Euclidean relations on an absolutely rigid disk because the length of its circumference decreased due to the Lorentz contraction, while the diameter remained unchanged. In the spring of 1911 Einstein meditated upon the paradox and even published a small note on the subject [30]. At the same time, he had been working at an article on the equivalence principle, so the behaviour of scales and clocks in accelerated frames of reference was of direct interest for him. In accordance with the general idea of that principle, some gravitational field (more complicated than homogeneous) should correspond to a uniformly rotating frame of reference. Therefore, the 'Ehrenfest disk' suggested an idea that the space geometry could be changed due to the action of gravitational field. In this way one might hope to overcome the difficulty that worried Einstein and consisted in the fact that the coordinates lost their direct metrical sense, when passing to accelerated frames of reference. In 1912 Einstein in his works concerning the scalar theory of gravitation recollects the "disk (or cylinder) of Ehrenfest" in connection with the remark that the use of absolutely rigid scales in accelerated frames of reference may lead to non-Euclidean relations [22].

Einstein met Ehrenfest in 1912. Undoubtedly, they discussed the problem of gravitation. In any case, the March paper by Einstein concerning the scalar theory of gravitation includes the Ehrenfest term "pocket measuring instruments" for local space-time and physical measurements [23]. The use of this concept indicated the preparation of the "infinitesimal" approach to the geometry on the physical, operational-measuring basis, rather typical of Einstein's ideas [12].

5. FOUR-DIMENSIONAL APPROACH UNDER THE CONDITIONS OF VIOLATION OF THE CONSTANCY OF THE VELOCITY OF LIGHT AND THE LORENTZ-COVARIANCE (MINKOWSKI, PLANCK, ABRAHAM)

As is known, Einstein was first rather sceptical about the 4-dimensional conception of the STR suggested by Minkowski. Only in 1910–1911 his attitude to the Minkowski formulation changed radically. In his Zurich report in January, 1911 Einstein spoke about "extremely interesting mathematical trend, for which the theory is mainly indebted to mathematician Minkowski"[2] – the trend which had led to a very clear interpretation of the theory of relativity, making its applications much easier [31].

It was the application of the 4-dimensional approach to the theory of static fields, outlined in the proof supplement to the March article (published at the end of May, 1912), that turned out to be the second, after the equivalence principle, most important step on the way to the TGCG [23]. For rather a long time Einstein, unlike Abraham, considered impossible the use of the 4-dimensional formalism in the theory of gravitation, because he thought this formalism to be the equivalent of the STR which, in his opinion, was violated even in the presence of homogeneous fields. Moreover, in 1912 Einstein admitted the possibility of deviation of the space geometry from Euclidean. Yet, the change in the attitude to the 4-dimensional formulation of the STR and meditations on the applicability of this formulation in the presence of gravitation, stimulated by the Abraham works, set Einstein on thinking about the possibility of extending the 4-dimensional interpretation of the inertia principle in the STR to the case of motion of a material point in the static gravitational field. In fact, he used here an original form of the correspondence principle – one of the basic methodological principles of physics. Einstein noticed that the equations of motion of a free particle in the STR (in their variational formulation) established by Planck in 1906 [32] and written in the Minkowski 4-dimensional form

$$\delta \int \mathrm{d}s = 0, \tag{1}$$

where $\mathrm{d}s^2 = c^2\,\mathrm{d}t^2 - \mathrm{d}x^2 - \mathrm{d}y^2 - \mathrm{d}z^2$ is the 4-dimensional interval of the STR, remain valid in the theory of static fields under the condition that the velocity of light is not constant, as in the STR, but is a function of spatial coordinates. This meant, in fact, that the 4-dimensional metric

$$\mathrm{d}s^2 = c^2(x, y, z)\,\mathrm{d}t^2 - \mathrm{d}x^2 - \mathrm{d}y^2 - \mathrm{d}z^2 \tag{2}$$

became of the greatest important in the theory of static fields. Minkowski and Laue have demonstrated the fruitfulness of this notion in the STR. Now, it found new life in the relativistic theory of gravitation, and prospects appeared to overcome the difficulty associated with the fact that the coordinates lost their real physical sense. The metric, but not the coordinates, acquired this sense. The metric (2) referring only to the case of static fields led, in fact, to the bending of space-time with non-zero curvature tensors [7]

$$R^{i}_{4j4} = -c \frac{\partial^2 c}{\partial x_i \, \partial x_j} \quad (i, j = 1, 2, 3) \tag{3}$$

$$R_{ij} = \frac{1}{c} \frac{\partial^2 c}{\partial x_i \, \partial x_j}, R_{i4} = 0, \qquad R_{44} = -c \, \nabla^2 c. \tag{4}$$

At this stage, however, Einstein had not been using the apparatus of the theory of curvature, though he intensively looked for the way to extend the 4-dimensional metric approach to arbitrary gravitational fields. After finishing his March paper on the static theory, Einstein wrote to a famous Polish physicist M. Smoluchowski in the letter dated March 24, 1912: "Es ist mir aber noch nicht gelungen, die dynamische Gesetze des Gravitationsfeldes aufzufinden. Das einfache Schema der gleichwertigen 4 Dimensionen gilt hier nicht in dieser Weise wie bei Minkowski" [45].

So, in the above-mentioned proof supplement Einstein found the decisive link promising the possibility of using the 4-dimensional approach in the case of arbitrary fields, the Mach ideas on the general relativity and mass as a dynamic characteristic of the action of bodies in the Universe being of an additional stimulating importance.

6. 'GENERAL RELATIVITY', INERTIA OF A BODY AS A CONSEQUENCE OF THE ACTION OF OTHER MASSES OF THE UNIVERSE (MACH, ABRAHAM)

Mach's criticism of the basic principles of the classical mechanics and his methodological instructions associated with the conception of "observability" were of considerable importance also for the development of the STR. No lesser role have they played in the development of the PE. Mach's ideas also proved very important for the creation of the TGCG. Of special importance here were the Mach criticism of the absolute space and time, the conception of general relativity, and his idea that inertia of bodies were caused by their interaction with all other masses of the Universe. Let us recollect that Mach was the first rector of the German University in Prague, where Einstein had been working in 1911–1912, and the leading professors of physics and mathematics were influenced by the ideas of the Austrian scientist (for instance, mathematician G. Pick and physicist A. Lampa who sponsored

Einstein's visit to Prague). According to Ph. Frank, who replaced Einstein at the Department of theoretical physics after he had gone to Zurich, Einstein was on especially friendly terms with Pick who interpreted many of Mach's arguments as anticipations of the theory of relativity [33]. In the obituary devoted to Mach [34] Einstein underlined in 1916 that the Mach interpretation of Newton's experiment with a bucket showed that he clearly understood the main idea of the PE and the necessity of extend the principle of relativity.

In May, 1912 Einstein finished his work, the main result of which was as follows: the shell K with an inertial mass M increases the inertial mass m of a material point inside the shell according to the formula

$$m' = m + \frac{kmM}{Rc_0^2}, \tag{5}$$

where R is the shell radius, k is the gravitational constant, c_0 is the velocity of light in vacuum. Just this result followed from the scalar theory of static fild, which could be considered only the first step towards the relativistic theory of arbitrary gravitational fields. Einstein noted in conclusion that fomula (5) gave an idea that the inertia of a material point was completely caused by all other masses through a certain interaction mechanism [27]. The Mach conception of inertia undermined the idea of space, or space-time, as being some 'absolute' governing the inertial motion of bodies and at the same time not affected by these bodies. This promoted, on the one hand, the development of the idea to generalize the special principle of relativity, and on the other hand – the formation of an idea that the geometry of space (or space-time) has to be dependent on bodies moving in it (first of all, on their masses). Works of Abraham concerning the scalar theory of gravitation could also be of the essential, stimulating importance. He had, for instance, a profound idea on the local validity of the STR in the presence of gravitation, and he bravely used the 4-dimensional approach to study the gravitational field despite the refusal from the condition of the Lorantz-covariance. Though Einstein originally criticized these ideas, he eventually adopted them when developing the tensor-geometrical field and, in particular, the 'infinitesimal' understanding of the 4-dimensional space-time. At the same time, the main point of disagreement between Einstein and Abraham – the attitude to the principle of relativity (Abraham was the opponent of Einstein's relativistic programme) – and the discussion about this point stimulated a deeper insight into the generalized understanding of this principle [28].

7. 'ABSOLUTE DIFFERENTIAL CALCULUS' – MATHEMATICAL APPARATUS OF GRAVITATION (PICK)

So, in the middle of the summer Einstein almost approached the ideas of the TGCG. The most essential step was the transition from the coordinates, as the main characteristic of space and time, to the metric of the 4-dimensional space-time ds^2, containing the variable velocity of light $c = c(x, y, z)$ and determining (by Equation (1)) the motion of a material point along geodesics of this space. Yet, this transition was only in contemplation. Even in the polemic note ("Reply to Abraham's remark") dated July the 4-th [28], Einstein went on to speak that the coordinates lost their simple physical sense, so the form of space-time transformations called upon to replace the Lorentz transformations could not be established.

But it is this small note that contains arguments suggesting an idea of the possibility of using a second-rank tensor as a mathematical characteristic of the gravitational potential. Not confirming these arguments by calculations, Einstein points that the scalar and vector potentials lead, in the framework of the Lorentz-covariant approach, to the contradiction with the equality betweein inertial and gravitational masses.

So, even in this last publication of the Prague period Einstein had not passed to the TGCG. But we clearly see the origins of this conception: transition to the 4-dimensional, and hence geometrical, inerpretation of the equivalence principle; recognition of the fundamental importance of the concept of metric and its non-Euclidean character in gravitational fields; representation of the principle of inertia by geodesics in the 4-dimensional space; conclusion on the unsatisfactoriness of the scalar and vector potentials; clear understanding of the necessity to generalize the principle of relativity and of impossiblity of restricting oneslef to the addition of only uniformly accelerated frames of reference; elements of the infinitesimal approach to the 4-dimensional geometry.

Eventually, all this had to lead to the use of the "absolute differential calculus" (i.e. the Riemannian geometry and tensor analysis) as the basic mathematical structure of gravitation. It is usual to associate the introduction of this structure into the theory of gravitation with the name of Einstein's Zürich friend, mathematician M. Grossmann, who joined in the development of mathematical aspects of the problem at the Einstein request, when he moved from Prague to Zürich at the end of the summer, 1912. There is, however, an evidence from Ph. Frank [33] that the Prague mathematician G. Pick, a pupil of Mach and the author of some papers on the differential geometry in non-Euclidean space [35], suggested to Einstein the idea of using the "absolute differential calculus" and the theory of differential quardatic forms for the construction of the relativistic theory of gravitation. In particular, when Einstein asked Pick to recommend him literature on the theory of determinants, Pick told him about the book by G. Kovalewski [36], which contained the chapter on the reduction of quadratic forms $\Sigma a_{ik} x_i x_k$ to the canonic form Σx_m^2. Einstein could encounter a similar problem when using the 4-dimensional 'metric' approach in a class of transformations wider than the Lorentz transformations [12]. Despite this fact the role of Pick still remains unclear, especially because Einstein himself has never mentioned the name of Pick in connection with the TGCG genesis, but repeatedly emphasized merits of M. Grossmann.

8. ON THE ROLE OF EXPERIMENT (E. FREUNDLICH)

Experimental incentives were very important for Einstein at that period. The return to the PE and to the development (on its basis) of the theory of gravitation in 1911 was associated, to a great extent, with the realization of the possibility to verify experimentally the effect of deflection of light in gravitational field, though Einstein, as before, considered the 'red shift' to be too small effect to measure it. A young Berlin astronomer E. Freundlich knew from L. V. Pollak, the demonstrator at the Institute of space physics in Prague, about Einstein's experiment on deflection of light in the solar gravitational field during the Sun eclipse, and that he wished to get in scientific contacts with astronomers. Soon they began correspondence the main point of which was the experimental corroboration of the equivalence principle. Freundlich vigorously supported the experimental ideas of Einstein. Hope for the success in these experiments stimulated further investigations based on the principle of equivalence.

From Einstein's letter to Freundlich dated September 1, 1911 we learn that Einstein attached so great importance to the experiment discussed that he even could give up the PE in the case of negative result: " . . . Dass deren Beantwortung durch die Erfahrung keine leichte Sache ist, dass weiss ich wohl, weil eben die Refraktion der Sonnenatmosphäre mitspielen mag. Aber eines kann immerhin mit Sicherheit gesagt werden: existiert keine solche Ablenkung, so sind die Voraussetzungen der Theorie nicht zutreffend." [5]

When substantiating the PE, which turned out to be the original point in the development of the TGCG, Einstein did not refer to famous experiments by Eötvös on the verification of the equality between inertial and gravitational masses either in 1907 or in 1911. But since he had learned about them (the name of Eötvös was first metnioned in this connection in 1913 in the joint paper by Einstein and Grossmann concerning the TGCG [37]) he always considered these experiments as the foundation of the TGCG.

9. EINSTEIN AND GROSSMANN

In October, 1912 Einstein moved from Prague to Zurich. The mathematician L. Kohlros who studied at the Zürich Federal Institute of Technology together with Einstein and Grossmann recollected that Einstein met Grossmann and told him about his concerns (in the field of gravitation – V. V.), saying once: "Grossmann, Du musst mir helfen sonst werd'Ich verrückt!". In 1912 Grossmann was the head of the Department VII at the Zürich Federal Institute of Technology for teachers in physics and mathematics. It was that Department where Einstein was invited to deliver lectures on theoretical physics. Grossmann who was the author of a number of papers on the non-Euclidean and differential geometry [39] was interested in the Einstein suggestion. Einstein recollected later that Grossmann conceived at once, though he was a little sceptical about physics [40]. In the letter to Sommerfeld dated October 29, 1912 Einstein wrote: "I am occupying myself exclusively with the problem of gravitation and believe that with the aid of a local mathematician who is a friend of mine I'll now be able to master all difficulties. But one thing is certain that in all my life I have never struggled as hard as that I have been infused with great respect for mathematics the subtler parts of which, in my simple-mindedness, I had considered pure luxury up to now! Compared to this problem the original relativity theory is child's play" [41]. Thus, since the autumn of 1912 Einstein and Grossmann began their frutiful cooperation. Einstein recollected that Grossmann agreed to work only at the mathematical part of the problem and discovered soon that the necessary mathematical apparatus, the "absolute differential calculus", had been developed at the very beginning of the XX century by Italian mathematicians G. Ricci and T. Levi-Civita on the basis of previous works of B. Riemann and E. B. Christoffel [40].

Einstein worked with Grossmann until he left for Berlin in 1914. The first version of the tensor-geometrical theory published in June of 1913 in the joint paper by Einstein and Grossmann [3] was known as the Einstein-Grossmann theory [37]. The main distinction of this theory from the GTR was the absence of the general-covariant equations of gravitational field in the former one. As for the field equations, this general-covariant, in its sense, theory used linearly-covariant equations, though with the help of purely mathematical arguments they could be made general-covariant equations containing tensor characteristics of the space curvature. The refusal from this general-covariant approach was associated with the fact that the corresponding field equations could not be made consistent with the principles of correspondence, causality and conservation of energy-momentum. Thus, the TGCG was not completely developed in the framework of the Einstein-Grossmann theory. Only two years and a half later Einstein solved the problem of field equations (without help of Grossman), thereby completing the construction of the general-covariant theory of gravitational field.[4]

10. THIRD LETTER OF EINSTEIN TO MACH

We have already mentioned the third (undated) letter of Einstein to Mach. F. Herneck, who published this Christmas letter, and believes that it could have been written either on the eve of 1912 or on the eve of 1913. However, the analysis of the works of 1911--1912 makes the first date impossible. If we admit that the letter was written on the eve of 1913, it might contain the first, though rather brief, consideration of the TGCG developed by Einstein with the help of Grossmann by the beginning of 1913: "Die Gesamtheit der Massen erzeugt ein $G_{\mu\nu}$-Feld (Gravitationsfeld), das seinerseits den Ablauf aller Vorgänge, such die Ausbreitung der Lichtstrahlen und das Verhalten der Masstabe und Uhren regiert". The main difficulty which, probably, delayed the publication of the new theory and was associated with the double covariance of the theory (in the general-covariant scheme of the theory it had to be restricted, then, by the linear covariance condition) was also clearly formulated in the letter: "Das Geschehen wird zunächst auf vier ganz willkürliche raum-zeitliche Variable bezogen. Diese mussen dann, wenn den Erhaltungssätzen des Impulses und der Energie Genuge geleistet werden soll, derart spezialisiert werden, dass nur (ganz) lineare Substitutionen von einem berechtigten Bezugssystem ist der bestehenden Welt mit Hilfe des Energiesatzes sozusagen angemessen und verliert sine nebulose apriorische Existenz" [11]. The Einstein letter also mentions considerable mathematical difficulties in the development of the new theory and insufficiency of the physical interpretation of the theory ("Aber is gelingt nur bei diesen abstrakten Dingen nicht recht, das Sachliche von der Form zu trennen" [11]).

Four letters, in all, of Einstein to Mach are known: two of them were written in 1909, and the fourth on July 25, 1913. In this, last letter Einstein writes about the completion of his work concerning relativity and gravitation which he "finished after endless efforts and agonizing doubts" and about the experiment on deflection of light in the gravitational field of the Sun during its eclipse, that was to take place in August, 1914. In all the four letters, written both in 1909 and in 1913, Einstein emphasizes the importance of Mach's ideas for the development of the theory of relativity, mentioning at the first place his 'Mechanics' and the work concerning the law of conservation of energy. "I am very glad that the deep importance of your studies on the basic principles of classical mechanics becomes obvious in the development of the theory"[5] – he wrote in the third letter [11]. "If so, – Einstein wrote in the fourth letter having in mind experimental evidence for the deflection of light – then your inspirational studies on the basic principles of mechanics will find – despite Planck's unfair critisism – brilliant confirmation" [11].

11. CONCLUSION

The principle of equivalence (PE), which established (since 1907) the undistinguishibility between homogeneous graviational field and uniformly accelerated frame of reference, proved to be the physical basis of the TGCG, its peculiar 'cell', "differential". In the 4-dimensional approach this means the identification of physical interaction with geometrical phenomenon. The PE, as such, have been established by virtue of the informal analysis of the empirical fact known from the time of Galileo and Newton (the equality between inertial and gravitational masses) from the standpoint of the relativivistic programme which grew out of the special theory of relativity. The PE required going beyond the framework of inertial frames of reference, thereby extending the original relativistic programme. In this case the coordinates lost their direct physical sense, so the determination of the limits of this extension became more difficult. Despite the 'progressive' character of the new approach (prediction of two gravitation-optical effects), the possibility of its extension to inhomogeneous fields remained rather problematical for a number of years.

The next important step was made with the use of the dependence of the velocity of light on the gravitational potential, stemming from the PE. Not restricting himself to the requirements of the theory of relativity, Abraham had constructed the 4-dimensional generalization of the Poisson equation for the scalar potential, according to this dependence. Stimulated by works of Abraham, Einstein first attempted at modifying (with due regard to the

PE) the scalar Poisson equation only for static fields, putting off, for a time, the solution of the problem on the class of permissible frames of reference. Aspiration for the natural consistency of the scalar theory with the PE and relativistic requirements, and for overcoming of the above difficulties associated with the extension of this principle to arbitrary fields made Einstein more and more disappointed in the scalar trend and evoked more and more doubts in the possibility of retaining geometrical concepts characteristic of the STR. Non-Euclidean type of metric relations in accelerated frames of reference (and, in accordance with the PE, in the presence of gravitational fields); infinitesimal character of the space geometry; realization of the deep connection of Mach's ideas on the absence of space-time absolutes and that the interaction with all masses in the Universe is the cause of the inertia of a body with the PE and the concept of general relativity – those were the main prerequisites of the TGCG that cropped up in the spring and summer of 1912 while Einstein was working in Prague on the scalar theory of gravitation and in the course of his intensive discussions with Abraham.

The change in Einstein's attitude to the Minkowski 4-dimensional formalism, which he understated previously, was of decisive importance in this situation. In the framework of this formalism the focus shifted from the coordinates to the metric and interval, ds. Einstein showed that in the STR the 4-dimensional metric variational formulation of free motion of a material point, established by Planck and Minkowski, could be extended naturally to the case of free motion in a static field if the metric, ds, were given a generalized sense when the velocity of light should be considered dependent on the space-time coordinates. In the light of the 4-dimensional, 'metric' approach the above prerequisites acquired a new, more exact sense and led directly to the "absolute differential calculus" as an adequate mathematical apparatus of the relativistic theory of gravitation and the TGCG.

A deeply physical approach of Einstein, giving the primary significance to interpretation, operational-measuring explanation of the concepts, experimental verification of theory, his devotion to the relativistic programme and heightened attention to Mach's criticial arguments on Newton's mechanics interpreted in the light of this programme, were the main Einstein's "weapon" on his way to the TGCG. Special emphasis should also be given to Einstein's brilliant use of methodological principles of physics, such as the principles of correspondence, symmetry, observability, simplicity, conservation and others.

A rather interesting feature of the TGCG genesis also leaps to the eye: almost any of the above prerequisites, almost any step towards the conception were associated with Einstein's scientific contacts. Abraham, Ehrenfest, Mach, Freundlich, Planck, Pick Grossmann and others contributed considerably, in some or other way, to the development of the TGCG – sometimes through an acute discussion, sometimes indirectly (like Planck or Minkowski), sometimes through the direct experimental (Freundlich) or mathematical (Pick, Grossmann) support.

The picture of the formation of the TGCG just described is confirmed, on the one hand, by the scheme of arguments used in the 'Physical part' of the 'Project' and, on the other hand, by Einstein's recollections on the event. The logical chain leading to the TGCG is in the 'Project' as follows: the PE → the dependence of the velocity of light on the gravitational potential → the scalar theory of static field → equations of motion of a material point in the field written in the form of the 4-dimensional variational principle: $\delta\int ds = 0$ with the metric containing the velocity of light considered as a function of space-time → arbitrary fields, in accordance with the PE, require the transition to arbitrary transformations and the Riemannian metric (the tensor components of the metric acquire, thus, a dual interpretation: they express the structure of the curved space-time and are, simultaneously, the gravitational potential).

Three times Einstein addressed the audience with his recollections and told, in particular, about his path to the TGCG (in 1933 [43], 1949 [44] and 1955 [40]). The corresponding chain of arguments in each of these papers is almost the same and agrees with the above-mentioned picture of the TGCG genesis. However, giving justly the primary importance to the key idea of the 4-dimensional metric formulation of the inertia principle, Einstein dos not mention other important points that played an essential role in the formation of this conception (arguments leading to the infinitesimalization of geometry and its non-Euclidean character; Mach's ideas on the nature of inertia; recognition of the lack of prospects in the vector and scalar approaches and so on).

Institute for History of Science

NOTES

[1] In the summer of 1913, soon after the appearance of the first publication concerning the tensor-geometrical theory of gravitation, Nordstrom modernized his theory so that it corresponded exactly to the equality between inertial and gravitational masses (the second theory of Nordstrom [26]). Later Einstein considered it to be the main alternative

for the tensor-geometrical theory, the final dispute between the two being solved by the experiment on the deflection of light in the gravitational field of the Sun.

[2] Translated by A. Talashev.

[3] The date of recept of this fundamental paper in the 'Zeitschr für Math. und Phys.' is not stated. Some indirect evidence shows that the paper was finished in April-May, 1913 and published in June of that year. In his letter to K. Laub dated 22/VII, 1913 [42] Einstein explained the delay with the reply by intensive work at the problem of gravitation which he had finished about two months before the letter was written. In his fourth letter to Mach dated 25/VI, 1913 there are the following words: "Dieser Tage haben Sie wohl meine neue Arbeit über Relativität und Gravitation erhalten, die nach unendlicher Mühe und quälendem Zweifel nun endlich fertig grworden ist [11]." In the article received in the 'Annalen der Physik' on 24/VII, 1913 Nordström refers to the "Project" [26].

[4] Modern specialists on the GTR, paying tribute to the contribution made by Grossmann to this theory, organized in Trieste, in July of 1975 the 'Grossmann Symposium' on the GTR. The programme of the Symposium contained the biography of Grossmann which, in particular, said: "Grossmann introduced Einstein to the absolute differential calculus, started by Elwin Christoffel (1864) and fully developed at the University of Padova by Gregorio Ricci Curbastro and Tullio Levi-Civita (1901). Allowing the encounter of the great mathematical achievements of the Italian geometers and profound physical insight of Einstein, Marsel Grossmann facilitated that most unique of mathematical and theoretical physics reached by Albert Einstein in what is still today considered the most elegant and powerful field theory of physics: The General Theory of Relativity". The author is grateful to A. A. Starobinsky for this information.

[5] Translated by A. Talashev.

[6] Translated by A. Talashev.

REFERENCES

[1] Pauli, W., *Relativitätstheorie,* Enc. d. math. Wiss. Bd. 5, T. 2, H. 4. Leipzig, Teubner, 1921, S. 543–775, S. 708.

[2] Guth, E., 'Contribution to the history of Einstein's geometry as a branch of physics', in *Relativity*, Proceedings of the Relativity Conference in the Midwest. Ohio, Cincinnatti, 1969, N. Y., London, 1970.

[3] Mehra, J., *Einstein, Hilbert and the Theory of Gravitation*, D. Reidel Publ. Co., Doredrecht, Holland, Boston, U.S.A., 1974.

[4] Pyenson, L. R., 'Einstein's early scientific collaboration', *Hist. Stud. Phys. Sci.* 7 (1976), 83–123.

[5] Pyenson, L. R., 'The Göttingen reception of Einstein's general theory of relativity', Dissertation, The Johns Hopkins Univ., Baltimore, Maryland, 1974.

[6] Earman, J. and Glymour, G., 'Lost in the tensors: Einstein's struggles with covariance principles, 1912–1916', *Stud. Hist. Phil. Sci.* 9 (1978), 251–278.

[7] Hoffmann, B., 'Einstein and tensors', *Tensor* 26, (1972), 157–162.

[8] Lanczos, K., 'Einstein's path from special to general relativity', in *General relativity*, Papers in honour of J. L. Synge, Oxford, 1972 p. 5.

[9] Lanczos, K., *Space Through the Ages*, London, N. Y., 1970.

[10] Lanczos, K., *The Einstein Decade, 1905–1915*, London, 1974.
[11] Herneck, F., *Die Beziehungen zwischen Einstein und Mach*, Wiss. Zeitschr. d. Friedrich-Schiller-Universität, Jena, 15 (1966), 1–14.
[12] Illy, J., 'Albert Einstein in Prague', *Isis* 70 (1979), 76–84.
[13] Stachel, J., *Einstein and the Rigidly Rotating Disk*, General Relativity a. Gravitation. One Hundred Years After the Birth of Albert Einstein, A. Held (ed.). N.Y., London, Plenum Press, 1980, p. 1.
[14] Einstein, A., *Ueber das Relativitätsprinzip und die aus demselben gezogenen Folgerungen*, Jahrb. d. Radioakt. u. Elektron., 4 (1907), 411–462.
[15] Lakatos, I., 'History of science and its rational reconstructions', in R. S. Buck and R. S. Cohen (eds.), *Boston Studies in the Philosophy of Science*, v. VIII. In Memory of Rudolf Carnap, D. Reidel Publ. Co., Dordrecht, Holland, 1970, pp. 91–136.
[16] McCormmach, R., 'Einstein, Lorentz and the electron theory', *Hist. Stud. Phys. Sci.* 2 (1970), 41–88.
[17] Einstein, A., 'Ueber den Einfluss der Schwerkraft auf die Ausbreitung des Lichtes', *Ann. Phys.* 35 (1911), 898–908.
[18] Abraham, M., 'Zur Theorie der Gravitation', *Phys. Zeitschr.* 13 (1912), 1–4.
[19] Abraham, M., 'Das Elementargesetz der Gravitation', *Phys. Zeitschr.* 13 (1912), 4–5.
[20] Abraham, M., 'Die Erhaltung der Energie und der Materie im Schwerkraftfeld', *Phys. Zeitschr.* Bd. 13, (1912), S. 311–314.
[21] Abraham, M., 'Neure Gravitationstheorien', *Jahrb. d. Radioakt. und Elektronik* 11, (1914), 470–520.
[22] Einstein, A., 'Lichtgeschwindkeit und Statik des Gravitationsfeldes', *Ann. Phys.* 38 (1912), 355–369.
[23] Einstein, A., 'Zur Theorie des statischen Gravitationsfelds', *Ann. Phys.* 38 (1912), 443–458.
[24] Nordström, G. 'Relativitätsprinzip und Gravitation', *Phys. Zeitschr.* 13 (1912), 1126–1129.
[25] Mie G. 'Die Grundlagen einer Theorie der Materie', *Ann. Phys.* 40 (1913), 1–66.
[26] Nordström, G., 'Zur Theorie der Gravitation vom Standpunkt der Relativitätsprinzips', *Ann. Phys.* 42 (1913), 533–554.
[27] Einstein, A. 'Gibt es eine Gravitationswirkung, die der elektrodynamischen Induktionswirkung analog ist?' *Vierteljahresschr. gerichtl., Med.* 44 (1912), 37–40.
[28] Einstein, A., 'Relativität und Gravitation. Erwiderung auf eine Bemerkung von M. Abraham', *Ann. Phys.* 38 (1912), 1059–1064.
[29] Ehrenfest, P., 'Gleichformige Rotation starrer Körper und Relativitätstheorie', *Phys. Zeitschr.* 10 (1909), 918.
[30] Einstein, A., 'Zum Ehrenfestschen Paradoxon', *Phys. Zeitschr.* 12 (1911), 509–510.
[31] Einstein, A., 'Die Relativitätstheorie', *Naturforsch. Ges., Vierteljahresschrift* 56 (1911), 1–14.
[32] Planck, M., 'Das Prinzip der Relativität und die Grundgleichungen der Mechanik', *Verhandl. Dtsch. Phys. Ges.* (1906), 136–141.
[33] Frank, Ph., *Einstein, His Life and Times*, N.Y., 1949.
[34] Einstein, A., 'Ernst Mach', *Phys. Zeitschr.* 17 (1916), 101–104.

[35] Pick, G. 'Sur les notions: droites parallèles, et la géometrie différentielle dans l'espace non-euclidien', *Compt. Rend.* **153** (1912), 1447–1449; **154**, p. 263; 'Zur nichteuklidische Geometrie', *Archiv d. Math. u. Phys.* **25** (1916), 135–137.
[36] Kowalewski, G., *Einführung in die Determinantentheorie*, Leipzig, 1909.
[37] Einstein, A. and Grossmann, M. 'Entwurf einer verallgemeinerten Relativitätstheorie und Theorie der Gravitation', *Zeitschr. Math. u. Phys.* **62** (1913), 225–261.
[38] Kohlros, L. 'Die Erinnerungen eines Kommilitonen', in *Helle Zeit-Dunkle Zeit.* Zürich, Stuttgart, Wien, 1956, 135–137.
[39] Grossmann, M., 'Die fundamentalen Konstruktionen der nichteuklidischen Geometrie', Pr. Thurgauische Kantonschule, Frauenfeld, 1904; Projective Konstruktionen in der hyperbolischen Geometrie, *Math. Ann.* **68** (1910), 141–144; Projectiver Beweis der absoluten Parallelkonstruktion von Lobatschevskij. *Verhandl. Schweiz. Naturforsch. Ges.* **35** (1911), 4.
[40] Einstein, A., 'Autobiographische Skizze', in *Helle Zeit – Dunkle Zeit*, Zürich, Stuttgart, Wien, 1956.
[41] Einstein, Albert, *Arnold Sommerfeld Briefwechsel*, Basel, 1968.
[42] Seelig, C., *Albert Einstein*, London, 1956, p. 142.
[43] Einstein, A., 'Einiges über Entstehung der allgemeinen Relativitätstheorie. G. A. Gibson Foundation Lecture, Glasgow', in *Mein Weltbild*, Franfurkt am Main, 1955, 170–175.
[44] Einstein, A., 'Autobiographisches (Autobiographical Notes)', in P. A. Schilip (ed.), *Albert Einstein – Philosopher-Scientist*, Evanston, 1945, p. 1–95.
[45] Teske, A., 'Z Einsteinovy Pražské korespondence'. Sbornik pro dějiny prirodnich věd a techniky, VIII, Praha, 1972, p. 228–231.
[46] McCormmach, R., 'Foreword', *Hist. Stud. Phys. Sci.* 7 (1976), p. XXVIII.
[47] Marsel Grossmann, 'Meeting on the Recent Progress of the Fundamentales of General Relativity'. Programme. University of Trieste, Trieste, Italy, 7–12 July, 1975, p. 2.

INDEX

THE UNIVERSITY OF WESTERN ONTARIO SERIES IN PHILOSOPHY OF SCIENCE

A Series of Books in Philosophy of Science, Methodology, Epistemology, Logic, History of Science, and Related Fields

1. J. Leach, R. Butts, and G. Pearce (eds.), *Science, Decision and Value*. Proceedings of the Fifth University of Western Ontario Philosophy Colloquium, 1969. 1973, vii+ 213 pp.
2. C. A. Hooker (ed.), *Contemporary Research in the Foundations and Philosophy of Quantum Theory*. Proceedings of a Conference held at the University of Western Ontario, London, Canada, 1973, xx+385 pp.
3. J. Bub, *The Interpretation of Quantum Mechanics*. 1974, ix+155 pp.
4. D. Hockney, W. Harper, and B. Freed (eds.), *Contemporary Research in Philosophical Logic and Linguistic Semantics*. Proceedings of a Conference held at the University of Western Ontario, London, Canada. 1975, vii+332 pp.
5. C. A. Hooker (ed.), *The Logico-Algebraic Approach to Quantum Mechanics*. 1975, xv+607 pp.
6. W. L. Harper and C. A. Hooker (eds.), *Foundations of Probability Theory, Statistical Inference, and Statistical Theories of Science*, 3 Volumes. Vol. I: *Foundations and Philosophy of Epistemic Applications of Probability Theory*. 1976, xi+308 pp. Vol. II: *Foundations and Philosophy of Statistical Inference*. 1976, xi+455 pp. Vol. III: *Foundations and Philosophy of Statistical Theories in the Physical Sciences*. 1976, xii+241 pp.
7. C. A. Hooker (ed.), *Physical Theory as Logico-Operational Structure*, 1979, xvii+ 334 pp.

8. J. M. Nicholas (ed.), *Images, Perception, and Knowledge.* Papers deriving from and related to the Philosophy of Science Workshop at Ontario, Canada, May 1974. 1977, ix+309 pp.
9. R. E. Butts and J. Hintikka (eds.), *Logic, Foundations of Mathematics, and Computability Theory.* Part One of the Proceedings of the Fifth International Congress of Logic, Methodology and Philosophy of Science, London, Ontario, Canada, 1975. 1977, x+406 pp.
10. R. E. Butts and J. Hintikka (eds.), *Foundational Problems in the Special Sciences.* Part Two of the Proceedings of the Fifth International Congress of Logic, Methodology and Philosophy of Science, London, Ontario, Canada, 1975. 1977, x+427 pp.
11. R. E. Butts and J. Hintikka (eds.), *Basic Problems in Methodology and Linguistics.* Part Three of the Proceedings of the Fifth International Congress of Logic, Methodology and Philosophy of Science, London, Ontario, Canada, 1975. 1977, x+321 pp.
12. R. E. Butts and J. Hintikka (eds.), *Historical and Philosophical Dimensions of Logic, Methodology and Philosophy of Science.* Part Four of the Proceedings of the Fifth International Congress of Logic, Methodology and Philosophy of Science, London, Ontario, Canada, 1975. 1977, x+336 pp.
13. C. A. Hooker (ed.), *Foundations and Applications of Decision Theory,* 2 volumes. Vol. I: *Theoretical Foundations.* 1978, xxiii+442 pp. Vol. II: *Epistemic and Social Applications.* 1978, xxiii+206 pp.
14. R. E. Butts and J. C. Pitt (eds.), *New Perspectives on Galileo.* Papers deriving from and related to a workshop on Galileo held at Virginia Polutechnic Institute and State University, 1975. 1978, xvi+262 pp.
15. W. L. Harper, R. Stalnaker, and G. Pearce (eds.), *Ifs. Conditionals, Belief, Decision, Chance, and Time.* 1980, ix+345 pp.
16. J. C. Pitt (ed.), *Philosophy in Economics.* Papers deriving from and related to a workshop on Testability and Explanation in Economics held at Virginia Poly-Technic Institute and State University, 1979. 1981.
17. Michael Ruse, *Is Science Sexist?* 1981, xix+299 pp.
18. Nicholas Rescher, *Leibniz's Metaphysics of Nature.* 1981, xiv+126 pp.
19. Larry Laudan, *Science and Hypothesis.* 1981, x+258 pp.
20. William R. Shea, *Nature Mathematized.* Papers deriving from the third international Conference on the history and philosophy of science, Montreal, Canada, 1981. Vol. I, 1983.
21. Michael Ruse, *Nature Animated.* Papers deriving from the third international Conference on the history and philosophy of science, Montreal, Canada, 1981. Vol. II, 1983.

GPSR Compliance

The European Union's (EU) General Product Safety Regulation (GPSR) is a set of rules that requires consumer products to be safe and our obligations to ensure this.

If you have any concerns about our products, you can contact us on ProductSafety@springernature.com

In case Publisher is established outside the EU, the EU authorized representative is:

Springer Nature Customer Service Center GmbH
Europaplatz 3
69115 Heidelberg, Germany

Batch number: 10823811

Printed by Printforce, the Netherlands